An Introductory Course in Science for Colleges

II
MAN AND THE NATURE OF HIS BIOLOGICAL WORLD

By FRANK COVERT JEAN
EZRA CLARENCE HARRAH
AND FRED LOUIS HERMAN
COLORADO STATE TEACHERS COLLEGE

WITH THE EDITORIAL COLLABORATION OF
SAMUEL RALPH POWERS
TEACHERS COLLEGE, COLUMBIA UNIVERSITY

GINN AND COMPANY
Boston · New York · Chicago · London · Atlanta · Dallas · Columbus · San Francisco

The Athenæum Press

GINN AND COMPANY · PRO-
PRIETORS · BOSTON · U.S.A.

PREFACE

There is a growing recognition of the importance of science. Almost every day the press carries reports of new discoveries and the further application of scientific principles to the needs of modern life. More progress has been made in this field during the past three quarters of a century than had been made in the whole previous eighteen centuries of the Christian Era. Within the life span of many now living man's knowledge of electricity and his ideas as to the constitution of matter and of radiant energy have been completely revolutionized; chemistry has developed a surprisingly long list of new products and has found thousands of applications in the processes of business and industry; biology has changed man's ideas as to the nutritional needs of the body, the nature and extent of hormone regulation, and the means adopted by the organism to combat diseases. Scientific knowledge has changed man's systems of thought and the manner in which he views the universe. In many respects it has altered his whole attitude toward life and the meaning of his existence. In these and a multitude of other ways science has entered into and affected man's everyday experiences.

Science has also profoundly influenced man's conception of dependable truth and the method of arriving at it. It has introduced a new method into rational processes and has laid down specific rules which man must follow if he is to reach valid conclusions. Indeed, in an increasingly complex society it now appears that the method of science as a factor in social progress may become quite as important as the purely material products of science and invention.

With this scientific advance coming partly from within and partly from without the schools, it was inevitable that its influence should affect in a fundamental manner both the content of the curriculum and the methods of instruction.

Educators readily recognized the importance of giving science a prominent place in the college program. Special courses of a general nature within each of the specific fields were formulated, and either designated courses or elections from groups of these courses came to be required.

At the outset high hopes as to the educational results of such a policy were entertained. But the actual outcomes of these traditional courses have proved rather disappointing. In most instances the subject-matter content was selected and organized by, and consequently from the point of view of, the subject-matter specialist. The result was too often a more or less detailed course that left the student in possession of certain specific facts but without an adequate conception of the general principles of that science, and more especially without an appreciation of the relation of that particular science to life needs. In other words, the students emerged from the course with a narrow specialized conception. It was comparable to that formed of a mountain range by a traveler who had proceeded only up the course of a stream lying in one of its deep cañons.

From the results of such a program, especially for the student who does not intend to specialize in the field of science, there has been a more or less pronounced revolt on the part of educational administrators. They are demanding courses which will interpret for the student the scientific phenomena of experience. Such courses need not be confined to any one particular field but may cut across the subject-matter content of several sciences. In short, administrators are demanding courses which are organized from the point of view of the life needs of the general student rather than of the student who wishes to become a specialist. They are insisting that the required general courses in science shall help the student to locate himself in his universe; that they shall help to free him from superstition and prejudice; that they shall teach him to rely upon established truth in ordering his own life and to value the leadership of the specialist; that they shall enable him to appreciate the careful, logical methods employed by the scientist in arriving at trustworthy conclusions; in short, that the generalizations of science shall be presented in such a way as to influence in a significant manner the beliefs, philosophy, and attitudes of the average student.

Toward the organization of such generalized introductory courses there is a distinct trend among the colleges and universities of America today. To this educational philosophy the authors of this text have subscribed. They do not believe, however, that an introductory course should omit the valuable element of actual contact with concrete materials on the part of the student. Nevertheless much of the time spent in traditional individual laboratory work can be more profitably spent in supplementary readings, classroom discussions, teacher demonstrations, and the employment of other modern visual aids. In the authors' judgment no course of this character should ever be attempted without the use of these valuable supplementary instructional agencies.

The two volumes comprising this series are the result of four years' experience in the classroom. Both the content and its organization have been greatly modified in the light of actual instructional use. Altogether something like two thousand students have been taught these courses. The books were first issued as a temporary class edition. The manuscript was then revised for the permanent edition in conformity with the composite judgment reached by the instructors in the use of the trial edition.

THE AUTHORS

ACKNOWLEDGMENTS

It is a privilege for the authors to acknowledge preëminently their indebtedness to the collaborating editor, Dr. S. R. Powers, and to Dr. G. W. Frasier, president of Colorado State Teachers College. The fundamental educational philosophy upon which these volumes are based came originally from Dr. Powers. The general outline of the field to be covered and the objectives to be attained were worked out with him in personal conferences. He has read all the manuscript, and his extensive criticisms and counsel have proved most helpful throughout the preparation of both books in the series. Dr. Frasier is thoroughly committed to the policy of broad, cultural courses and has manifested the deepest interest in the preparation of these texts. He has given the authors invaluable assistance both by way of suggestion and encouragement.

It is a pleasure to acknowledge our indebtedness to the many authors and other private individuals, to the publishers, and to the institutions whose coöperation and courtesies in the matter of supplying pictures and giving permission to use illustrations and quotations have been of inestimable service in preparing the manuscripts.

The authors desire to acknowledge in a special manner the assistance of the following persons who have read and offered criticisms on certain chapters of the manuscript: Mr. L. B. Aldrich, of the Smithsonian Institution; Mr. Herbert J. Arnold, of Teachers College, Columbia University; Mr. Clarence Birdseye, of the General Foods Corporation; Dr. L. L. Burlingame, of Leland Stanford University; Dr. Frederick L. Fitzpatrick, of Teachers College, Columbia University; Dr. R. G. Hoskins, of Harvard University; Dr. M. L. Huggins, of Leland Stanford University (now of Johns Hopkins University); Dr. E. V. McCollum, of Johns Hopkins University; Dr. R. J. Pool, of the University of Nebraska; Dr. Paul Popenoe, of the Human Betterment Foundation; Professor G. D. Swezey, of the University of Nebraska; Mr. C. F. Talman, Librarian of the United States Weather Bureau; Dr. Hugh K. Ward, of Harvard University; and Dr. H. B. Ward, of the University of Illinois.

While the coöperation and suggestions of these readers are acknowledged with deep gratitude, yet they must not in any sense be held responsible for errors that may occur. The authors themselves assume full responsibility for factual defects of any character.

CONTENTS

UNIT V

Mendel made Fundamental Discoveries which have made Possible Improved Plants and Animals

UNIT VI

The Knowledge of Man's Origin and of the Laws of Heredity gives Promise of an Improved Society

UNIT VII

Man has discovered Much as to the Nature and Control of Disease

UNIT VIII

Man's Cultural Development moved Slowly at the Outset but has been greatly accelerated by Science and Invention

AN INTRODUCTORY COURSE IN SCIENCE FOR COLLEGES

II. Man and the Nature of his Biological World

UNIT I

Life took the Form of Plants and Animals and became adapted to live under Many Conditions

LIFE as it resides on the earth is a profound mystery. No one has been able to say with assurance where it came from originally; neither has he been able to explain just what it is. Once life had established itself on the earth, however, it proved to be very persistent and plastic. It has assumed an almost endless variety of forms and has acquired the ability to live successfully in widely separated and most diverse places.

One of the most remarkable of all its achievements is the fact that living things have solved the whole question of vital energy. Green plants have acquired the power to arrest the rays of sunlight and, from the energy supplied in this way, combine inorganic substances into life-sustaining foods. Animals, with a few minor exceptions, have not been able to make food themselves, but have solved their problem in another way: they have turned to the plant as a source of nourishment, and in so doing have become dependent upon the sun also.

Perhaps the most striking feature of the earth is its varied forms of plants and animals and their adaptations to live successfully. What these organisms do because they are alive and how they provide for their continued existence makes a most interesting story.

Chapter I · Life is always found Associated with Organized Protoplasm

Introduction. When did life first appear on the earth? That is an interesting question, but unfortunately one that cannot be answered. It must have been a long time ago, and yet it obviously was millions of years after the infant earth first came into existence. Regardless of whether the view as to the gaseous origin of the earth proves true, or whether the planetesimal hypothesis wins out, or whether both must be supplanted by a still more plausible theory, eons of time must have elapsed before the first simple form of life arrived on the scene. This conclusion is forced upon us by a consideration of the preliminary changes that must have taken place in order to furnish a set of conditions under which life could have survived: a hydrosphere must have been formed, an atmosphere must have accumulated, and parts of the earth materials must have become soluble. For, regardless of what form primeval life took, it would have had to have oxygen, water, and food-making materials to continue its existence. To have fulfilled these prerequisite conditions necessary to life would have required ages of slow geological change. So, in terms of man's comprehension, the earth had attained a considerable age before the first simple forms of protoplasm took up their earthly abode.

There are many theories as to the origin of life. The question as to where life first came from has been a much-mooted one. It still is an important problem for many people who are philosophically inclined and like to indulge in mental speculations. A multitude of opinions is found concerning this question, just as in the case of all great problems connected with the universe. Some persons formulate an answer out of pure fancy, with scarcely a thought as to whether their

5

conclusions are in harmony with known facts or not; others attempt to gather all available facts, and upon these as a basis build up a hypothesis. It will be recognized that the latter approach is the only reliable one; yet it must be admitted that, so far as the solution of the problem is concerned, results to date are far from conclusive. Only the leading theories proposed will be considered, and the reader will be left to draw his own conclusions as to the plausibility of each.

Perhaps the most widely accepted idea as to the origin of life is that of special creation — the view that living forms appeared here in response to Divine command. This, it will be recognized, is not a scientific conclusion in any sense. It rests wholly upon the basis of authority and is accepted as an act of faith. Though this view cannot be disproved, it does not appeal to the minds of many scientifically inclined persons. Many of them readily admit the probability that the universe is directed by an infinite intelligence, but they insist that for that very reason the whole process of creation must have come about in a regular, orderly fashion and in accordance with the law of cause and effect. The interests of scientists, then, lie not so much in finding out whether there was a Divinity back of the phenomenon as in trying to discover the scientific forces involved and the chain of events followed in producing life. From this general point of view several hypotheses have been proposed.

One of the earliest theories was that living organisms sprang from nonliving matter. This idea has been called the theory of spontaneous generation. The ancients accepted it almost to a man. Even the illustrious Aristotle taught, apparently without question, that worms, insects, and fishes sprang spontaneously from mud. What could be more plausible than the belief in spontaneous generation? Doesn't all stagnant mud and slimy ooze literally swarm with life and aren't all putrefying bodies full of maggots and other organisms? What, then, could be more logical than the conclusion that these living things came from the inanimate matter which they infest? For centuries this fallacy was accepted by many intelligent people. As late as the seventeenth century Alexander Ross, discussing Sir Thomas Browne's doubt as to whether mice may be bred through putrefaction, flays his antagonist thus:

So may we doubt whether in cheese and timber worms are generated, or if beetles and wasps in cow-dung, or if butterflies, locusts, shell-fish, snails, eels, and such life be procreated of putrefied matter, which is to receive the form of that creature to which it is by formative power disposed. To question this is to question reason, sense, and experience. If he doubts this, let him go to Egypt, and there he will find the fields swarming with mice begot of the mud of Nylus [the Nile], to the great calamity of the inhabitants.

Strange to say, too, this old myth, instead of being silenced by the advent of the microscope, was for a time bolstered up by it; for every drop of water examined and every tiny object scrutinized under the microscope swarmed with microörganisms. Was this not further proof that these minute organisms had arisen from inorganic matter? The evidence seemed conclusive. What could be more reasonable? All of which goes to show how much everyone needs the questioning habit of mind, the scientific spirit, toward the things he observes.

Francesco Redi, an Italian physician, in 1668 dealt this fallacy of spontaneous generation, already gray with age, its first blow. By placing beef in a wide-mouthed bottle and tying a piece of fine veiling over the mouth of the bottle, he proved that maggots do not develop in putrefying meat. It is only when flies can get to the meat to deposit their eggs that these fly larvæ appear. Later Lazaro Spallanzani, another Italian, dealt the whole theory a smashing blow. Although trained for the clergy, Spallanzani became so interested in natural science that he never engaged in churchly offices. Following his bent of mind he became a university professor instead and came to enjoy the reputation of being one of the most eminent men of his time. In the course of his studies he became very skeptical of the whole idea of spontaneous generation. He believed that putrefying substances abound with tiny organisms because the original parents get into the substance from some exterior source and then multiply there. To prove his point he took scrupulously clean flasks of meat and vege-

2

table juices, hermetically sealed the necks of the flasks, and then immersed them in boiling water for three quarters of an hour for the purpose of destroying all living organisms that might be contained in them. The flasks were then set away for observation. All the flasks remained clear, not a sign of life appeared, and no putrefaction took place in the broths. From these results this bold investigator concluded that living organisms do not arise spontaneously within nutrient fluids.

Prejudiced minds brought all sorts of objections against Spallanzani's results. Among other things, they claimed that boiling would destroy the generative power of the broth. These claims were refuted when Spallanzani uncorked the flasks in the air and showed that the juice cultures later became filled with microörganisms. For almost a hundred years thereafter questions of doubt and controversy regarding the whole matter continued to arise — another evidence that when error once becomes intrenched, it is hard to rout. Finally, in 1860, Louis Pasteur of France entered the lists and took up his lance against the theory. We shall hear more of Pasteur later. Suffice it to say here that he was a most painstaking investigator. During his study of fermentation he became thoroughly convinced that substances do not become filled with organisms and ferment spontaneously. Minute forms of life appear in decaying matter only when they enter from the outside and multiply there, he declared. To prove his point he prepared a battery of flasks containing suitable broths, each of which was provided with a tight stopper through which was inserted a long capillary glass tube bent in the form of a swan's neck; in other words, the tube upon leaving the cork bent down, then curved up again, and was left open at the outer end. When these flasks of soup were thoroughly sterilized by heating and were set out into the air they did not ferment, as soups in all other kinds of containers did when exposed to the air. The reason was that when the germs entered the outer orifice of the tube, they dropped to the lowest point in the curve and could not "climb" uphill and drop down into the broth. Instead they collected at the bottom of the curve. To demonstrate this fact, Pasteur first skillfully tipped the flasks so as to run a small portion of the soup out into the curve of the capillary neck and then tipped them in the opposite direction to permit the soup to flow back again into the flasks. The flasks were now replaced in warm incubators. The next day the cultures were found to be teeming with minute organisms.

By this experiment and many others which Pasteur conducted he dealt the final blow to the then tottering theory of spontaneous generation. For a short time a few conservatives tried to revive the belief, but they were always overwhelmed with evidence against it. Finally, it succumbed and is now as permanently interred as the ancient Egyptian mummies. Like the mummies, too, when the theory is brought out to the light these days it is exhibited only as a relic of the past — as the desiccated remains of a giant fallacy, so vigorous in its day as to require the growing scientific spirit two thousand years to conquer and dispatch.

Another idea as to the origin of life is that it came to the earth from some other planet. This suggestion was made by Lord Kelvin in 1871. The idea has never gained much credence, however, because it yet remains to be explained how life could have reached the earth from some other celestial body. About the only way imaginable is that it rode in on a meteorite; and this does not seem very plausible, considering the baptism of fire it would have had to undergo on its way in through the atmosphere.

Preyer reversed the point of view and held that lifeless matter came from the living substance. He believed the inanimate to have come from the animate, much as slag separates off when ores are being refined in the smelter. This was a purely philosophical conception, with little if any evidence to support it, and has never been regarded seriously by scientists.

Ernst Haeckel, a German biologist, advanced the idea that life had its origin in the primeval sea. There appears to be some evidence to support this theory. In the first place, living forms seem certainly to have originated in marine water. All evidence gathered from fossil remains supports this conclusion.

In the second place, practically all the forms of simple life today reside in water. Even the individual cells of higher animals are surrounded by lymph, which is a highly watery fluid.

The adherents of this theory believe that in the evolutionary changes accompanying the earth's development such a combination of chemical and physical conditions arose as to produce life; in other words, under these primeval conditions, life by chemical and physical processes sprang from nonliving matter. Whether it emerged in more than one place or not, or whether it only rose once and the original protoplasm subsequently evolved into all the multitudinous, diverse, and complex forms found inhabiting the globe today, the adherents do not pretend to know. In fact, this theory, along with all others relating to the origin of life, must be taken as a hypothesis with very little definite evidence to support it. All scientists do agree, however, that, regardless of how life first originated on the earth, it is not coming into being anew at the present time. Prehistoric conditions may have favored a spontaneous generation of living protoplasm, but that age is long past. No life springs from inanimate matter today.

It is probable that more scientists favor the view of Haeckel than that of any other, yet when they give assent to this idea they admittedly do so as an act of intellectual faith rather than a settled conviction. If pressed for a definite answer regarding this interesting but baffling question, most scientists honestly and frankly turn agnostic and freely admit that they do not know. Many of them believe that there was a great guiding intelligence behind this vital event, but how it was done none would pretend to say. The solution of this mystery may be too difficult for mankind to solve. Nevertheless, one thing is certain: that if the problem is ever solved, it will be through careful scientific investigation and experimentation.

Life has not been adequately explained. This question is another deep biological mystery that the mind of man has not been able to penetrate. The problem seems to have presented itself as soon as man's intelligence arrived at the point where he began to wonder about the nature of the cosmos and himself. The ancient Greeks reflected on the matter, and every cultured people since that time have discussed it. Numerous theories have been advanced, some of which have appealed to the intellect and others have been so fantastic as to be entirely incredible.

During the centuries of discussion, however, the two theories that have challenged the attention of scholars most are those of vitalism and mechanism.

The vitalists believe that the bodies of plants and animals are made up of different kinds of materials and that these must be constantly renewed and supplied with energy to maintain life. But, they contend, this does not account for the whole of life. Something more is necessary. So they believe that to the material body a vital principle of some kind has been superadded. What this principle is or where it came from they do not attempt to explain. Perhaps it could be said that they regard it as an attribute of nature which resides in and manifests itself through the living body.

The mechanists, on the other hand, are just what the word indicates. They believe that the body is a machine and nothing more. This machine must be kept in constant repair and be furnished with energy-supplying foods. When this is done the physical and chemical reactions taking place within the body constitute life. They believe that motion, digestion, growth, reproduction, nervous reactions, and even intelligence itself can all be explained on this basis.

The reader will readily surmise that between two views as diverse as these regarding the nature of life little harmony can exist. The mechanists demand of the vitalists that they show them the mysterious vital principle which they hold to be the real basic cause of life. Failing in this the vitalists are then asked to point out a single action, a single physiological process, or a single response, even thought itself, that is not accompanied by a chain of chemical and physical processes in the body. This, of course, they cannot do; for physical and chemical reactions seem intimately connected with every manifestation of life. Driven into close quarters now, the vitalists are compelled to acknowledge that although they cannot demonstrate a vital principle they feel certain of its existence. The only reason why they cannot prove their position, so they reiterate, is because they do not yet have sufficient knowledge of this vital force.

About this time the vitalists, with some feeling, turn on the mechanists and fire a broadside of disconcerting questions at them. Let us grant, they say, that the bodies of living things, including man, are machines. How are you to explain choice of action? To use an illustration of Will Durant's: Take a toy engine propelled by a spring. Wind up the spring and set the engine down on the floor a foot or two from the wall toward which its course is set. Immediately the engine hits the wall and rebounds. When the force of the rebound is spent, the engine again heads into the wall with the same result as before. Alternate collision and rebound, unless the engine's course is accidentally deflected, will continue indefinitely or until the power stored up in the spring is exhausted. That is the way a machine acts. In contrast take a long aquarium containing paramecia and, midway of its length, set a glass partition part way across. Now place a light at one end and observe the paramecia. Some of them strike the glass plate on their way toward the light. They back up and try again with the same results, just as the toy engine did. But, unlike the engine, after a try or two with the same painful results the paramecia begin to deflect their course and finally approach at such an angle as to pass around the glass obstruction and continue their movement toward the light unimpeded. The vitalists complacently ask their mechanist friends whether they think the engine could have ever done that. They ask other vexatious questions. They want to know when a machine ever started *of its own accord* to do a piece of work as a man does. They ask whether a machine was ever known to reproduce another of the same species. Finally they ask to be shown any machine, regardless of how complicated it may be and how delicately adjusted, that manifests the state of consciousness; that can think, and feel, and will, and act. The only rejoinder the mechanists can make in this dilemma is about the same as that made by the vitalists under similar circumstances. The mechanist replies that the only reason he cannot explain the whole phenomenon of life mechanically is because he does not yet know enough about the complex physical and chemical processes taking place in the bodies of living organisms. If he did, he avers, he could explain all. The vitalist replies that if he knew all about the vital force, he too could explain. In the final analysis it does not seem that the vitalist is relying upon faith to a much greater extent than is the mechanist.

It must be conceded that all that man knows about the bodies of living things, both structurally and physiologically, he has learned through the avenue of physical and chemical processes. Yet there are many things about living organisms that cannot, in any sense, be explained on that basis; indeed, to believe that all physiological processes can be explained on the basis of physicochemical reactions (many of which are so complex that they are not yet understood) requires just about as much faith as it does to believe that some vital principle will be discovered.

For this reason many thoughtful scientists these days are becoming dissatisfied with the extreme views held by both of these schools. They believe that life must be completely explained by some other hypothesis. Each of the old theories may have some elements of truth in it, but neither one contains the whole truth. Emergent evolution, which will be discussed later, has been proposed as one possible alternative.

Life resides in the protoplasm. The vital point in the body where life was supposed to reside was a disputed question among the ancients. For example, in regard to man it was believed by some that the liver was the seat of life. However, as man's knowledge of the body grew this idea became untenable.

Another belief sprang up, to the effect that life resides in the heart. Doubtless the advocates of this theory had noted that the heart was a very vital organ and that when it was pierced by any kind of missile death followed promptly. This idea led these ancients to the conclusion that the emotional characteristics of persons are also connected with the heart.

Relics of this old belief come down to us in expressions such as "big-hearted" and "good-hearted" people.

This position was superseded by the theory that life resides in the brain. It seems that this view rose from the growing knowledge that the brain assumes such importance in the control and direction of the bodily activities. Indeed, the conviction that the brain is the seat of life became so deep-seated with some of the older scientists that they named a part of this organ the "tree of life." Nevertheless, as man's knowledge of human physiology continued to grow it was soon found that there are other vital points in the body as well, and he became skeptical of the proposition that life resides in any one particular part of the body.

Subsequent investigation and the development of physiological and anatomical sciences have proved that this skepticism was well founded, for we know today that life cannot be located in any one organ in the bodies of either plants or animals. Instead, life has finally been found to reside all over the body within little compartments called cells. Inside these cells life seems to be intimately and inseparably connected with a peculiar viscid substance called protoplasm. Since protoplasm and cells (the tiny structures into which protoplasm is organized) are so important in any discussion of life, we shall pause for a time to consider them.

In 1835 Félix Dujardin, a French zoölogist, noticed in his microscopic study of lower animal forms a semifluid, jelly-like substance which he said was endowed with the qualities of life. He recognized a resemblance between it and flesh, and called it *sarcode*, which means flesh-like. This substance had been observed earlier by other biologists, but no one until then had considered it vitally connected with life.

In 1846 Hugo von Mohl, a German botanist, during his study of plant cells, noted this jelly-like substance, or plant *schleim*, as he designated it, and finally called it *protoplasm*. By 1850 there had come to be a feeling among biologists that the sarcode of the zoölogists and the protoplasm of the botanists were the same thing. Eleven years later (in 1861) Max Schultze, after an extensive study of the whole matter, announced that this protoplasm is the living substance and that it is practically identical in both animals and plants. From that time on the term "sarcode" lost ground, and the living contents of the cell became permanently designated as protoplasm. Thomas H. Huxley later clinched the matter by saying that "protoplasm is the physical basis of life."

Protoplasm has definite physical and chemical characteristics. If one wishes to examine protoplasm to determine its appearance and how it acts, there are certain plants that offer special advantages in this respect. For instance, the colorless hairs that are found thickly scattered over petunia stems or the purple-colored hairs that cover the filaments of the spiderwort plant are excellent for this purpose. When these hairs are properly mounted under the microscope, the protoplasm may be found to occupy characteristic parts of each tiny cell.

One who was taking his first look at this material expression of life would naturally expect to see something rather striking in appearance — something entirely different from anything he had ever seen before. He might even experience a sense of the uncanny. If the observer had feelings of this sort, one look through the microscope would disillusion him. For protoplasm is not at all imposing in appearance. On the contrary, it looks very ordinary, resembling the white of an egg, although it is not quite so clear and homogeneous. To use another figure, it looks very much like a soft, colorless jelly with granular products, or food materials, interspersed through it. It is elastic, like a gelatinous substance, and, if deformed by exterior means, tends to resume its former shape when the distorting force is removed.

Chemically protoplasm is said to be a colloid, the name coming from the Greek word *kolla*, "glue." Colloids of the type to which protoplasm belongs consist of liquid globules of one kind suspended in another liquid. Cream, with its butterfat globules interspersed through the milk, is a colloid of this type. Colloids of this kind are often called emulsoids. Instead of being a simple emulsoid, like tiny drops of oil thickly dispersed through water, protoplasm is a very complex one; indeed, it is so complicated in its composition and behavior that our best biochemists are still baffled when it comes to the matter of explaining its real structure and just how it carries on its activities. We do know, however, that protoplasm has droplets of carbohydrates, proteins, fats, and often other substances suspended in its more liquid portion, the whole forming an amazingly intricate colloidal system.

One of the strange facts about protoplasm, as has already been stated, is that all protoplasm, regardless of its source, seems to be very similar in chemical composition and appearance. While there are doubtless always some differences between protoplasm from different organisms, yet that which comes from an orange, for example, appears very much like that which comes from a piece of beefsteak. It is difficult to get at the chemistry of protoplasm, because it always has to be killed before it can be analyzed, and dead protoplasm is certainly very different in many ways from that which is living. For instance, we are told that the biceps muscle of a man's arm can sustain a much heavier weight when living than it can after death.

Although chemistry so far cannot give us definite information as to the structure of protoplasm, it can give us rather reliable data as to the particular elements combined to make up its basic part. All analyses show carbon, hydrogen, oxygen, and nitrogen to be present. In addition sulfur and phosphorus are almost always found. Besides these six elements chemical analysis always reveals the presence of others. But there are no others that seem indispensable.

Formerly, in connection with the question of life, much more importance was attached to the exact composition and structure of protoplasm than is accorded them today. For it used to be believed that life depended largely upon what protoplasm is made of and how its constituents are put together. Now life is looked upon as being not so much the result of a chemical state as of chemical and physical activity. The cell exhibits life not so much because of what it is made of as because of what it does. In other words, life is the expression of a set of dynamic changes going on in the protoplasm rather than the result of a static composition.

In connection with the activities of protoplasm, water is always an important constituent. Extract all the water from protoplasm, and it dies immediately. As plants pass into the dormant state parts of them dry down greatly in order to meet the rigors of winter without injury. Seeds and buds always do this. But they never lose all their water. Kernels taken from an ear of corn that had hung in a dry laboratory for two years were still found to have 12 per cent of water in them. When protoplasm is in the active state it always has a much higher percentage of moisture. It has been found that seeds cannot become active in germination until the water content of their growing parts has reached about 70 per cent. This explains the delay in growth when seeds are first planted. Before they can sprout they must absorb enough water from the soil to permit the protoplasm to become active. In some plants the moisture content is much higher, reaching from 90 to 95 per cent. A luscious watermelon may have 98 per cent of its weight composed of water. Active animal bodies always have a high water content. Muscles are about 75 per cent water, and the very aristocrat of tissues — the gray brain cells — is 84 per cent water. Even living bone tissue is about 25 per cent water. Dr. D. T. MacDougal, an eminent student of plant life, believes that something like 98 per cent of the increase in volume due to growth in plants is produced by the absorption of water.

Protoplasm manifests definite physiological properties. Although the chemical and physical properties of protoplasm are important, the aspect of protoplasm that holds most interest is what it does because it is living; in other words, its physiological activities. It is at this point that the mysteries of life present themselves in their most perplexing form.

Fundamental activities of protoplasm are growth and repair. It is the nature of living organisms to begin life as infants and, with proper nourishment, grow until they reach the adult state. Even one-celled plants and animals grow to some extent after they are formed by cell division. Living things also wear out the protoplasm that comprise their tissues, and this has to be replaced to keep the body in a healthy, vigorous condition. Sometimes injuries occur and wounds have to be healed. In some instances whole organs even, such as a leg or the tail, are lost by violence and have to be replaced by regeneration. Though these processes all have elements of difference, yet they all have one feature in common: that living protoplasm has to be produced. How this is done by the organism is an unfathomed secret of nature. But with a certainty and skill that excite our profound admiration, living cells go about their daily task of building up protoplasm from lifeless food and keeping the body repaired and functioning like a smoothly running machine.

It is the protoplasm also that liberates the energy from food to keep the body going. Every living thing uses some energy. Warm-blooded animals use relatively large quantities of it; for they not only must be supplied with energy to do their work, but must also be kept warm. Now to get energy from coal or wood by placing it in the fire box of an engine and burning it is an easy matter. The greater part of our industrial power is derived in this way. But to release energy from liquid food within the cells and to do it at a temperature so low as not to damage the delicate tissues of the organism is quite another thing. Yet protoplasm can do this. In some way, by employing certain chemical substances called enzymes, it can produce chemical changes at a temperature so low as not to injure the tissues and still keep flowing a sufficient stream of energy to meet all needs of the organism.

In this connection we must pause to introduce two or three terms with the meaning of which everyone should be acquainted. Plants take the elements of earth and air and elaborate them into foods to sustain their bodies. From these foods animals as well as plants build up their bodies and derive their energy. The processes involved in all these transformations and changes are many and complex. Nevertheless physiologists employ a term to cover them all. This term is *metabolism*. All building-up processes, such as food-making, tissue repair, and growth are more specifically designated as *anabolism*, and the destructive organic activities concerned in the "wear and tear" of the body and the chemical decomposition of food to secure energy are called *katabolism*. From what has been stated it is evident that *metabolism* is the more general term and includes the meaning of both the others.

Another remarkable power of protoplasm is sensitivity, or irritability. A signpost may stand in the earth for years or a bowlder may lie on a hillside for centuries without the least indication of a single response to its environment except the chemical and physical changes incident to weathering and decay. These objects may be pounded, illuminated, warmed, chilled, and drenched with rain, yet they remain perfectly inactive. There is not a single reaction to the environmental forces in any way. It is entirely different with a living organism. Allow light to strike the protoplasm of a plant, and its leaves bend toward the source of illumination; warm the chilled and stiffened larva of a mosquito, and it becomes a lively wriggler; place fertilizer at certain places in sterile soil occupied by plants, and their roots become impacted at these points to secure the additional nutriment; let a brass band strike up a lively air, and small boys follow it in procession along the streets. All forces in nature that can affect living things become stimuli to living protoplasm, and the protoplasm in turn has the power to respond to them. All higher animals have especially sensitive portions of their bodies to enable them to receive these incoming stimuli with ease and rapidity. These specialized structures are called sense organs. Living things respond to their environment in a thousand different ways every day, depending upon the kind of stimuli received. It is only organisms whose protoplasm can respond appropriately that have the ability to survive and to succeed.

What is true of organisms in general regarding the sensitivity of protoplasm is doubly important in the case of man. Not only must he make suitable physiological responses to his environment to live, but the whole of his mental processes also depends primarily upon this power. One cannot think unless he has something to think with. He sees an orange, and certain stimuli, or perceptions, such as its color and shape, come to him; he feels the orange and gets another impression

— this time as to its texture; and so on for its smell and taste. Now, with these raw materials for the brain cells to react upon, man is able to connect them and form a concept as to the nature of an orange. In this way he gets the concrete idea of all other objects in his world. Later he learns to read, and from the stimuli furnished by the printed page he gets still other concepts. He also learns to understand the spoken word and, through his auditory sense, gets materials of thought in this way. After he has accumulated all these protoplasmic stimuli from various sources, he learns to put them together and to associate them in a multitude of ways. So it is these sense impressions, these protoplasmic stimuli, to be explicit, together with their derived products, that form the basic materials from which long trains of complicated thought may be built up. It is for this reason that a wide and varied experience is so important in the life of the well-educated man. The broader his experience through travel, reading, and conversation, the richer stock of stimuli he has to be used in elaborating his thought. Teachers and parents who wish to achieve the most in developing the mental power of the boys and girls under their care should never overlook this fundamental biological principle.

Reproduction is a very important physiological function of protoplasm. Age, senility, and finally death is the fate of every individual living thing. The only organisms that can evade it are the unicellular ones, and they must do it through cell division. So the one way that either plants or animals can escape physical extinction is through reproduction.

Reproduction is of two kinds: asexual and sexual. In asexual reproduction a piece of the original parent's body, such as a cell, a spore, or a bud, becomes detached and grows under favorable conditions into another individual of the same kind. It is akin to the process of cell division involved in growth in multicellular organisms, except that in asexual reproduction the cells or other reproductive structures become separated from the parent and thereafter lead an independent existence. In sexual reproduction a simple-looking bit of protoplasm, the egg, receives material from a different bit of protoplasm, the sperm, and then, if properly nourished, grows into a new organism like the parents.

This ability to reproduce itself is one of the most remarkable of all the characteristics possessed by protoplasm. It is an extraordinary thing for living bodies to be able to take inanimate food and transform it into living, active stuff to promote growth and repair; but when one individual can evolve a microscopic bit of protoplasm and endow it with such powers as to make it capable of being separated from the parent's body and of beginning a separate existence of its own, the wonder grows. There is no greater achievement of nature in the whole biological field. Besides this, another striking aspect of reproduction is the fact that like begets like. One may take two eggs, one of a rose and the other of a human being, and compare them. The first is microscopic; the latter, under the most favorable circumstances, can just be seen with the unaided eye. They are both naked masses of protoplasm and, when magnified, look very similar. Yet innately how different they are! for one, true to its inherent qualities, always produces a rosebush, while the other normally never fails to produce a human being possessed of all his high physical, mental, and spiritual powers.

Protoplasm is organized into cells. It was not until after the invention of the microscope that biologists suspected the existence of living cells. About 1665 Robert Hooke, an Englishman, made a microscope and was greatly elated over its power to magnify objects. One day while testing it out he was examining all sorts of things to see how large his instrument made them appear. It happened that he picked up a piece of cork, cut off a thin shaving, and placed it under the objective. He was surprised to see a group of little compartments, which he called "cells" because of their fancied resemblance to the cells of honeycomb. As everyone knows, cork tissue as used for corks is dead, and what Hooke really saw was the walls of the cells instead of the much more important protoplasmic contents. Today we know that all the vital processes are connected with the protoplasm and that the cell wall's chief function is to provide a protection for the more delicate parts within. To call one of these units of structure a cell, then, is a misnomer. Nevertheless Hooke's term caught the fancy of men at that time, and the name became fixed through tacit consent and custom.

It was not until much later that the full importance of Hooke's discovery became apparent. In 1838 a German botanist by the name of Schleiden became interested in plant structures and made a rather extensive study of them. He was impressed by the fact that, no matter what kind of plant he used or what part he took, the tissues always presented a series of tiny cells. About the same time a German zoölogist named Schwann had noted in his studies that the same conditions of cellular structure held for animal tissues. It happened that these two men became acquainted and were dining together one October day in 1838 when they fell to comparing notes in regard to the observations they had each made in the course of their researches. They were both struck with the similarity of the tiny compartment structure which each reported for the kind of tissues he had been studying. So impressed were they that after dinner, according to Locy, they went to Schwann's laboratory to examine the section of animal tissue which the latter had been working on. Schleiden immediately recognized the striking similarity of the nuclei in this section of the dorsal cord and in the plant tissues he had been studying. Schwann came to the conclusion that the elements of structure in animal tissue were practically the same as those in plants. Subsequent study confirmed their ideas, and in 1838 Schleiden published the cell theory for plants. His statement was to the effect that the living parts of all plants are composed of tiny structures called cells. In 1839 Schwann, who was the calmer and more retiring of the two but also, as often happens, much the better scientist, published the cell theory as applied to animals. This discovery as to the cell structure of living organisms was one of the most important ever made in the whole field of biology. One writer has gone so far as to say that, excepting Darwin's theory of evolution, it was the greatest biological discovery made in the nineteenth century. Today we know the cellular structure of living organisms to be a fact. It is no longer a theory. Yet custom often prompts us still to refer to the idea as the cell theory.

Cells vary widely in different organisms and also from part to part of the same organism. For this reason it is practically impossible to describe any one cell and say that it is typical of all cells. For our purpose it will be sufficient to portray a particular type of plant cell and leave the many exceptions to more specialized discussions. Let us take a cell from the hair of a petunia stem.

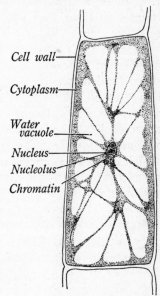

FIG. 1. Cell from a Hair on a Petunia Stem

Because this cell was not subjected to distorting pressures during growth, as cells within stems are, it is observed to be more or less cylindrical in shape. It has a wall which upon examination is found to be composed of a non-living substance called cellulose. The paper upon which this book is printed is chiefly cellulose. The principal function of the cell wall is to incase and protect the delicate protoplasm within, much as the shell of an oyster houses and preserves the oyster's soft, tender body.

Inside the cell wall the protoplasm is differentiated into two main parts: the cytoplasm and the nucleus. The cytoplasm comprises all the protoplasm between the nucleus and the cell wall. It completely lines the inside of the wall, much as the lining of a garment does, and if the cell is vigorous, it usually forms strands extending in from the protoplasmic lining to the nucleus near the center. Under favorable conditions this cytoplasm is active and may be seen streaming in currents about the cell wall or along the nuclear strands. As one watches it under the microscope he experiences a strange feel-

2

ing; for here is stuff almost as simple in appearance as the white of an egg, yet exhibiting that most mysterious and baffling of all attributes — life. The question as to how it does it insistently arises. Yet, with all man's wisdom, there is no answer.

Within the cytoplasm granules of various kinds exist and cavities of different sizes may be found. The granules are often food or waste particles. The cavities at first were believed to be little vacant spaces, and from this mistaken idea were given the diminutive term *vacuole*. Later investigation revealed that the original idea as to the nature of a vacuole was entirely wrong. Instead of being a vacuum it is a cavity which is filled with a watery fluid called cell sap. The vacuoles seem to be very important structures connected with the metabolic processes of the plant. In addition to the water, which is all-important, the vacuoles may contain mineral matter from the soil which the plant has not yet used, they may contain elaborated food materials, and they doubtless often include waste products from the surrounding protoplasm. One eats an orange chiefly for the abundant supply of water, acid, sugar, and other organic products stored within the large vacuoles found in the cells of this fruit. In older cells the protoplasmic strands supporting the nucleus in its central position often disappear, and the nucleus settles down to one side of the cell. In this case, then, the whole central part of the cell may be occupied by a single large vacuole.

Though many important processes are doubtless carried on by the cytoplasm, the nucleus seems to supersede it as a center of activity. Someone has called the nucleus the headquarters of the cell, which, considering the rôle it takes, is not an inappropriate term. The nucleus in higher plants is more or less spherical and has three main parts: the nuclear membrane, the nucleolus, and the chromatin network. The nuclear membrane is composed of protoplasm and incloses the rest of the nucleus. Within the nuclear membrane is found a structure somewhat round in appearance, which readily takes a nuclear stain. This tiny body is called the nucleolus. Its function in all cases is not positively known. Embryonic cells or growing plant-tissue cells may have two or more nucleoli within the same nucleus. Apparently the most striking feature of the nucleus, however, is the chromatin network. This structure in prepared, stained cells, as its name indicates, appears to take the form of a net which is stretched out through the periphery of the nucleus and consists of delicate fibers of linin to which are attached granules of chromatin. This nuclear net is not distinctly visible until after the cell is stained with a dye, such as safranine. In the latter case the granules take on a beautiful red color; it is because of this characteristic that they are called chromatin (Greek *chroma*, "color"). The chromatin assumes its chief importance because it is now known to carry the inheritance factors; in other words, the hereditary characteristics possessed by any organism depend wholly upon what kind of chromatin was transmitted to it in the reproductive cells received from the parents. In addition to the nucleolus and the chromatin network the rest of the nucleus is occupied by a watery fluid called the nuclear sap.

Aside from carrying the inheritance determiners, the nucleus seems to exercise a directive influence over the whole cell and its activities. When cells divide, the nucleus assumes the most important rôle; in fertilization the nuclei of the fusing cells again take the leading part; and, with very few exceptions, if cells are deprived of their nuclear contents they die immediately. There is a certain protozoan called the *stentor* with a long chainlike nucleus. Gruber found that by mounting these one-celled animals on a microscope slide and gently rocking the cover glass back and forth, he could often break the cell up into pieces some of which contained nuclear fragments and some of which did not. Many of the cell portions containing nuclear matter would each round up and regenerate into a new individual. But all the pieces wholly deprived of nuclear fragments, although they might be relatively large, promptly perished. The mature, tubular, elongated cells which, placed end to end, form the food-carrying tubes of higher plants normally have no nuclei, but they are believed by some botanists to receive the necessary nuclear stimuli from certain companion cells which are always found accompanying them. Some unicellular organisms, such as bacteria, have no organized nuclei, but they are all believed to have nuclear matter diffused through the cell.

Before closing this general discussion of cells we must again emphasize the fact that there is no such thing as a typical cell. Some cells are long and some are short; most of them have walls, but a few do not; many have chlorophyll, but the greater number of them are devoid of it; some of them have heavy, resistant walls, and others have walls so thin and delicate that they can scarcely be detected; most animal cells have centrosomes, and most plant cells do not; and so the enumeration of variation runs. Yet all living cells have one thing in common, and this common character is more important than the differences that exist: they all have protoplasm, which is the seat of life and the center of all their activities.

Tissues may live outside the body. In closing this chapter it will be of interest to trace briefly what appeals to the authors as being one of the most extraordinary experiments ever carried on with living tissue. It is being carried out at the Rockefeller Institute in New York City in connection with a study of that dread disease, cancer, and is throwing much light on the nature of life and protoplasm.

So far as present knowledge goes it appears that cancer is caused by an abnormal growth activity of what is otherwise more or less normal tissue. During the period of youth, and when healing an injury, growth seems to be a normal activity of embryonic tissue. But under these conditions the rate of growth is temperate. The enlargement of the tissue is restrained in some way compatible with the supply of food available and the welfare of the surrounding tissues. In the case of cancer, however, the situation is different. The tissue at the site of the cancer begins to grow in an abnormal manner. All restraint appears to be off, and the cancer tissue grows at an inordinate rate. It does not confine itself to the available food supply, but attacks the surrounding tissues and consumes their substance much after the manner of a parasite. As

Paul de Kruif has said, in his breezy but vivid style, the cancer cells seem to be tissue that has gone on a "joy ride."

At the Rockefeller Institute it was conceived that since growth is such a striking characteristic of cancer tissue, it might be well to learn all that could be learned about growth as a normal physiological function. With this object in mind Dr. Alexis Carrel, on January 17, 1912, deftly took a piece of living tissue from the heart of an embryonic chicken and placed it, together with a supply of nutritive juice, in a warm, moist incubator for observation. For a time the heart pulsation continued, and the fleshy mass exhibited behavior perfectly normal for excised tissue. Then it began to slow down and beat more feebly. Just before it stopped it occurred to Dr. Carrel and his assistant, Dr. Ebeling, that it might be the accumulation of waste products which seemed to enfeeble it and slow it down. To remove this difficulty, they gave the piece of tissue a bath in a saline solution and immediately replaced it in the incubator with a generous supply of nutritive material. To their great delight the mass again began to beat vigorously and exhibit all the characteristics of young, vigorous chicken-heart tissue. But a more interesting thing was to follow. The tissue actually began to grow by producing new cells and enlarging them to normal size.

The experiment continued. Whenever senility began to manifest itself and the tissue slowed down in activity and growth, it was again washed and fed, which restored its normal vigor. It was found that the tissue could be kept in an active state by cleansing and replenishing the nutritive material every forty-eight hours. Growth was so rapid, too, that it was found necessary to cut off and discard half the whole mass at the end of each forty-eight-hour period. Not only did the tissue thrive, but it exhibited a remarkable longevity. The average life of a chicken is said to be two or three years and the extreme age limit to be seven years. But this famous piece of chicken-heart tissue has beat all records for longevity; for although more than twenty years have passed, it still remains as active and vigorous as when removed from the chicken that memorable January morning in 1912. Dr. Carrel says that with continued care there is reason to believe that it would live a thousand years or, for that matter, be immortal. Had it been possible to leave the tissue intact and to keep it growing at its normal rate, its volume now would have been as large as the earth. How long the experiment will be continued is not known. It requires absolute faithfulness and regular attention on the part of the attendant. For this reason it is probable that it will be discontinued whenever Dr. Carrel feels that he has learned all of value that can be derived from the study. Excised tissue of other organisms, such as rats, guinea pigs, and even man, has been tested and found to behave in a similar manner.

Dr. Carrel's investigation seems to indicate that life and protoplasm will remain indissolubly associated for an indefinite period of time unless the tissue suffers an accident, or becomes diseased, or is unable to throw off the toxic waste substances that accumulate with age as a result of normal metabolism. If man could learn to stave off all these eventualities he might not only extend his period of vigorous physical and mental activity but even discover a way to prolong life itself.

QUESTIONS FOR STUDY

1. Make a list of the theories as to the origin of life and give the reasons why you believe or disbelieve them.

2. Contrast the vitalistic theory with the mechanistic theory.

3. Give in outline form the attributes of life.

4. In what substance of the body does life reside?

5. Make a list of the men who were responsible for our present ideas of protoplasm and state the contribution made by each.

6. What is the cell theory? Of what value has it been to man?

7. Name the parts of a representative cell and give the function of each.

8. Of what value to man may the experiment with the embryonic chicken heart tissue prove to be?

9. Why is it important that man should have some knowledge as to the nature of life and where it resides?

10. What contributions did Redi and Spallanzani make to the growth of biological knowledge?

11. Give a brief account of the experimental work of Pasteur in overthrowing the theory of spontaneous generation.

12. Discuss the reaction of protoplasm to different stimuli.

REFERENCES

BURLINGAME, L. L., HEATH, H., MARTIN, E. G., PEIRCE, G. J. General Biology (second edition), pp. 13–27.

CRAMPTON, H. E. The Coming and Evolution of Life, pp. 1–14.

HERBERT, S. The First Principles of Evolution, pp. 43–48.

LOCY, W. A. Biology and its Makers, pp. 237–252, 276–293.

WELLS, H. G., HUXLEY, J. S., and WELLS, C. P. The Science of Life, Vol. I, pp. 1–23.

CHAPTER II · Plants are the Food-Makers of the World and, as Such, are Adapted to Succeed in their Environment

Introduction. There are probably upward of two hundred and fifty thousand different species of plants now definitely known by botanists. The number is constantly growing as new species in the less-frequented and more remote parts of the earth are being discovered. Plants live everywhere that they can get a foothold — on rocks, in the water, under dense shade, in open fields, and on the sandy, wind-swept desert. Naturally, under these conditions, different plants have an almost endless variety of structure and life habits. In complexity they vary from the simplest threadlike forms that grow in water to the most highly differentiated composite, such as the dandelion. Fundamentally they all do the same things: they must obtain food, assimilate it, grow, and reproduce. However, they differ widely as to how they carry on these functions and as to how they are built to achieve them.

So, then, in talking about the plant as a living, working thing we shall have no particular plant in mind. Rather, what we shall say might apply to any plant of normal habits, such as those that make up field crops, the flowers that adorn lawns, or the vegetables that grow in gardens.

Plants and animals differ as to mobility. The two general groups, plants and animals, differ widely with respect to mobility. There are a few aquatic plants, such as the simple diatoms and other algæ, that move about from place to place. There are a few animals which are not capable of locomotion. These are represented by such forms as the troublesome barnacles that cement their shells to ship bottoms, or the polyps that build up coral reefs. But as a rule plants are stationary and animals are mobile. This contrast in habit makes a wide difference in their general form and appearance. The more

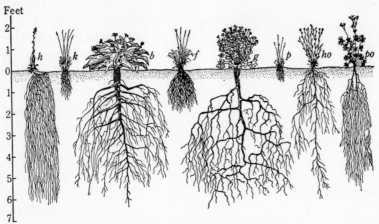

Fig. 2. Schematic Representation of the Size Relation between the Roots and Tops of Certain Prairie Plants of Southwest Washington[1]

complex animals have legs or wings or both, with muscles to move these appendages and nerves to control them. The higher types of plants have no such organs. On the contrary, once established, most plants stand stock-still all their lives in the same spot. But they must have food-making materials from both the soil and the air. So, not being able to walk or fly, they must have some means of obtaining these supplies. In this necessity we find the secret of the general form which plants assume.

Below ground they divide their bodies up into roots, which ramify outward and downward in all directions. To make contact with all particles of available soil, these roots divide and subdivide until their ultimate extremities are almost microscopic in size. The distance penetrated and the volume of soil reached in this way are surprising. Few people have any idea of the extent of the ordinary plant's root system. Weaver, of the University of Nebraska, has given us in Fig. 2 an idea of

[1] From "The Ecological Relations of Roots." *Bulletin No. 286*, Carnegie Institution of Washington, D. C. By permission of the Carnegie Institution.

the comparative size of tops and root systems of some of the prairie plants of southeastern Washington. Fig. 3 is the picture of a mature corn-plant root system excavated in eastern Nebraska by one of the authors and drawn to scale. Note that the squares represent square feet.

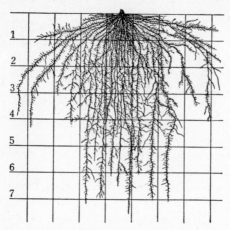

FIG. 3. Corn Root System excavated when Mature and drawn to Scale[1]

The necessity of obtaining the gases of the air and the sunshine is the main reason that the above-ground parts of plants take the form they do. By developing branches and twigs, much more room is provided for the attachment of leaves. In this way too the leaves themselves are brought into contact with the largest possible volume of the required aërial gases as their molecules are either blown or slowly diffuse through the air.

The roots of plants are structurally adapted to perform their functions. The lowest types of stationary plants have no roots at all; they simply cling to rocks by gelatinous walls or holdfasts. One who has visited the seashore has doubtless found the big kelps attached so firmly to the rocks by tough holdfasts that he was unable to pull them loose. Mosses have tiny threadlike structures that penetrate down into the humus. They doubtless help the plants to get some moisture and, perhaps, a little mineral matter, but structurally they are not true roots in any sense. In fact, botanists call them *rhizoids*, which means "rootlike." It is not until we come to the ferns that we find true roots; and from that group through the

[1] From "The Development and Activities of the Roots of Crop Plants." *Bulletin No. 316*, Carnegie Institution of Washington, D. C. By permission of the Carnegie Institution.

conifers and flowering plants, all species, with the exception of some of the parasites, possess them.

Plant roots have more than one duty to perform. One of their chief functions is that of anchorage. It would be quite impossible for most plants to maintain their upright position were it not for the fact that they are securely anchored to the soil and braced in all directions by lateral roots. It is doubtful whether an engineer ever supported a tall chimney more effectively with guy-wires than roots anchor trees to the earth. To gain an idea of the efficiency of roots as anchorage organs among field plants, one has but to watch a full-grown grain-field or a cornfield as it sways back and forth during a wind storm.

Some plants use their roots as reproductive organs. Sweet potatoes are nothing more than the fleshy roots of this plant, designed for reproduction. Dahlias also are propagated by their thick, fleshy roots.

Roots may be effective storage organs too. Practically all plants that live more than one year store large quantities of food in their roots and other below-ground parts to be used early the following spring to promote growth. Annuals like-wise often store food temporarily in their roots. Radishes are edible only because of this temporary storage practice. As soon as their substance is removed to promote develop-ment of the above-ground part of the plant, the root becomes pithy and unpalatable. Beets, carrots, turnips, parsnips, and all other root vegetables owe their value to this habit of root storage.

By far the most universal work of roots, however, is that of absorption and conduction. The true roots of all plants have this work to do, and for this reason we wish to call special attention to this function. The main roots of plants divide and subdivide as they grow out in all directions through the soil. In this manner they obtain water and mineral matter. The fine root endings are hairlike in dimensions. A short way back from the tip, these fine termini are furnished with a dense growth of very short, woolly-looking root hairs to add to their absorbing surface. The walls of these root hairs are very

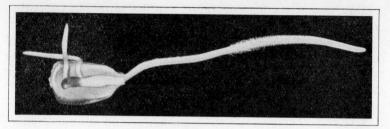

FIG. 4. Corn Seedling showing Root Hairs

thin, — much thinner than tissue paper, — and they absorb the available water and dissolved minerals from among the soil particles with great ease. Plants such as field crops and garden vegetables could not thrive if they were not furnished with root hairs. They could not obtain sufficient water and inorganic food-making materials. It is estimated that field corn has its absorbing area increased from five to twelve times by the production of these root hairs.

Land plants get all their chemical salts from the soil. The principal inorganic salts required by the plant are those containing nitrogen, phosphorus, sulfur, potassium, magnesium, calcium, and iron. Under certain conditions any one of these may become too deficient for good plant production, but the two of which the soil is most likely to become depleted are nitrogen and phosphorus. To keep the supply of these two elements adequate in soil which has long been subjected to cropping requires scientific soil management in which the crops are properly rotated. A legume, such as one of the clovers or field peas, is included in the rotation to maintain the nitrogen supply. When the phosphorus supply in soil is exhausted, there is no other way to replenish it except by the application of a fertilizer containing this element. This may be done by adding barnyard fertilizer, bone meal, phosphatic rock, or some other form of phosphoric compounds. Nitrogen also may be added in the form of a fertilizer.

In the Eastern states thousands of acres of agricultural lands that grew profitable crops during colonial times have become so exhausted of phosphorus and nitrogen that they cannot be farmed profitably at present and have been abandoned as farm lands by their owners.

A most interesting aspect of the behavior of roots is their sensitivity to the force of gravity. When a seed, such as a bean, germinates in the soil, the first sprout that pushes out immediately turns downward into the earth. This response of the root to the force of gravity is called *geotropism* and means "to turn toward the earth." Considering the work that roots must do, the advantage of this habit is obvious. This downward course of the primary root cannot be easily prevented. Should one dig up a newly sprouted bean, turn it so as to leave the root pointed upward, and again cover it with earth, the root will not grow upward through the soil. On the contrary, it will bend within a few hours so as to reverse its direction of growth completely and continue again to penetrate downward into the soil. This seemingly insignificant characteristic of roots usually escapes our attention altogether, yet it is of the greatest importance. If roots did not exhibit geotropism, agricultural production, on anything like the scale we know it now, would be impossible; for every seed, regardless of how small it might be, would have to be placed in the soil "right side up with care" so that when it germinated, its roots would be headed downward into the earth.

After the dilute soil solutions are absorbed from the soil by the root hairs, they are passed from cell to cell across the fleshy part of the root to special water-carrying tubes, called tracheary tubes, which are located near the center of the root. The solutions are conducted upward in these tubes, supplying the cells along the way with water, but most of them finally reach the leaves, where they are used by the plant. The stems are specially fitted by reason of their structure to form this connecting link between the roots and the leaves.

Stems are structurally adapted to serve the plant in many ways. The first real stems, like roots, were introduced into the plant world by the ferns and their relatives. The mosses manifest some tendencies in this respect, but their upright parts are very simple in structure and do not possess many of the characteristics found in true stems.

When it comes to the functions of stems, they are first of all supporting organs. They spread the leaves out and lift

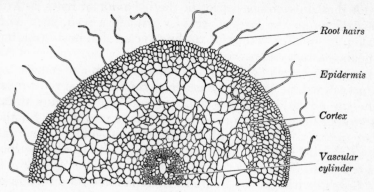

FIG. 5. Cross Section of Root to show Root Hairs, Epidermis, Cortex, and Vascular Cylinder

them to the air and sunshine so that they may carry on their metabolic processes to the best advantage. Stems also support the flowers above the ground, where they may attract pollinating insects or where the pollen, if it is dry and dusty, may be caught up by the wind and scattered broadcast to other plants of the same kind. Stems likewise hold up the fruit to the sunshine and near the leaves, from which the fruit receives its abundant supply of food. In this elevated position, fruits when mature are borne where they may most advantageously attract the eye of birds, catch into the hair or fur of animals, or be wafted about by gusts of wind.

But if stems are supporting organs they are also separating organs; for it is obviously impossible for stems to hold the leaves, flowers, and fruits above the ground without at the same time carrying them a corresponding distance away from the roots, which supply the necessary water and mineral matter to sustain them. For this reason stems must also be conducting organs through which all supplies from the soil pass upward to the more peripheral parts of the plant.

Stems are also very important storage organs. During the summer months, when food is being manufactured rapidly by plants, much of it is deposited in the bark and storage tissues of wood, pith, and medullary rays. This will either be used later that same season, as in the case of annual plants, or will be drawn upon to support early growth the next spring. For example, in corn plants it is a surprisingly short time after shoots and silk appear until the kernels are in the roasting-ear stage. This rapid development is made possible only because a great quantity of sugary sap is stored temporarily, during the earlier summer months, in the watery pith of the stem. As soon as the kernels form, this reserve food is quickly withdrawn from the stem to supplement that going to the ear from the leaves, and the rapid maturity of the seed is thus accomplished.

Sugar and sirup are secured from the sweet sap expressed from the stems of sugar cane and the sorghums. The whole maple-sugar industry is based on the fact that most of the maples store large quantities of carbohydrate foods to be used in growth early the following year. When this food is changed to soluble sugars and is flowing through the stem in the early spring, the sap containing the sugar is secured by tapping the tracheary tubes in the trunk and letting it flow out into containers. The sap is then gathered and concentrated by evaporation to either the sirup state or the sugar state by the application of heat. All trees store large quantities of food temporarily in their stems and twigs. It would astonish the average person to see the large quantity of starch stored at the end of the season within the body of any ordinary deciduous tree.

Plants vary widely as to stem structure; so as we discuss them we shall have in mind only such types of stems as are found in common, familiar plants. If one were to cut a cross section of a mature cornstalk or almost any other similar monocotyledonous stem, he would find a hard outer covering of fibrous tissue containing a pithy portion within. Throughout the pith he would find a number of what appear to be denser spots scattered over the face of the section. These scattered organs are the cut ends of brownish threadlike structures that run up and down through the stem, coalescing

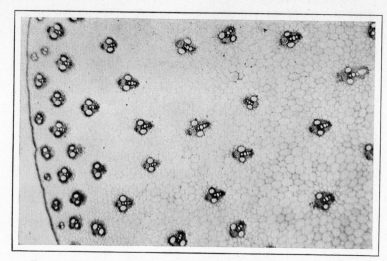

Fig. 6. Cross Section of a Corn Stem showing the Arrangement of
Fibrovascular Bundles and Pith

freely at the joints, many of them ultimately leaving the stem
to pass out into the leaves, where we recognize them as veins.
These longitudinal threads are the principal organs of conduction, both of food and of water, and are also very important in helping to form the skeleton of the plant. They are
called *fibrovascular bundles.* If the leafstalk of the great plantain be pulled apart carefully, the fibrovascular bundles may
be pulled out intact as long "strings." Celery sometimes is
tough and stringy. In such cases the "strings" pulled out of
the celery stalk are fibrovascular bundles that have become
tough and fibrous because of unfavorable weather conditions
or an insufficient supply of water.

When the inner side of these bundles is examined more
minutely, it is found to contain two, three, or more large
tracheary tubes which, as we have stated, are the chief
water-carrying tubes. These tracheary tubes are embedded in
and surrounded by tough fibrous tissue. The part of the bundle
toward the outside of the stem has a light-looking area composed mostly of sieve tissue, whose chief function, as many

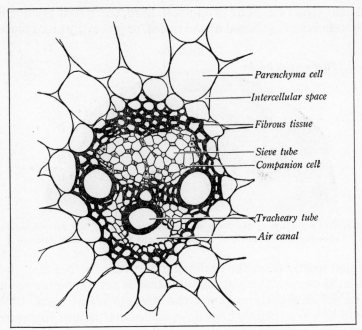

Parenchyma cell

Intercellular space

Fibrous tissue

Sieve tube
Companion cell

Tracheary tube

Air canal

FIG. 7. Fibrovascular Bundle of a Corn Stem
Highly magnified

believe, is to carry elaborated food lengthwise through the stem. The sieve-tube region, like the tracheary tubes, is also inclosed by fibrous tissue. These threadlike strands, then, are composed chiefly of elements for conducting water and food, embedded in and surrounded by tough, strength-giving fibrous tissue. It is for this reason that the whole organ is called a fibrovascular bundle — a term which will be recognized as being most appropriate. The portion of the bundle lying toward the inside of the stem and containing the tracheary tubes is known as the *xylem*; the part facing the outside of the stem and containing the sieve tubes is called the *phloëm*. The pith of the stem, as has been found, is chiefly for storage; it also serves to some extent as conducting tissue, especially when food is to be carried crosswise through the stem.

It may now be understood that the corn stem is a very

2

efficient structure. The cylindrical, hard, outer bast portion of
the stem is strengthened at the nodes, or joints, by cross plates

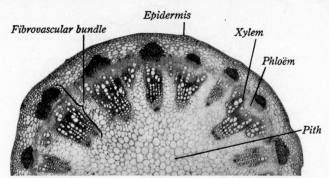

FIG. 8. Cross Section of the Stem of the Ragweed, *Ambrosia artemisiœfolia*,
showing the Structure of the Herbaceous Dicotyledonous Stem

formed mostly from the coalescing fibrovascular bundles. Between the nodes the long bundle strands act as so many taut
wires which greatly augment the strength of the stem. It is
a question as to whether our most skilled mechanical engineers,
with the same amount of material at their disposal, could con-
struct a stem as efficiently as nature has done it.

Dicotyledonous plants, almost all of which have netted-
veined leaves, have a stem structure that is different from
that of the monocotyledonous plants. In the herbaceous
dicotyledonous stems, represented by such plants as the bean,
the pea, the sunflower, and the cotton plant, the fibrovascular
bundles are arranged in a vascular cylinder. A great part of
the stem is occupied by a pulpy tissue with large, thin-walled,
highly vacuolated cells called *parenchyma*. In stems of this
type the pith, the tissue between the bundles, and the cortex
are all composed of parenchyma. Other familiar examples of
parenchyma are the pulpy parts of all fruits, such as the apple,
the banana, the watermelon, and the softer parts of all leaves,
stems, and roots. By referring to Fig. 8 it will be seen that
the cut ends of the bundles form a ring in the cross section of
this type of stem and have their xylem and phloëm elements
in the same relative positions as in the monocotyledonous stem.

In woody dicotyledonous stems of trees and shrubs the arrangement of the bundles is similar to that found in the

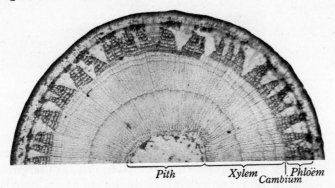

Pith   Xylem   Phloëm
Cambium

FIG. 9. Cross Section of the Stem of the American Linden, *Tilia americana*, showing the Woody Dicotyledonous Structure

herbaceous stem except that the bundles are more fibrous and woody, are squeezed closer together, and form a relatively larger portion of the stem. The pith is small, the bundles are elongated radially, and the parenchyma tissue between the bundles has been reduced in width until it forms what appears in cross section to be mere lines extending out, like rays, from the pith; indeed, these radial structures in stems of this type are called *medullary rays*. Their chief functions are storage and the conduction of food crosswise in the stem. The woody part of these stems, extending from the pith out to the cambium, which lies between the bark and wood, is composed almost wholly of the xylem of the bundle, which has numerous tracheary tubes and relatively large amounts of fibrous tissue, together with the medullary rays. The bark includes the phloëm of the fibrovascular bundles with its sieve tissue and fibers, together with some parenchyma and corky, insulating tissue.

Between the bark and the wood is found the cylinder of tender, growing tissue called the *cambium ring* because it appears as a ring in cross section. This cambium layer neither stores nor conducts food, as many people erroneously suppose;

FIG. 10. Section of a Redwood Tree showing Bark, Cambium Layer, Sapwood, and Heartwood

its sole function is growth. During the spring and early summer months it grows rapidly both by increasing the number of cells and by enlarging their size. After this it differentiates a new layer of wood, or xylem, on the inside and a new layer of phloëm on the outside. Relatively few people know that trees and shrubs add a ring of bark on the outside of the cambium each time they add a ring of wood on the inside. But they do; and these outer rings may be seen as evidence of the fact. The cambium layer makes it possible to strip the bark off the more solid woody part of the tree trunk or branch. The cells of this region are so soft and delicate that it is an easy matter to tear them apart and, in this manner, separate the bark from the wood. The slimy substance left on the outside of the wood after the bark is removed is the protoplasmic and watery content of the disrupted cambium cells. The rapid expansion of the cambium is what makes it possible for boys to slip the bark off the willow stems in the spring and fashion them into whistles. The newly formed rings, or layers, of wood and bark are the most active in carrying water and food, respectively, and as both the wood and the bark become more distant from the cambium ring they grow less active and finally cease to function in this capacity altogether. These remote parts then die and become dark in color. The inner dead heart of the tree is called *heartwood* to distinguish it from the outer, light-colored, functioning *sapwood*. In old trees the outer bark, being dead, cannot increase its circumference. Consequently, as the tree trunk expands in growth, heavy pressure is brought to bear upon the dead bark, which finally bursts, presenting great cracks that extend longitudinally up and down the stem.

Plants produce the food for living things. The life span of plants and animals may vary from a day or two to many years, but regardless of whether they survive briefly or for a long period of time, work must constantly be done within their bodies. Each individual cell has living, functioning protoplasm with its specific work to do, and the doing of work always involves the expenditure of energy. This brings us, then, to the question of the supply of energy for living things. It is a very important one because plants and animals can no more live and function without a constant supply of energy-giving materials than a steam engine can run and furnish power without a supply of coal to heat its boiler. In the case of living organisms the sole source of their energy supply is the food which they consume. The only exception to this is the energy received by chloroplasts which we shall note more in detail in the following paragraphs.

With respect to their source of food, there is a fundamental difference between plants and animals which should always be kept in mind. There are some plants that get their food supply from other plants or animals and there are also a few animals that have the power to make their own food, but, in general, animals always get their food from plants. Many of them take it directly by eating the leaves, stems, roots, and fruits of the plants upon which they forage. A smaller number get it indirectly by eating the bodies of other animals, which in turn have fed upon plants. With the exception of the parasites that occur in this group, however, plants are much more independent than animals. Instead of relying upon other organisms the plant resourcefully makes an abundance of food — not only that which is required to meet its own needs but also a great supply of nourishing materials stored up in stems, leaves, seeds, roots, tubers, and bulbs, which animals have learned to appropriate and use. The main difference between animals and plants, then, on the nutritional side, is that animals, with very few exceptions, are dependent upon plants for their food supply, but plants have the power to take the inorganic materials of the soil and air and combine them into food. In reality, then, anyone who wishes to observe the real food factories of the world should not make an inspection tour of the great flour mills, the meat-packing houses, or the large preserving establishments, but should go into the fields on a warm summer morning and look at the growing plants. There, standing in unbroken ranks, silent and unostentatious, are the real food factories of the world. Let us see, then, how these highly important and most interesting food-making processes are carried on by the green plants of garden, field, and orchard.

Leaves are structurally adapted to carry on certain functions. While some food may be elaborated and changed in the stems, twigs, and roots of plants, the greater part of it by far is made in the leaves, which are peculiarly adapted to this and other important functions. We shall now consider how the structure of leaves fits them for their work.

Leaves are normally joined directly to the stems and twigs; indeed, leaves are nothing more nor less than highly modified extensions of the twigs themselves which have become very much flattened in order to facilitate the absorption of light rays. If one examines a cross section of an ordinary horizontal leaf, such as that of an elm tree, a sunflower, or a bean plant, it will be found that its upper surface is covered by a colorless, transparent, cutinized layer of epidermal cells. These cells are for protection against mechanical injury and excessive loss of water. They are transparent, and so the light rays pass directly through them to the green cells beneath. Immediately below the epidermis and perpendicular to it is found a layer of elongated parenchyma cells. The cells of this layer are dark and green because they are abundantly supplied with tiny green bun-shaped protoplasmic bodies called *chloroplasts*. This layer of elongated cells is known by botanists as the palisade layer. In some regions, especially those subjected to brilliant sunlight, there may be two or three layers of palisade cells. Directly below the palisade area is a region of parenchyma cells which have a great deal of space between them. Because of the presence of these intercellular spaces, this part of the leaf is very porous, and is called the *spongy parenchyma*. The intercellular spaces also extend up between the palisade cells.

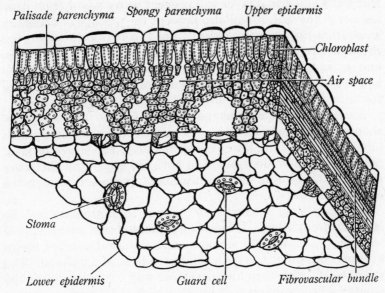

Fig. 11. Diagram of a Cross Section of a Sunflower Leaf showing Structure

The chloroplasts show the characteristic green color because they are stained with a coloring matter called chlorophyll. Grass stains on clothing are produced by the crushed contents of these chloroplasts.

Extending through this spongy parenchyma, more or less in the form of a network, are the fibrovascular bundles which we recognize as the veins and veinlets of the leaf. These slender bundles, like the streams at the head waters of a river, unite to form larger ones that extend back through the leaf-stalk into the stem and, in a more or less coalesced fashion, on down into the roots. Covering the underside of the leaf in a protective capacity is another layer of epidermal cells similar to the one on the upper surface. This is called the lower epidermis.

We see, then, that the whole leaf is inclosed in a transparent epidermal layer which is largely for the purpose of protecting the delicate tissues within from the excessive and disastrous loss of water which might occur on a warm summer day. But if the epidermal layer will prevent water vapor from passing out of the leaf, it will also prevent the passage of necessary gases from the air into the leaf. For this reason the plant has through adaptation modified the character of its epidermis, and instead of this layer's being continuous and unbroken, it is generously interrupted by pores of a special form called stomata (Greek *stoma*, "mouth"; plural, *stomata*).

Stomata are oval in shape and are formed by two lunar-shaped cells called *guard cells*. While the leaves of some plants have stomata in the lower epidermis only, the leaves of most plants have them distributed in both the upper and the lower sides. In this way a better diffusion of the gases through the epidermis results.

It is surprising to know how many of these stomata may be found in a single leaf. It has been estimated that a single apple leaf may have 400,000 stomata, and the total number on an average corn plant has been computed to be 104,057,850. The guard cells function in an interesting way to open or close the stoma. The mechanism through which they work is turgor pressure, or the pressure of water within the cell. Whenever the cell is abundantly supplied with water so that its walls are distended, the stoma is open; but if, for any reason, the pressure within the guard cells is reduced, the cells shrink, tending to close the stomatal opening.

Food-making is the chief function of leaves. Leaves participate in many living functions of the plant, but their chief business is that of making food. Considerable food is made in other parts of the plants; but because of their favorable position and the abundant supply of food-making substances which they receive, leaves are by far the plant's most important food-making organs. In food manufacture there are three important processes known as photosynthesis, protein synthesis, and fat synthesis. The leaves are concerned chiefly with the first, although they, as well as other parts of the plant, form the other two kinds of food also.

Photosynthesis. Photosynthesis is the first step in the manufacture of food. In this process the green cells may be regarded as the sugar-making factories, and the chloroplasts themselves as the workers. Since sugar can be made in green cells only, it is now evident why plants usually produce such a large number of leaves. An acre of average-size corn plants in eastern Nebraska has been estimated by Kiesselbach to have about two acres of leaf surface.

The materials used by plants in photosynthesis are water and carbon dioxide. The former is absorbed by the roots, is transported up the stem through the fibrovascular bundles out into the veins, and finally passes into the leaf cells. The carbon dioxide is obtained mostly from the air, which under normal conditions contains about 0.03 per cent of it. The chief sources of this atmospheric carbon dioxide are combustion, decay, volcanic action, and the respiration of animals and plants. It is estimated that people alone liberate as much carbon dioxide annually through respiration as would be produced by the combustion of two hundred and thirty million tons of coal. The constant use of it by green plants, however, keeps the concentration of carbon dioxide in the air very low and fairly constant. Diffusing from the air through the stomata into the intercellular spaces of the leaf, the carbon dioxide comes in contact with the damp walls of the spongy parenchyma and palisade cells, dissolves, and, like the water and minerals from the soil, enters the cells by osmosis.

Under proper temperature and light conditions the chloroplasts become active in combining the carbon dioxide into a photosynthate which is believed by most botanists to be some form of sugar. But, strange to say, the chloroplasts will not work unless they are activated by light. In this respect the chloroplasts of higher plants are the most remarkable organs in the world. Automobile engines, to do work, must be supplied with gas, steam engines derive their power from the coal burned under their boilers, and beasts of burden and man obtain the energy they use from the food which they consume. But chloroplasts are energized in an entirely different manner, and therein lies their uniqueness.

We already know that light is a form of energy coming to the earth from the sun and that, although we receive but one two-billionth of the sun's radiation, if all these rays could be caught and utilized they would furnish energy continuously at the rate of two hundred and thirty trillion horse power. Man, despite many earnest attempts, has never been able to capture this energy on a large scale for the purposes of industrial power. Chloroplasts, or other chlorophyll-bearing parts of plants, and a few animals are the only living organisms that have the ability to do this. So, extracting the energy from the light rays in some way, chloroplasts work all day and are apparently just as efficient workers at the close of the day as when they began in the morning. To use a figure: if human field hands could put on green coats and work all day supplied only with energy from the rays of light absorbed by these green coats and then return from the fields in the evening as energetic as when they went out, we should have a situation analogous to the action of chloroplasts. Just how chloroplasts effect this capture of energy is not known, but it is of the greatest importance in the economy of nature. The carbohydrate foods made by the chloroplasts are the original source of all energy used by plants themselves as well as that used by animals. Moreover, as has already been made clear, the coal, gas, and petroleum supplies derived from the earth came from plant tissues whose energy was stored originally in the plant by the action of chloroplasts. It is from these natural resources too that by far the greater part of our industrial power comes. With this knowledge before us we are in a position to see just how important chloroplasts are. They are the only agents, animate or inanimate, that have the ability to arrest the sunlight, to capture its energy, and to fix it in a form available to living organisms and industry. If chloroplasts did not have this power, life and civilization would pass speedily from the earth.

Just how the compounds water and carbon dioxide are torn apart and built up to form photosynthate, or sugar, is not known. Several theories have been advanced, but none of them have been proved. Botanists have not been able as yet to develop a technique by which they can look down into the cells of leaves and see what goes on there. Neither is it known just what the first food product is. It may be glucose,[1] or grape sugar, or it may be cane sugar (called sucrose by the chemists). The product may even be different in different plants. Let us assume that it is glucose, as is believed by most botanists, and let us also disregard the intermediate steps involved. If we do this we may write the reaction thus:

$$6\ CO_2 + 6\ H_2O + energy \longrightarrow C_6H_{12}O_6\ (glucose) + 6\ O_2$$

This indicates that six molecules of carbon dioxide and six molecules of water have been converted into a molecule of glucose as a product and into six molecules of oxygen as a by-product. Work has been done on this sugar molecule by the chloroplasts to form it, and, by virtue of this work, potential energy from the sun has been stored in it.

From this initial and comparatively simple sugar, plants build up other carbohydrate products. This apparently is done mostly by combining the molecules and extracting a part of the hydrogen and oxygen in the proportion that they occur in water. Thus:

$$C_6H_{12}O_6 + C_6H_{12}O_6 \longrightarrow C_{12}H_{22}O_{11}\ (sucrose) + H_2O$$

The sugar we buy at the grocery for family use is sucrose. Other forms of sugar may be made also. For purposes of storage the plant may make the insoluble starch form of carbohydrate, which we conceive to be derived in this way:

$$n(C_6H_{12}O_6) \longrightarrow (C_6H_{10}O_5)n\ (starch) + n(H_2O)$$

We use the symbol n here because the starch molecule is a very complex one. Chemists know that carbon, hydrogen, and oxygen are combined in the given proportions, but they do not know just how many proportional parts are built into the

---

[1] The term "glucose" as used in this book may include various forms of simple six-carbon sugars.

starch molecule. Consequently they use the indefinite symbol n to express this fact. Or the plant may make cellulose with which to build cell walls instead of starch. If it does, the reaction is believed to be something like this:

$$n(C_6H_{12}O_6) \longrightarrow (C_6H_{10}O_5)n \text{ (cellulose)} + n(H_2O)$$

Thus it is seen that the direct or indirect products of photosynthesis may be varied, depending upon the needs of the plant. However, regardless of what photosynthetic product is formed, they all have some characteristics in common. As used in this connection they all consist of carbon, hydrogen, and oxygen, and the hydrogen atoms are always twice those of oxygen, or in the same proportion that they are found in water. For these reasons the products of photosynthesis are all called *carbohydrates*. Starches and sugars are the principal carbohydrate foods used by man, and the production of those plants that yield large quantities of this type of food, such as sugar cane, sugar beets, potatoes, wheat, rice, corn, and the other cereals, form the very basis of agriculture.

We must now call attention to the oxygen which results as a by-product of photosynthesis. It was seen that every time a chloroplast makes a molecule of glucose it consumes six molecules of carbon dioxide from the air and, at the same time, tends to liberate six molecules of oxygen through the stomata back into the air. This exchange of oxygen for carbon dioxide between the plant and the air on the basis of equal volumes is also of indispensable value to animal life. It tends to keep the ratio of the atmospheric gases constant. Moreover, at the same time it removes the carbon dioxide, which would vitiate the air for animal use, and replaces it with oxygen, the very gas which all animals must have to support respiration. It has been estimated to require about a hundred and sixty square yards of leaf surface working through the summer to balance a man's respiration for a year. Thus in photosynthesis the plant performs a twofold service: it makes a class of food which is the source of most of the energy and power in the world and it replenishes the air with life-giving oxygen at the same time. In this connection we may see what a fortunate thing it is for animal life, including man, that plants are so persistent in their growth. They crowd into every available space where conditions make it at all possible and cover the earth with a rich verdure. In addition to beautifying the earth and supplying energy-containing foods, they are constantly renewing the life-supporting gas of the air. If people always realized the indispensable service plants are constantly performing in the world, they would appreciate them more. In fact, they would then be much less inclined to complain about the weeds which preëmpt the roadside and insistently attempt to edge their way into field and garden whenever the least opportunity presents itself.

FIG. 12. A Geranium Leaf with the Chlorophyll Extracted

It has then been stained with iodine to show the presence of starch. The diamond-shaped area which has been shaded from sunlight has no starch present

Protein synthesis. Protein synthesis does not occur solely in the leaves; but because of the abundance of carbohydrates made there and the liberal supply of mineral matter carried up from the roots by the transpiration current, it is probable that more proteins are made in the leaves than in any other part of the plant.

Protein synthesis is a complicated process, and for this reason it is imperfectly understood. Looking at the matter simply, however, it seems probable that starting with carbohydrates as a base, plants add to them by chemical union certain elements which are constituents of proteins. Among these are nitrogen, sulfur, and phosphorus. Proteins always contain carbon, hydrogen, oxygen, nitrogen, and sometimes sulfur. The proteins of cell nuclei always contain phosphorus also. These minerals come from the soil in the soluble form and are carried to the leaves by the ascending stream of water from the roots. The first direct use of these minerals by the plant in metabolic processes seems to occur in protein-making.

The proteins are the only foods which the plant can utilize to build up protoplasm. For this reason proteins are found to be very abundant in certain storage organs where rapid growth is to take place later. Especially is this true in seeds such as the bean and pea, as well as in other legumes. In many cases the principal food stored in the seed is some form of carbohydrate or fat, perhaps most often starch. We find starch particularly abundant in the seeds of cereals, such as wheat, corn, and rice. With this abundant supply of carbohydrate food made in the leaves of the parent plant and stored in the seed, the young plantlet is placed in an especially advantageous position. Immediately upon germinating, indeed, the very first thing the embryo does is to push out roots into the moist soil to absorb the necessary nitrogen, sulfur, and phosphorus and conduct them back into the seed. Here the growing plantlet uses these minerals to combine with its store of carbohydrates to make proteins for further growth of the roots and leafy shoots. In this case the proteins are made by the cells in and adjacent to the points of growth. Perhaps no better example is afforded of how nature provides for her children. Buried deep down in the dark earth, the one kind of food which the young plantlet cannot make is carbohydrates; but in this respect the tiny embryo need have "no thought for the morrow" because the mother plant took care of that need long before by storing an abundant supply of carbohydrates inside the seed coat.

Proteins are also formed abundantly in the growing cells when vegetative reproductive organs, such as bulbs, tubers, corms, and rhizomes, begin active growth. Here, just as in the seed, the supply of carbohydrates was made and stored in these underground parts by the parent plant the previous year. These protoplasm-building foods are likewise made extensively at growing points in the stem and other aboveground parts. Here, though, the carbohydrates may come directly from the leaves.

Fat synthesis. Since fats are insoluble in water, they too must be made in practically all parts of the plant as well as the leaves. Just how they are made is even more obscure than the synthesis of proteins. It seems probable that fats are also evolved from the carbohydrates as a base. For instance, when the castor bean starts to ripen, it has been found to contain relatively large quantities of carbohydrates but little oil. When it is mature, the opposite relation exists. It may be that some fats are derived from protein constituents also.

Fats are the most condensed form of energy-containing food produced by plants. They contain proportionally more of the carbon and hydrogen atoms and less of the oxygen than do the carbohydrates. It is the possibility of increasing the oxygen content through oxidation that gives fats their energy-producing power.

All plants make some fats, but some make much more than others. Among the abundant fat-producers are nuts, olives, the cotton seed, the castor bean, and conifers. In the tallow tree (*Sapium sebiferum*) of eastern Asia fat is so abundant in the waxy-coated seeds that it is used for making candles. The extracted oils of cotton seed and corn, in their prepared and refined form, have come to be used extensively as cooking compounds and butter substitutes.

Our knowledge of the food-making materials used by plants grew up gradually. Today, with our extended knowledge of the materials used by plants in the production of food, we are likely to take it for granted that men have always possessed this knowledge. We are inclined to overlook the fact that our acquaintance with the needs of plants in this respect, as well as our knowledge of many other phases of plant physiology, has followed the course of all scientific development. Men have achieved what they know of plant requirements as the result of insistent seeking and hard work. Sometimes they proceeded by trial and error and sometimes on the basis of the most carefully thought-out inferences derived from previous experiments. But gradually their knowledge grew, and in general they came to know the materials that plants need. What men have learned of the food requirements of plant life is of inestimable value, although there is much detailed knowledge yet to be gained. It forms the very foundation upon which all intelligent systems of crop production are built and is also the basis of our whole science of soil management. What crops shall be grown on different kinds of land, what methods of tillage shall be used to produce the best results, and what fertilizers shall be used to make the soil most productive are questions which require that men should know what the plant needs to grow and thrive.

For almost two thousand years men believed the Aristotelian dictum that plants get their foods from the soil. All larger plants are anchored to the earth, their roots forming extensive absorbing organs that ramify in all directions through the soil. What conclusion could be more logical, then, than that their foods come from this source? But in the early part of the seventeenth century Van Helmont, a Belgian chemist, with a commendable skepticism questioned this belief and attacked it with the sure weapon of experiment. He placed two hundred pounds of thoroughly dried soil in a pot in which he later planted a healthy willow cutting weighing five pounds. The pot was covered to exclude dust and was watered daily with rain water. In the true spirit of the scientist, for five long years Van Helmont cared for the cutting, which grew readily. At the end of the period the tree weighed one hundred and sixty-nine pounds, but when the soil in which it had grown was thoroughly desiccated, it was found to have lost but two ounces. From this result Van Helmont, failing to take the air into consideration, concluded that the material of the plant is formed from water — a classic evidence that scientists are likely to secure but incomplete facts or, indeed, to arrive at wholly erroneous conclusions when for any reason they fail to take into account all the factors involved in an experiment. Later Priestley, who discovered oxygen, noted in 1774 "the restoration of air, in which a candle has burnt out, by vegetation." He did this by placing a plant in the air under a glass container in which a candle had burned to extinguishment. In 1796 Jan Ingen-Housz, a Dutch physician who lived in England, wrote a noted book on the nutrition of plants.

Ingen-Housz was a physician who seems to have derived a great deal of pleasure from experimenting with plants in his garden. As a result of his studies he decided that plants get carbon dioxide both from the air through the leaves and from the soil through the roots and, by combining this with oxygen, make "acids, oils, mucilage, etc." The idea that plants get carbon dioxide to any great extent through the roots was, of course, a mistake, but it was several years before the error was disproved. In 1804 De Saussure, another physician who turned to plant science, wrote his *Chemical Researches on Vegetation*. He used plants inclosed in glass containers and, by careful work, showed that plants secure from the air practically all the carbon dioxide they use, and took the position that little or none of it comes from the soil. He also showed that water, along with the carbon dioxide, is decomposed in the leaves when the plant is illuminated. Moreover, by careful experimentation, he was able to show that when photosynthesis takes place the plant liberates into the air a volume of oxygen approximately equal to the amount of carbon dioxide consumed. Finally, by a long system of balance sheets, he arrived at a fairly clear idea of what mineral elements were necessary and how much of each is essential for healthy plant nutrition. Harvey-Gibson, in his *Two Thousand Years of Science*, tells us that the work of De Saussure "constitutes a landmark in the history of botany." From his time on men may be considered to have had a fairly intelligent idea of what raw materials plants use in their metabolic activities and what one comes mostly from the air and what ones come from the soil.

Plants lose water through transpiration. Turning now from food-making, we find that the liberation of large amounts of water into the air in the form of vapor is also an important physiological process in plants. The sun's power vaporizes the water, and in this respect the plant is no more active than a dish would be if the water content of it was being evaporated by the hot summer sun. But the plant does respond to the situation by taking certain measures to reduce the excessive loss of water whenever that loss is so great as to endanger its life.

2

By far the most of the water vapor escapes through the stomata of the leaves, but some of it escapes from the epidermal surface of the leaves, fruit, stems, and twigs also. The structure of leaves, as the reader will recall, is especially conducive to the loss of water. The leaf must have large intercellular spaces to admit an adequate supply of carbon dioxide to the photosynthetic cells within. It must also have stomata in order to permit this gas to get by the barrier interposed by the epidermal layer against the drying effect of the air. Moreover, the stomata of the leaf must permit the passage of gas both ways, for oxygen escapes and carbon dioxide must get in.

Fig. 13. Apparatus used to demonstrate Transpiration through the Leaves of a Plant

A, geranium in Ganong shell, loses weight by transpiration through the leaves. *B*, control in Ganong shell, with the plant cut off just above the surface of the soil, remains practically constant in weight

For these reasons, then, it can be seen that the porous leaf with its thousands of stomata would provide an almost ideal structural condition for the loss of water vapor.

Transpiration is also speeded up by another condition. It has been observed in the experience of everyone that when a vessel containing water is warmed, it hastens the evaporation of the liquid. Leaves, under normal conditions, are always so arranged as to intercept a maximum of the impinging light rays. Strange to say, however, only a very small part of the light energy arrested by the leaf is actually used in photosynthesis. It has been estimated that although about 70 per cent of the direct light rays are absorbed by the leaf, only something like from 2 to 3 per cent of the light energy is actually utilized in the manufacture of sugar. All the other

66 or 67 per cent of the light energy is transformed into heat, which, if not removed effectively, would tend to raise the internal temperature of the leaf and the surrounding air to a point that would be highly dangerous to its protoplasm. How, then, can the plant manage to dispose of this excess heat? It is done through transpiration; for every time a gram of water is evaporated, as we have found, five hundred and thirty-six calories of heat disappear and take the form of latent heat of vaporization.

Because of the large quantities of absorbed heat to be disposed of, the amount of water transpired by plants is surprising. One experiment revealed that a sunflower leaf with a surface area of one square meter lost water at the rate of 275 cubic centimeters per hour, which would require 166,800 calories of heat. Kiesselbach found that at Lincoln, Nebraska, a large corn plant, under conditions favorable to transpiration, might lose as much as $1\frac{1}{4}$ gallons of water per day. A grown birch tree with 200,000 leaves has been estimated to give off from 660 to 880 pounds of water on a hot summer day. This would make about $1\frac{1}{2}$ to 2 large barrels of water. It has been computed that from 90 to 99 per cent of all the water absorbed by plants goes directly through their stems and out into the air again by transpiration. So heavy is this loss that Briggs and Shantz, as a result of careful experimentation, have found that it may take 368 pounds of water to produce 1 pound of water-free corn-plant tissue and that the production of 1 pound of dry wheat-plant tissue may require 513 pounds of water.

In the light of these statements it now becomes clear why it is important to prune back the leaf area of plants when transplanting them. As the plants are taken up from the soil a large portion of the water-absorbing root system is broken off and left behind. Then, when they are reset, it is necessary to cut back the aboveground parts to reëstablish more nearly a balance between the absorbing and transpiring surfaces, else serious consequences to the plant might follow.

Transpiration is believed by many to benefit the plant in another way. It is thought that the transpiration stream carries large quantities of mineral matter upward through the stem to points of usage at a much more rapid rate than they otherwise could be transported. This speeds up anabolism, which permits the plant to grow faster and to be more fruitful.

Such a rapid loss of water from the surfaces of the plant cannot always be considered a direct advantage. On the contrary, it may become a positive menace. Particularly is this true when, because of a scarcity of water in the soil or because of severe atmospheric conditions, the water tends to be lost more rapidly than it can be absorbed. This sometimes happens in the best crop-producing regions and often occurs in arid and semiarid climates.

Because of this danger, which may become imminent at any time, the plant is compelled to adopt structural and functional adaptations to check transpiration. This may be done in many different ways. It may be accomplished by a thick, waxy cuticle, as in the prairie rose; by more than one layer of epidermal cells, as in the cactus; by placing the stomata at the bottom of pits in the epidermis, as in pine needles; by covering the surface with hairs that reflect much of the light, as in the mullein; by rolling the leaves to reduce the transpiring surface, as in corn and many of the grasses; by producing hydrophylous, or water-holding, colloids that tenaciously hold the water back, as in some of the cactuses and other desert plants. Finally, if the plant is driven into a corner, figuratively speaking, and the water loss becomes too excessive, the turgor of the guard cells is reduced, the plant wilts, and the stomata are tightly closed. This results in a heavy reduction in the rate of water loss and often saves the life of the plant.

Respiration is an important physiological process of plants. Since plants are living things they must work constantly during the active season and expend energy. Even in the dormant condition and in dry seeds some energy is required. Aside from part of the energy used by chloroplasts, this energy must come from the carbohydrate component of the food. Once the energy from the sunlight is locked up in the carbohydrate molecule, the only way the plant has of liberating it for use is through chemical change, just as man must release energy from coal and oil by oxidation for use in the industries. In some ways the chemical action that takes place in a plant cell, however, is quite different from that which is produced in a furnace. For this reason physiological oxidation is called *respiration*.

Respiration in both plants and animals is fundamentally the same. It takes place in every living cell of the body and results in the release of energy to support vital functions. The mechanics of the process, however, are very different in the two types of organisms. Higher animals have lungs or their equivalent to gather in oxygen, and an

Fig. 14. Birch Bark showing Lenticels

elaborate circulatory system to transport this dissolved gas about through the body so that every living cell may be adequately supplied. Plants have neither lungs nor a circulatory system to carry the oxygen. Oxygen must enter the plant either through the stomata or through special organs designed for gas exchange, called *lenticels*. Lenticels may be found thickly distributed over the surface of young twigs, stems, and even over the surface of roots and tubers. They appear as tiny, corky-looking spots that interrupt the epidermis and outer bark. They may be plainly seen even in a potato tuber. In older stems the lenticels and epidermis are pushed out by the growing tissues beneath and slough off. It is probable that in such cases the deeply furrowed cracks admit oxygen to the inner tissues. In some trees, as in the birch, the lenticels are very large, and provision is made for their continued enlargement through special growth.

After the oxygen of the air has entered the plant from the outside, nature has provided no way by which it may be propelled about through the plant body. Instead, the force that brings it ultimately in contact with the wall of each living cell is that of diffusion. To promote this action, however, the plant has produced a remarkable system of intercellular spaces. As the oxygen slowly diffuses through these intercellular spaces, it comes in contact with the moist cell walls, dissolves, and passes into the cell by osmotic action. This action reduces the oxygen pressure in the intercellular spaces, and more oxygen then diffuses in from the outside to restore the oxygen-pressure equilibrium. Some plant stems that have a hard outer shell, such as corn, have special air-conducting canals that extend down from the leaves through the fibrovascular bundles. Plants like the bulrush and the water lily often have the stem or the leaf petiole with a very porous interior through which an exchange of gases may take place between the air and the roots embedded in water-logged soil.

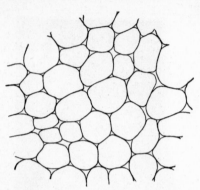

FIG. 15. Cross Section of a Corn Stem showing the Intercellular Spaces

The modern view of plant respiration held by most careful students of the problem is that there are two steps in the process: one is called anaërobic (without air) respiration, and the other is known as aërobic (with air) respiration. In anaërobic respiration no oxygen from the air is required, and the food substances, perhaps most often simple sugars, are broken down chemically to liberate energy to be used by the organism. Certain intermediate by-products, such as alcohol, organic acids, and carbon dioxide, result. Now many of these intermediate substances, such as alcohol and the acids, are either toxic to the plant or interfere in other ways with the life functions of the protoplasm. Consequently they must be promptly removed. Their removal is accomplished through the second step, or aërobic respiration. In this process oxygen from the air is used to oxidize the intermediate waste by-products into the harmless end products, water and carbon dioxide. Heat, of course, is generated, but is not necessarily needed. It may be a by-product just as the carbon dioxide and water are waste by-products.

If this view of respiration is correct, it is evident that the energy supply comes from the anaërobic respiration and that the aërobic, or oxidation, phase is merely a purification process to rid the plant body of intermediate waste by-products. The waste by-products may be used again immediately in photosynthesis during the daytime or they may escape into the air by the very same channels through which the oxygen entered. Because of this it is evident that under favorable conditions oxygen escapes from plants during the daytime and that carbon dioxide escapes during the night.

Plants carry on digestion. All the food which plants make is

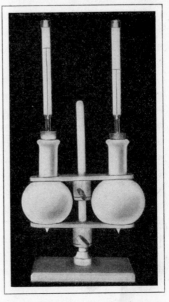

FIG. 16. Dewar Bulbs used to determine the Heat of Respiration in Seeds

Note the difference in temperature recorded by the two thermometers. The difference in temperature rise may be as much as 20° C.

as a rule not used immediately. Some is utilized as it is made, to promote growth and repair, and some is used to furnish energy; but during the most active period, when conditions are favorable, each day there is a constant excess of food produced. This is stored in the roots and stems as fats, carbohydrates, and proteins to be used at some future time by the plant itself or else it is sent to the developing fruits to be deposited in the seeds for subsequent use by the young embryos. For the sake of economy this reserve food is most often stored as an indiffusible fat or as an insoluble carbohydrate or protein. Space within the plant cells is thus conserved, and there is less danger from injury by frost. But when either the parent plant or the young offspring is ready to use this reserve food, it must be changed back into the soluble form so that it can be transported and utilized. This process of making indiffusible foods diffusible and making insoluble foods soluble and liquefying them is called *digestion*.

If there were no way to speed up this digestive process in plants, it would take place at a rate far too slow for the needs of the organism. The weak acids and alkalies within the cells would tend to liquefy some of the food; but since the general temperature of plants is but little if any higher than the surrounding air, the process would be an infinitely slow one. Even if it had a way to raise the temperature to the point where the food would dissolve rapidly, the plant could not do it; for such a temperature would endanger the very life of the protoplasm itself, causing it to coagulate and die. Consequently, if rapid growth is to be promoted at all, the plant must in some way change food rather rapidly from the insoluble to the soluble form at the low temperatures prevailing in the plant body. This is made possible through the production and use of organic catalyzers (Greek *katalusis*, "to loosen down" or, more freely, "to unloosen").

Catalyzers, as the reader may remember, are substances that when employed in but very small quantities will speed up the rate of chemical reactions often to a remarkable degree. For instance, by the use of platinum black a mixture of hydrogen and oxygen may be made to combine with explosive force. Under ordinary conditions, with no catalyzer present, the mixture will not unite at all unless heated to the ignition point. Someone has put the matter tersely but vividly when he said, "A catalyst is like a tip to a waiter in that it accelerates a process which otherwise would proceed with infinite slowness." Although they are employed widely in all kinds of chemical reactions, just how catalyzers effect this acceleration is imperfectly understood.

Organic catalyzers are called *enzymes* (Greek *en + zumē*, "in leaven"). This name was applied by Buchner, a German scientist, who first isolated an enzyme from yeast. He ground up yeast cells with sand and diatomaceous earth and, after subjecting the whole to a hydraulic pressure of from three hundred to four hundred atmospheres, succeeded in obtaining a transparent liquid that caused sugar to ferment readily without the intervention of yeast cells themselves.

It is now believed that practically all physiological activities within the bodies of both plants and animals are promoted by enzymatic action. Their number may run into the hundreds. Enzymes occur in such small quantities within organisms that, with perhaps one or two exceptions, they have so far escaped chemical analysis. For our purpose we shall note specifically but four of the main groups of enzymes.

1. *Hydrolases*. Hydrolyzing enzymes promote digestion by the addition of water and the subsequent splitting of the molecules acted upon into simpler substances. Of these we may note:
   a. Amylases, which convert starches into sugar.
   b. Cellulase, which digests food stored in the form of hard, cellulose walls, as in the date seed.
   c. Invertase, which forms fermentable sugars, such as glucose, from cane sugar.
   d. Lipases, which aid in the breaking down of fats into glycerine and fatty acids. Such enzymes are active when hickory nuts, cotton seed, corn, and castor beans germinate.
   e. Proteases, which promote the digestion of all kinds of proteins into amino-acids and other soluble products.
2. *Oxidases*. These enzymes accelerate oxidation within the living cells and are active in respiration.
3. *Coagulases*. These promote the coagulation of certain substances into the insoluble form.
4. *Fermenting enzymes*. These are active in fermentations, as when yeast cells digest sugar into alcohol and carbon dioxide.

In animals, as we shall find later, the enzymes and the digestive juices are secreted mostly by special digestive glands. With a few exceptions plants have no such organs; instead, the protoplasm of each cell where the food is stored secretes its own enzymes directly.

Reproduction is an important physiological function of plants.
Many plants, like the fragile algæ of pond and stream, are
short-lived; others live much longer; and when we come to
the giant Sequoias of California we have what possibly may
be the oldest living things in existence. There is but one living
thing that is thought by some to be older. That is the Big
Tree of Thule standing about two hundred and fifty miles
southeast of Mexico City. It is a cypress, fifty feet through
at the base, and has been estimated to be at least five thousand
years old. But whether plants live a short time or a long time,
they must provide for progeny if their own kind is to continue
on the earth. In fact, reproduction seems to be the chief
business of plants. As we see them in the form of perennials
standing year after year, we are likely to underestimate the
importance of reproduction. Plants are always struggling
toward the one crowning achievement in their existence;
that is, the production of offspring.

Reproduction on the part of plants is an extremely ex-
hausting process; for not only must the young embryos in
most cases be nourished to the point of complete development,
but they must be sent out into the world supplied with an
adequate ration of food and provided with protective seed
coats. This occasions a heavy drain on the energies and physiological activities of the parent. Especially is this evident
when we remember with what prodigality many plants reproduce. Annuals exhaust most of their energy in this supreme
effort and die directly after the seeds have been matured;
many perennials, such as fruit trees and nut-bearing trees,
after one prolific yield, are so drained of food and vitality
that they do not bear at all the next year or, if they do, produce but a few seeds.

Just as there is no typical cell or leaf, so there is no one
particular way of reproduction. On the contrary, plants
live in so many different kinds of places and have to adapt
themselves to such wide divergencies of habitat that their
variations in reproductive practices are almost legion. Yet,
fundamentally, reproduction of whatever kind involves certain principles and processes in common. It is these common

elements in which we are interested and to which we shall direct our attention.

Asexual reproduction. The simplest type of reproduction is asexual reproduction, which requires no fertilization. A piece of the parent body simply becomes separated and begins an individual existence of its own. Usually this vegetative reproductive structure is either supplied with a store of food to nourish it until it can become established or else it maintains a connection with the parent body, through which it is fed until it becomes rooted and produces leaves. Tubers produced by artichokes and potatoes, bulbs developed by lilies and onions, rhizomes employed by rhubarb and blue grass, and corms grown by the gladiolus are all examples of familiar and important asexual reproductive structures. The runners of strawberry plants are also typical asexual reproductive organs that keep a connection with the parent until the young plant gets a start of its own.

The chief advantage gained by plants in reproducing by one of these asexual means is to insure the growth of the offspring even if they have to struggle against heavy odds. Young potato plants spring up quickly, with the heavy reserve of food in the tuber to support them. Paper-white narcissus plants shoot up and bloom profusely within a few weeks because of the abundant food supply obtained from the bulbs at their base. Scattered tufts of blue grass within a year or two spread out and thoroughly occupy the lawn with a continuous rich green carpet because of the rhizomes that burrow out in all directions from the bases of the scattered plants. Each rhizome at its outer extremity sends up a new bunch of shoots. Many shrubs, such as sumac, coral berry, and hazelnut, push out to occupy grassy plots in which their seed cannot get a start. They do this by sending out long rhizomes, or runners, which strike root, develop stems, and grow a crown of leaves that shades and finally kills the grass.

Sexual reproduction. Sexual reproduction always involves fertilization and is carried on by higher plants through the medium of a structure which we call the flower. Flowers vary widely in size, parts, color, and form, but all of them are

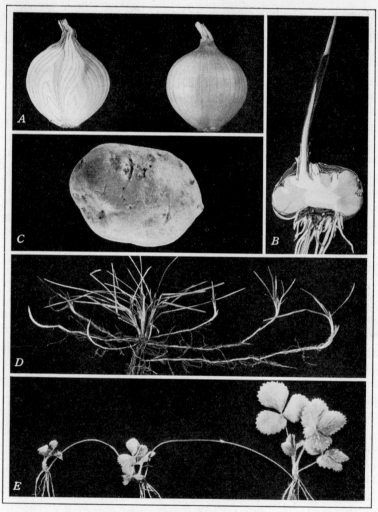

Fig. 17. Vegetative Structures produced by Plants for the Purpose
of Reproduction

A, bulb of the onion; B, corm of the gladiolus; C, tuber of the potato;
D, rhizome of the blue grass; E, runner of the strawberry

concerned, in one way or another, with the production of the
essential organs called the stamens and pistils. Most often
these organs are borne in the
same flower, as in the rose,
but they may be grown in
different flowers, as represented by the corn plant.
Indeed, in some species, the
stamen-bearing flowers and
pistil-bearing flowers are produced on entirely different
plants, as in the case of the
willow, the aspen, and the
cottonwood. It is for this
reason that many cotton-
wood trees grown for shade
in the towns and villages of
arid regions of the West are

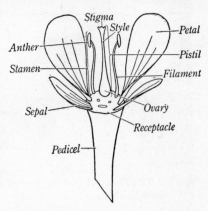

Fig. 18. Diagram of a Flower showing
the Arrangement of its Parts

a nuisance because of the cottony fruits they produce, while
other trees of this species are entirely free from this objection.
Flowers, in addition to these essential stamens and pistils, may
have a more or less showy envelope consisting of the calyx and
corolla, or they may be devoid of these parts.

There are four important steps between the formation of the
flower and the completion of seed dispersion. *Pollination.* Pollination is the process of transferring the pollen from the
stamen to the pistil. Some plants are self-pollinated, as the to-
bacco and oats, but most of them succeed better when cross-
pollinated. In fact, cross-pollination is so important in many
cases that all kinds of devices have been developed by the
plant to insure its success. The corn plant bears great quantities of dry, dusty pollen that blows about profusely in the wind.
A medium-sized corn plant has been estimated to produce
fifty million pollen grains. Most of them are lost. Doubtless
for every pollen grain that finds the stigma of a corn silk,
thousands of others perish on the dry ground. But the comparatively few grains that do reach their destination are all-
important in the production of a corn crop; for if any silk

fails of pollination, the ovary at its base does not develop but shrivels and dies, leaving a vacant place on the cob at harvest time. Most forest trees as well as grasses and many other plants depend upon the wind also for pollen distribution. It is surprising how far pollen may be carried by the wind. A pistillate date palm at Otranto, Italy, after having been barren for several years, unexpectedly set some seed. Investigation showed that the only possible source of pollen was a staminate tree located at Brindisi, over forty miles northwest of Otranto. Riley tells us that at St. Louis, after a thunderstorm in March, he saw the ground covered in places with pine pollen which, because of its characteristics, was judged to have been blown from the long-leaved pines in the Southern states, at least four hundred miles away.

U. S. Bureau of Plant Industry

FIG. 19. The Effect of Self-Pollination and Cross-Pollination in Blueberries

Left, self-pollinated ; right, cross-pollinated

By far the greater specialization of plants to insure cross-pollination is exhibited in connection with insects. Some flowers, like those of the buckthorn, produce a remarkably fragrant secretion that attracts insects to their nectar pockets. The hairy heads and bodies of insects, such as bees, are thoroughly dusted with pollen as they enter the flowers. When they visit the next flower the pistil is so located that the stigma scrapes off some of this pollen, and the insect gathers another load of pollen from that flower to be carried to the next. Other plants, such as irises, orchids, trumpet creepers, and scarlet sage, produce blossoms with gorgeous colors that attract pollinating insects. In many cases the insect seems to be made for some particular flower, and that flower, in turn, seems to be fashioned for that special insect. As Sir Francis Darwin has said, "The co-ordination between a flower and

Fig. 20. Corn Ear showing Unfertilized Kernels

the particular insect which fertilizes it may be as delicate as that between a lock and its key." It has been reported that an early attempt to introduce scarlet-runner beans into Nicaragua was unsuccessful because the right species of bumblebee was not there to pollinate it. Smyrna-fig culture in the United States failed entirely until the wild pollen-bearing Capri fig was imported, together with its pollen-bearing wasp, from the east-Mediterranean district. Many flowers of the tropics and some in the temperate zones are adapted to pollination by the long, slender-billed humming birds. The yucca, or soap weed, of the Western plains is pollinated in a unique manner. This can be accomplished only by a special moth, the pronuba. The beautiful white female, about a half-inch long, inserts her ovipositor into the ovary of the flower and deposits her eggs. Immediately afterwards she ascends to the top of the flower and pokes pollen which she has gathered from the stamens into the canal running down through the pistil. The stigmatic surface lines the interior of the canal, and from this surface pollen tubes grow down to effect fertilization. Seeds develop from the fertilized ovules,

and when the young larvæ are hatched out they feed on these developing seeds. No other food can be substituted. There are so many seeds produced that a few uninjured ones always survive to perpetuate the plants. When fully grown the larvæ bore holes through the seed-capsule wall and lower themselves to the earth by a slender thread which they spin. Here they bore into the ground, spin a cocoon about themselves, and pass the winter.

FIG. 21. *Pronuba* Moth (Twice Natural Size)

This moth pollinates the Yucca flower

The next spring the moths, after pupating, emerge just at the flowering season of the yucca.

Some plants are pollinated by water. Among the most interesting of these is the common eelgrass. This is a submerged plant rooted in the mud beneath shallow water and having the stamen-bearing flowers and pistil-bearing flowers on different plants. Plants of this type are called diœcious to distinguish them from the monœcious plants, like the corn and castor bean, that produce both the staminate and the pistillate flowers on the same individual plant. When the flowers of the eelgrass are mature the stamen-bearing ones become detached from the parent, rise to the surface, open, and

FIG. 22. Eelgrass

Its remarkable adaptation for pollination has been described [1]

[1] From Stevens's *Introduction to Botany*. By permission of D. C. Heath and Company, publishers.

float about on the water, using the floral envelope as a sort of raft. At the same time the stalk that bears the pistillate flower elongates rapidly until it reaches the surface, where the flower opens. The little staminate flowers, impelled around by wind and water currents, float up to the female flower, and the protruding stamens are brought into contact with the stigma to effect pollination. Soon after pollination the long stalk of the pistillate flower shortens by forming coils that pull the flower beneath the surface of the water, where it may mature unmolested by wading birds, aquatic mammals, or devouring insects.

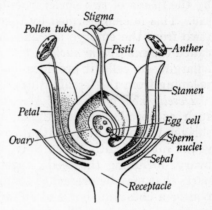

Fig. 23. Diagram of Flower showing Structures concerned in the Formation of the Seed

Fertilization. In most plants fertilization follows pollination directly and results in the fusion of the sperm and egg cells in the ovule, or immature seed, within the ovary. To accomplish this the pollen grain, after being caught by the stigma of the pistil, germinates, producing a long, slender pollen tube that penetrates the entire length of the style. This may be a very short distance, as in the cottonwood, or a much longer one, as in the Easter lily or field corn. The pollen tube secretes an enzyme that dissolves the tissue ahead of it, and these digested cells in turn are believed to supply the needed nourishment to the growing tube. On reaching the ovary the pollen tube usually penetrates the opening, or micropyle, in the ovule and delivers two sperm cells to the embryo sac. Many pollen tubes may grow down through the style, but one alone succeeds in delivering the sperms to fertilize the egg. In ovaries containing many ovules each ovule requires fertilization by the sperms from different pollen tubes. One of the two sperms in each case fertilizes the egg,

2

which will subsequently develop into the small plantlet, or embryo. The other fuses with a nuclear body already formed by the fusion of two nuclei from opposite ends of the embryo sac. This nucleus formed by the fusion of these three nuclei (two from the embryo sac and one from the pollen tube) is called the *endosperm nucleus,* and as it grows, multiplying and enlarging its cells, it provides food for the young embryo developing from the fertilized egg.

Development of the seed. After fertilization the flower parts, such as corolla, calyx, stigma, and style, usually die and shrivel, while the ovary with its ovules, now developing seeds, grows into a fruit. Botanically, fruits consist of the mature ovary, including the seeds, together with certain accessory parts that may be present in some cases. In this sense a bean pod or a chestnut, bur and all, are just as much a fruit as a peach or an orange. There are many different kinds of fruits.

The principal feature of a fruit in which we are interested is the seed which contains the young plantlet, or embryo. In some seeds the embryo remains small, but, packed around it, is an abundant supply of stored food in the form of the mature endosperm. The endosperm, it will be remembered, came from the so-called triple-fusion nucleus. The whole structure, both embryo and endosperm, is securely packed away in a tough skin consisting of two seed coats and sometimes the ovary wall securely grown together. Wheat, barley, and corn are good examples of such a seed. In other plants the young embryo consumes the whole endosperm while the seed is developing. In such cases the embryo becomes relatively large. The cotyledons, or seed leaves, become greatly enlarged, and while the seed is germinating and pushing its shoots aboveground the young plant feeds upon the sustenance stored in these swollen seed leaves. Beans and peanuts are good examples of this latter type of seeds.

Seed dissemination. Plants are confronted with the problem of dissemination if the species is to survive. This necessity has been met in many ways by different plants, and the adaptations exhibited at this point are quite as striking as those manifested in connection with pollination. The degree of success attained is shown whenever a plot of ground is plowed up and left unmolested or whenever an area is denuded by fire, ice, or other destructive agents. Soon plants begin to spring up over the bare ground, and within a very few years the area is as densely populated as the surrounding region. Let us see how plants accomplish this.

Many seeds and fruits are provided with appendages of some sort that catch the wind and cause them to be borne about for long distances. The fruits or seeds of maple, box elder, elm, cottonwood, and catalpa have devices of this sort that are very effective. Forest burns are usually soon covered with young aspen trees, whose seeds have blown in from other forest areas, often miles away, that have not been burned over. One reason why the pestilential dandelion is so successful in establishing itself as an unwelcome guest on our lawns is because of the fine parachute with which its tiny fruits are provided. Every country boy has doubtless

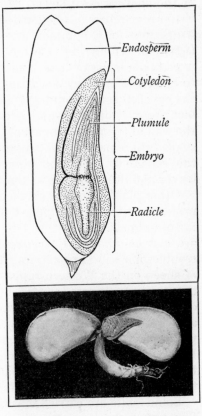

FIG. 24. Seeds showing the Embryos

Above, longitudinal section of a kernel of corn showing embryo, endosperm, and seed coats. Below, lima bean laid open to show the embryo with its large cotyledons and the seed coats

amused himself by striking a mature milkweed plant to see the great crowd of seeds released from the dry pod, which, with their cottony appendages, go drifting about like a cloud of snow.

Some plants discharge their seeds with explosive force. Beans and peas do this to a limited extent. Purslane and some of the spurges are likewise very active in this respect. The writer remembers having gathered some snow-on-the-mountain fruits once for a botanical friend who was making investigations concerning the possible sources of rubber. The fruits were taken into the house and spread out on the kitchen table to be dried. As they dried tiny "snaps" could be heard, and the expelled seeds would go rolling across the floor six or eight feet away. Touch-me-nots get their name from the fact that the wall of the mature ovary when touched or jarred suddenly snaps into a mass of twisted and contorted segments, causing the seeds to fly about in all directions, rattling among the dead leaves and other objects as they go.

Other plants produce on their fruits or seeds spiny, hooked projections that are adapted to being caught in the hair or fur of animals and, in this way, these reproductive structures are carried about. Everyone has had experience with the troublesome sand bur and its habits in this respect. The wool of sheep on the Western range may become filled literally to the point of stiffness by sand burs. Horses on winter pasture often have their manes and tails so matted with cockleburs and burdock fruits that they become stiff and boardlike. Cattle that are pastured in the lowlands in the late fall may have the hair of their heads and necks so filled with the small bootjack-like fruits of the "Spanish needle" that they take on the brown appearance of these fruits.

Some fruits are carried effectively by water. Black-walnut trees often grow on the lowlands along rivers and other streams. In the autumn when the fruits, surrounded by their large but light and spongy hull, fall to the ground they are washed into the streams by freshets and, floating, are transported long distances. The seeds of many water plants are carried by this agent.

One of the most interesting adaptations through which plants effect seed dissemination is that of attractive, pulpy, edible fruits. This feature especially appeals to our sense of fitness because the animals (usually birds) are compensated for the disseminating services which they render. In this class may be found the wild berries of all kinds, herbs with pulpy fruits, and many shrubs. On a farm in one of the central states a home was established in a cultivated field. The new house was surrounded by a grove of young cottonwood and maple trees. At that time the tilled soil was bare of everything except the young trees and a few annual weeds that had escaped the cultivator. Now, after about forty-five years, except where persistent effort keeps it clear, the undergrowth in the grove is a tangled mass of shrubs, vines, and briers. Wild raspberries, blackberries, woodbine, poison ivy, grapes, bittersweet, dogwood, prickly ash, and sumac grow in great profusion, although the nearest groves in which these plants could have previously grown are from one-fourth to one-half mile away. Such plants usually have hard, bony seeds that successfully resist the effect of animal digestive juices. The birds ate fruits which grew in the older surrounding groves and then flew to the grove in question, where the seeds were cast to the ground in the birds' excreta. Subsequently these seeds grew into the tangled undergrowth found in the grove today. Pine seeds and nuts of various kinds are carried by squirrels and other rodents, although in this case the only seeds to grow are the ones from the animal caches that escape being eaten.

When one sees all these fine adaptations for pollination and seed dissemination which work out so successfully for plants, if he is of a reflective turn of mind at all he is likely to stop and ask himself how all these adjustments came about. Some people take them as sure evidence of a definite design and coördination between plants and animals and other forces in nature. Others take the opposite view and hold that these successful and highly coöperative adaptations are not the result of intervening intelligence at all, but are the results of long ages of variation and change — change which resulted in survival of the better variations. Most botanists are inclined to take the latter view rather than that of definite design. At this point, however, we are departing from the precincts of science and entering the field of philosophy, where we shall leave the reader to formulate his own opinion.

QUESTIONS FOR STUDY

1. What is the difference between plants and animals with respect to food?

2. Make a list of the functions of the roots of plants.

3. What are the functions of the stems of plants?

4. Name the parts of the stem and describe the work of each.

5. What is the meaning of the statement that most of the energy on the earth comes either directly or indirectly from the sun?

6. Describe the process of photosynthesis in detail.

7. Where does a plant obtain the materials from which food is manufactured?

8. How do the various classes of food differ from one another?

9. Why are plants such heavy water users?

10. What purposes does transpiration serve in the plant?

11. How is the process of respiration carried on in the plant?

12. Describe the process of digestion in the plant.

13. Describe the process of pollination.

14. What is the function of the seed?

15. How may seeds and fruits be dispersed?

16. Why do plants produce so many seeds?

17. What is the value of plants to man?

REFERENCES

GAGER, C. S. General Botany.
HOLMAN, R. M., and ROBBINS, W. W. Textbook of General Botany.
POOL, R. J., and EVANS, A. T. First Course in Botany.
ROBBINS, W. J., and RICKETT, H. W. Botany.
SMITH, G. M., OVERTON, J. B., GILBERT, E. M., DENNISTON, R. H., BRYAN, G. S., and ALLEN, C. E. A Textbook of General Botany.

CHAPTER III · The Animal Body is Much Like a Machine

Introduction. Animals in some ways are similar to plants. They must take food, they must digest and assimilate it, they must throw off waste, and they must reproduce. Yet the manner in which these functions are accomplished and many other details of their life are very different from those of plants and often more interesting. Since our purpose in these books is to acquaint ourselves with the universe in which we live, our story would not be complete if we left out animals. Moreover, man himself is an animal, and that fact adds interest to the study.

It is evident that we cannot discuss all animals or even all the various groups of animals. That would require many volumes. We are interested not so much in how all animals are built and how they live as in how the course of animal life, in general, runs. What we need to know is how a typical animal is constructed and how its body functions. Then, from this knowledge we may gain some ideas, broadly speaking, as to how animals are made and how they succeed. For this purpose we might select the rabbit or the dog, if we chose, either one serving very well as an example. But certainly the most interesting animal in all the world is man himself. From time to time we shall need to make little excursions out among the other animals to see how certain structures and functions developed. But we shall do this not so much for the sake of the other animal characteristics themselves, but rather that we may better understand man's relation to other organisms and better comprehend his place in nature.

The skeleton is the structural framework of man's body. All larger animals must have a frame, or a skeleton. If they did not have one they would be shapeless, unsightly objects

77

spread out like a mass of jelly, and unable either to move or to secure food and protection. Strange to say, many animals when they produced a skeleton formed it on the outside of their bodies. It resembled the coat of mail put on by the old knights before they engaged in the contests of the tournament. In a large measure too these *exoskeletons*, as they are called, serve the same purpose. For this type of skeleton, possessed by such animals as the oyster, the crawfish, and the insect, not only furnishes rigid structures for the attachment of muscles but also forms a singularly effective organ of protection. Any boy who has molested a turtle and has seen it draw its legs, tail, and head up under its hard, rooflike back, or who has tried to remove a clam from its tightly closed shell, has first-hand knowledge as to how efficient this type of skeleton is as a protective structure. But just as the knights found their armor clumsy and unwieldy to move about in, so were these animals encumbered with an exoskeleton unsuited to free movement. Nature, then, just as she has done thousands of times in the past, but just how nobody knows, seems to have introduced another general line of development. She provided her children with an inside skeleton called an *endoskeleton* — one that would permit of more agile movement, a more varied response, and, consequently, a better adaptation. So animals, from the most primitive fishlike organisms on up to and including man, have an endoskeleton.

In man the skeleton is an admirably fashioned structure consisting of about two hundred and six bones. A few persons have an extra pair of floating ribs to make two more bones. Once in a while a person is found with an extra head bone too, caused by the failure of the cranial bones at the top of the head to coalesce properly in infancy. The human skeleton is articulated into a system of levers and points of muscular attachment so as to combine the whole into a surprisingly efficient machine. So, then, the bony structure of man, and, in truth, of all higher animals, not only holds the body upright, but its individual bones, as levers, form part of the machine that aids in moving the body about.

So important is the skeleton to the organism that it is

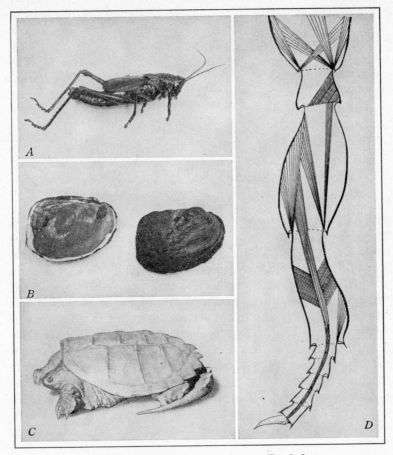

FIG. 25. Series of Photographs showing Exoskeleton

A, grasshopper; B, mollusk; C, turtle; D, leg of insect showing
attachment of muscles

practically formed by the end of the second month of the
embryological period. In this early state, however, the framework of the body is not true bone but either a tough, elastic
gristle, technically known as cartilage, or bone-forming membranes. Almost immediately thereafter the cartilage and the
membranes begin to take on lime and phosphorus and pass
over into true bony tissue. This change is slow, nevertheless,

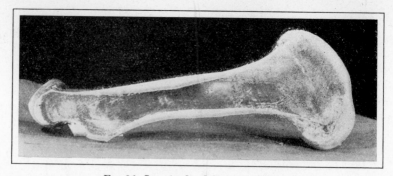

FIG. 26. Longitudinal Section of Bone

Note the marrow in the central part and the cancellous bone near the ends

and requires years for its completion. At birth and all through early childhood and youth a portion of each bone remains soft and cartilaginous so that growth can take place in it. In the shaftlike bones of arms and legs and other parts of the body the cartilaginous region is near the extremities, just back of the enlarged ends. Outside of the possibility of growth, the soft bones, with their cartilaginous sections, are of the greatest importance to children and other young animals. It enables them to fall and to tumble about violently with perfect impunity as they always do in early age when learning to walk or to perform other highly coördinated movements. If older people took such falls, bone fractures of the most disastrous character would result. By the end of about the twenty-fifth year the bones have all reached their full size, and ossification is complete.

The structure of the main skeletal bones is interesting. The most skilled mechanical engineer could not excel the efficiency of nature in building up these lever-like supports. Cylindrical in contour, with a relatively small, hard central shaft and enlarged porous ends, these bones are admirably fitted for their work. The highly ossified dense shaft fits the bone to endure great strain without breaking, while the enlarged ends furnish a splendid base for the similar end of the bone above. Within, the large bone extremities are not solid but are highly porous, being filled with what is known

as cancellous bone. This cancellous bone affords considerable
strength, yet is comparatively light. In older persons these
cancellous regions may dis-
integrate, leaving the bone
hollow and greatly weak-
ened. The more solid
parts of the bone tend also
to become brittle with age
and disease. For this reason falls by old people are
much more likely to pro-
duce a fracture than when
incurred by persons of
middle age. The bone
canal is filled with a fatty
substance called marrow,
which, strange to relate,
has very little to do with
the skeletal system di-
rectly; rather, it is related
to the circulatory system,
as we shall see later.

The movable joints of
bones are highly efficient
and smoothly working
structures. The articu-

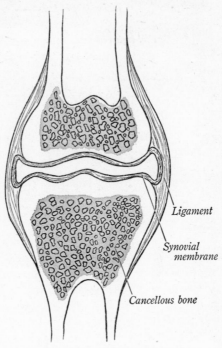

FIG. 27. Section of a True Joint

lating extremities are so enlarged as to give ample bearing surface, and each extremity is then padded with a smooth, elastic
layer of cartilage. These cartilage-covered ends are further inclosed by a membrane (the synovial membrane) which secretes
a fluid that keeps the joint lubricated so as to insure easy,
smooth, and ready movement. Then, to hold the bones in
place and to keep each contour of one bone bearing upon its
counterpart on the face of the other, the whole joint is lashed
together by tough, strong sheets of connective tissue called
ligaments. The muscles, too, with their stout tendons
stretched from bone to bone across the joint, assist in holding
the joints intact and ready to function properly.

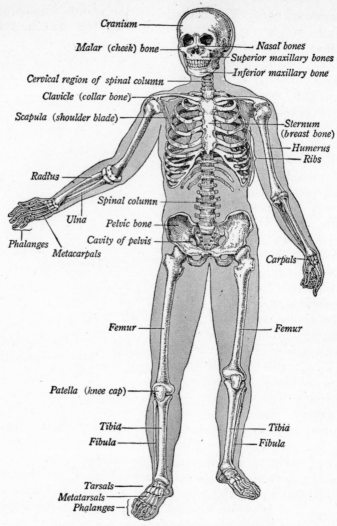

Cranium

Malar (cheek) bone

Nasal bones
Superior maxillary bones
Inferior maxillary bone

Cervical region of spinal column

Clavicle (collar bone)

Scapula (shoulder blade)

Sternum
(breast bone)

Humerus
Ribs

Radius

Spinal column

Ulna

Pelvic bone

Phalanges

Cavity of pelvis

Metacarpals

Carpals

Femur

Femur

Patella (knee cap)

Tibia

Tibia

Fibula

Fibula

Tarsals
Metatarsals
Phalanges

Fig. 28. The Human Skeleton (Front View)

In youth the cartilaginous pads of joints are highly elastic and the young person walks with a "spring in his heel." As age comes on lime is deposited in the cartilaginous tissue and the step then becomes less elastic.

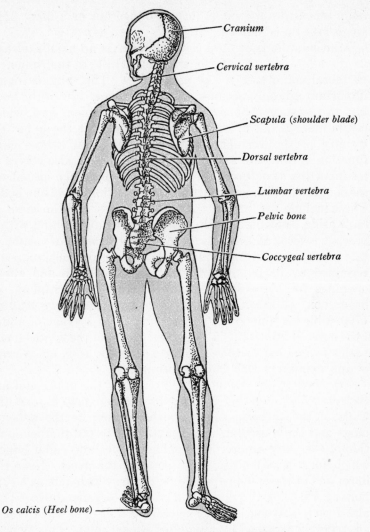

Cranium

Cervical vertebra

Scapula (shoulder blade)

Dorsal vertebra

Lumbar vertebra

Pelvic bone

Coccygeal vertebra

Os calcis (Heel bone)

FIG. 29. The Human Skeleton (Back View)

Because the freely moving joints are such intricately constructed organs, an injury of any kind to them is not a matter to be taken lightly. Indeed, a joint infection once incurred is a most serious matter. Dr. Clendening says that "a dirty knife stuck into a joint is just as dangerous as a dirty knife stuck into the abdomen." [1]

Man has the type of skeleton known as an axial skeleton, as do all other vertebrate animals. Of this the spine, or vertebral column, is the principal part. The spine is highly important not only because it forms the chief axis of the body but also because it houses and protects the spinal cord, which is the main trunk of the nervous system. At the upper extremity of the vertebral column is the skull, whose chief function is to incase and protect the brain. It also serves as a supporting structure for a whole group of highly specialized sense organs which are located at the anterior end of the body. These include the tongue, the nose, the eyes, and the ears. In fact, aside from the mouth, which affords an opening with a passage leading to the interior of the body, and jaws to facilitate mastication of food, the whole contour of the skull is determined by its adaptations to house the brain and afford openings for these special sense organs. The cranium, or brain box, of the normal white male ranges from 1220 to 1790 cubic centimeters in capacity, with an average of about 1450 cubic centimeters. The cranial cavity of the white female varies from 1090 to 1550 cubic centimeters, with an average of approximately 1300 cubic centimeters.

Attached to the vertebral column near the upper end are the ribs which, together with the breastbone, help to form the walls of the thoracic cavity, within which lie the delicate lungs and the muscular heart. Attached to the thoracic portion of the skeleton are the shoulder blades and collar bones, which form a sort of girdle to support the arms. Near the lower end of the spine is the pelvic girdle, which acts as a supporting base for the spine and visceral organs above. This also forms a strong, stable skeletal portion to which the legs are securely attached by a ball-and-socket joint. This type of joint affords the maximum degree of strength and movement.

The bones of the body, as compared with other tissues, are poorly supplied with blood vessels. For this reason all bone injuries, especially compound fractures, in which the danger of

---

[1] Logan Clendening, *The Human Body*, p. 58.

infection is imminent, should receive the best of surgical attention. Infections such as tuberculosis of the bones should also receive immediate attention; for the limited flow of blood to these parts makes it difficult for the system to overcome the invading germs and heal the afflicted part. Fractures should be reduced (set) with great care, and the broken ends of the bones should be held securely together in order that the slow, bone-healing processes of nature may continue until the union of the broken members is again complete.

Man's skeleton is held in place and moved by muscles. The skeleton of the body is like the mechanical parts of a steam engine. The engine may be built with the greatest care and its parts may be capable of smooth and efficient performance; but if it has no power behind the pistons, no steam in the cylinders, it remains a passive, inactive machine. So it is with the human skeleton. The bony levers form a potential machine capable of ready action, but of themselves these bones can do nothing. They must be moved by a force from without. This force is furnished by the muscles.

The muscular parts of man's body form the more solid fleshy portions. In the carcasses of animals killed for food the muscles comprise the red meat. The highly nutritive round steak of the beef is practically all composed of the powerful hind-leg muscles of the animal, together with the thin sheets of connective tissue that hold the individual muscle strands together. The more muscle the cut contains and the less connective tissue it has, the tenderer the steak will be.

The power of contraction to produce movement is possessed by all protoplasm to a greater or less degree. Even the simplest animal imaginable, the amœba, is believed to thrust out its pseudopodia by protoplasmic contractions in other parts of the cell. In worms like the earthworm the walls of the body are composed of two layers of smooth muscle. In one of these the cells extend lengthwise, and in the other they extend in a circular fashion. The worm progresses by alternately contracting the longitudinal muscles, which shorten and thicken the body, and then by contracting the circular muscles, which stretch the body out again.

In higher animals, such as man, the muscular tissue is of three types. The main skeletal muscles are *striated*. They are composed of elongated cylindrical cells that contain many fine contractile cytoplasmic fibrils. These fibrils show alternate dark and light bands, and it is for this reason that they are said to be striated. The cells are aggregated into muscular strands and bundles that are held together by sheaths of connective tissue. The action of the striated muscles is quick and powerful, and it is because of this that they are so efficient in producing movements in the more massive and rapidly bending parts of the body,

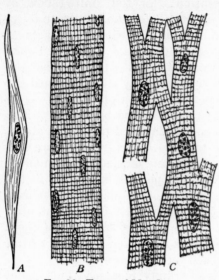

FIG. 30. Types of Muscles

A, smooth muscle; *B*, striated muscle; *C*, heart muscle

such as the arms, legs, back, and head. It is estimated that the stress on the biceps muscle of the arm when the average man is "chinning" himself is at least a thousand pounds. The skeletal muscles are attached to the bones by means of tough, strong endings of connective tissue called tendons. An excellent example is the large hard cord, the tendon of Achilles, just above the heel. This tendon is named after the supposed vulnerable spot in the heel of the Greek hero Achilles.

The *smooth muscles* are made up of spindle-shaped cells laid down in sheets. This type of muscle was introduced into the animal kingdom much earlier than the striated form and is the kind of tissue found in the contractile parts of many lower animals, such as the worms, clams, and snails. The action of the smooth muscle is relatively slow, and though it is a strong tissue it is not nearly so strong as the striated muscles. Animals equipped with smooth muscles only are always slow and sluggish in their movements. In the human body a remnant of the smooth-muscle structure is retained in the more slowly responding parts: for example, the walls of the stomach, intestines, bladder, blood vessels, Fallopian tubes, and a few other parts not under the control of the will. It is interesting to note that in most portions of the body where this type of muscle persists, it forms the walls of tubes — structures that in form are similar to those of the worms, in which the smooth-muscle organization is supreme. The structures in the human body composed of smooth muscles are usually made of two or three layers of cells. It is their combined action that produces the rhythmical peristaltic contractions so prominent in the normal functioning of stomach and intestines.

The *cardiac* (or *heart*) *muscles* are peculiar in that they resemble both the smooth and the striated muscles. They form the heavy walls of this powerful blood pump and are further unique in that they seem to possess the ability to initiate contraction within themselves. The contraction of all other muscles, as we shall find later, is initiated in response to a nerve stimulus. But although the contractions of the heart muscles are not initiated by the nervous system, the rate and frequency of heartbeat are partially controlled by it.

When the muscles in the human body are taken separately and named, their number seems almost legion, as students of human anatomy have found. The number probably varies slightly in different individuals, but normally the number of skeletal muscles alone has been given at approximately five hundred and fifty-five. The accompanying picture will give some idea of how intricately and effectively the muscular system is built up and arranged. Most assuredly one of the sacred writers was expressing a profound biological truth when he said that man was "fearfully and wonderfully made."

All the details of muscular action are not completely known. It has been found, however, that as a muscle contracts, the energy-containing substance, glycogen, decreases. Muscular energy and heat are liberated and lactic and phosphoric acids are produced. The source of energy, then, is the glycogen,

2

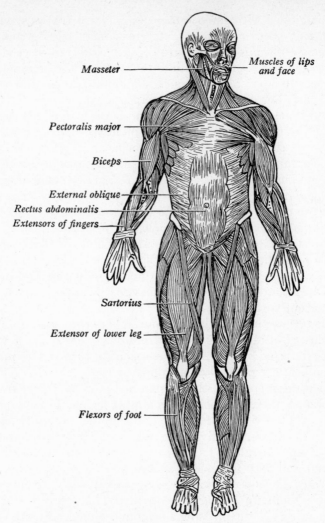

Masseter

Muscles of lips and face

Pectoralis major

Biceps

External oblique
Rectus abdominalis
Extensors of fingers

Sartorius

Extensor of lower leg

Flexors of foot

FIG. 31. Arrangement of the Skeletal Muscles of Man

and the energy is liberated for muscular use through chemical change. Either carbohydrate or a portion of the lactic acid is oxidized for the reconversion of the lactic acid into a hexose phosphate complex. The latter chemical change restores the

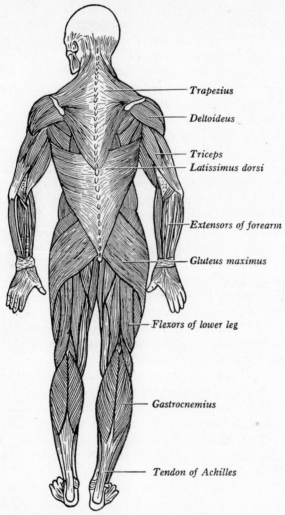

Trapezius

Deltoideus

Triceps
Latissimus dorsi

Extensors of forearm

Gluteus maximus

Flexors of lower leg

Gastrocnemius

Tendon of Achilles

Fig. 32. Arrangement of the Skeletal Muscles of Man

muscle to its original condition. The oxygen involved comes from the lungs and is carried to the muscle cells by the circulation of blood and lymph. As the glycogen is depleted in the muscle, it is replenished from the reserve supply stored in the liver. In reality, then, muscles are a sort of chemical engine which converts the chemical energy stored in the glycogen into a form of mechanical energy to be used in moving the parts of the body.

Muscular action is an interesting phenomenon. When the muscle begins to act, the small amount of carbon dioxide and lactic acid generated at the outset serves as a stimulant to muscular action. This increases the irritability of the muscle cells, which in turn enables them to respond to the nerve stimuli more vigorously and consequently to do more work than they could when contraction first began. This is one reason that race horses and baseball pitchers are "warmed up" by moderate preliminary exercise before they are sent into the strenuous work of the race or game. Considering the greater ease and facility with which one accomplishes mental tasks after a few minutes of effort, it may be that brain cells too need a preliminary "warming up" to do their best work.

As strenuous exercise continues, however, the lactic acid of muscular activity accumulates. After a while the consequent accumulation of this waste begins to interfere with the contraction of the muscle. A chemical change has taken place which appears to affect the muscle and motor end-organ and which enervates the muscle. Under these conditions the muscle may be whipped up by the action of a stimulant; but the results are only temporary, and in the end it is in a worse condition than when the stimulant was administered, because through increased action even more lactic acid has accumulated. The only source of permanent relief is rest. The inactivity of the rest period gives the blood stream an opportunity both to replenish the glycogen supply in the muscle and to dispose of the waste products. In case of muscular fatigue, then, sleep is the best restorer because all body processes fall to a low level. To effect these desirable changes, then, everyone needs adequate sleep. Massage of the fatigued member also facilitates recovery, because the increased circulation induced carries the waste products away more rapidly.

Often, after prolonged periods of comparative inactivity, especially in the case of middle-aged and older people, marked muscular soreness may follow violent exercise. This condition may be the result of two causes. It may be produced by an excessive amount of chemical change resulting from too prolonged exercise. In such cases sleep, the application of heat, and massage may all be of value as restoratives. The soreness may also be due to numerous little tears in the muscle cells brought about by too strenuous activity. For this condition rest is the best prescription. Athletes going into training, unless they "ease in" with great care, are likely to suffer severe muscular soreness from both these causes.

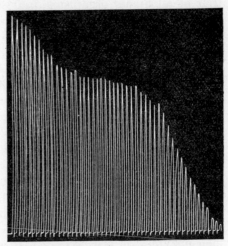

FIG. 33. Normal Fatigue Curve of the Flexors of the Middle Finger of the Right Hand

Lifting a weight of three kilograms. Contractions at intervals of two seconds (Maggiora) [1]

Animals utilize food as a source of energy. The bodies of higher animals are very much like machines in another respect. They must have a constant supply of energy to keep them going. To secure this energy, living things must take food, digest it, transport it about through the tissues, and there assimilate it. Waste products of assimilation too are bound to be formed which, if allowed to accumulate, would interfere with the working of the animal body as surely as dirt and mineral by-products interfere with the efficient operation of a gas engine. So these by-products of metabolism must be continuously removed if the animal machine is to run smoothly. As J. A. Thompson has said, the living organism is "a self-stoking, self-repairing, self-preservative, self-adjusting, self-increasing, self-reproducing engine."

[1] From Williams's *Text Book of Anatomy and Physiology.* Published by W. B. Saunders Company.

To study this whole energizing process, let us again turn to man as our example. We shall find that in man's body, as in that of all higher organisms, four services must be provided if this energy supply is to be continuous.

Food is prepared for absorption in the alimentary canal. The preparation of the food is accomplished within a long, winding tube called the alimentary canal (Latin *alimentum*, "nourishment") which traverses the whole length of the body cavity. In the average man it is approximately thirty feet long. The food is taken into this tube, and efficient provision for the digestion of each kind is made at appropriate points. The food products are moved along through the digestive tract by peristaltic movements which follow one another in rhythmic succession.

Relatively large quantities of digestive juices are added to the food to soften and help to liquefy it as it progresses along through this tortuous tract. These juices secreted by different organs consist of copious quantities of water, together with such enzymes as are necessary to effect digestion. We can perhaps gain a clearer understanding of the food-preparing service by tracing food through the digestive tract.

The work of the mouth in digestion. In the mouth the food is masticated and mixed with saliva to soften it in order that it may be swallowed. The mastication breaks the food into small particles not only that it may pass through the pharynx and esophagus with ease, but also that it may present a large surface for the action of the digestive juices. The inability of the enzymes to attack with facility the large particles of food swallowed in hasty eating is one source of the discomfort so often experienced shortly afterwards. Poorly masticated food lies longer in the stomach, too, and this in turn is conducive to fermentation that may add to the distress. One constituent of the alkaline saliva added to the food in the mouth is ptyalin. This is a starch-digesting enzyme that tends to break the starch into maltose, a complex sugar which has to be further broken down before it can be used by the body. When the food is ready to be swallowed, deglutition takes place, and the softened, masticated bolus of food passes quickly through the pharynx and esophagus into the stomach.

The action of the stomach in digestion. The stomach of the average adult holds in the neighborhood of a quart. In

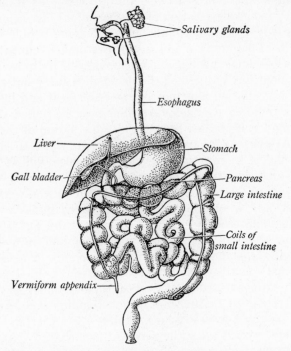

Fig. 34. Alimentary Canal of Man showing **Different Parts** in their Respective Positions

reality it is nothing more than a specialized swollen portion of the alimentary tract. It serves two functions: it is a temporary storehouse of great importance, and it advances the digestion of the food another step.

Our knowledge of the gastric (Greek *gastēr*, "stomach") functions was greatly enlarged through the accidental shooting of a young Indian by the name of Alexis St. Martin at Mackinac, Michigan, in 1822. The accidental discharge of a shotgun tore off the skin and muscles of the abdominal wall just over the stomach and also carried away a portion of the stomach wall itself. Dr. William Beaumont, a young army surgeon, called in to attend him, experimented by stitching the edges of the stomach to the torn abdominal wall. Surgery in those days was not nearly so far advanced as it is today. Contrary to the expectation of Dr. Beaumont his patient lived, and the edges of the wound healed, leaving a hole into the stomach. Beaumont could see the stomach contract. He could also see the gastric juice ooze out on its inner surface. Being scientifically inclined, he decided to carry out some experiments. He tied a string to a piece of meat and inserted it into the Indian's stomach. An hour later he pulled the meat out, to find it frayed at the edges by the action of the gastric juice. At the end of two hours the meat was half digested. At the end of three hours the meat had been completely dissolved, and the string could be withdrawn with the empty loop intact. Gastric juice was poured out by the walls of the stomach only when food was being chewed in the mouth or when it had entered the stomach. Using a tube, Dr. Beaumont withdrew some of the gastric juice and had it chemically analyzed. This analysis revealed hydrochloric acid as one of the normal constituents of the stomach secretion. Beaumont, after an extensive study of St. Martin's stomach, prepared a written report of his findings which remains a classic in the field of gastric physiology even today. St. Martin lived for almost sixty years after the accident occurred.

In the stomach vigorous peristaltic movements thoroughly mix the food with a heavy flow of gastric juice. Like other digestive fluids, this juice is largely water. However, in addition to the water, it contains two important enzymes and the acid already mentioned. The principal enzyme is pepsin, which attacks protein foods and hastens their partial decomposition. The products formed are peptones and proteoses, which are intermediate products in protein digestion and which themselves must be further broken down before they can be absorbed and assimilated. Another enzyme, rennin, assists in the coagulation of soluble milk proteins, and the curds so formed are then acted upon by the pepsin. The hydrochloric acid stimulates gastric digestion by acidifying the food, which is necessary if the pepsin is to work normally. This acid is usually found in the proportion of about 0.2 per cent. It is believed by many to function also as a disinfectant to kill bacteria that enter the stomach with the food. Gastric action and secretion are inhibited by stimulation of the sympathetic centers of the autonomic system. For this reason anger, pain, worry, solicitude, and distaste for food may delay digestion. Active exercise also retards digestion by deflecting a heavy blood supply away from the visceral organs to the skeletal muscles. When gastric digestion is greatly delayed, the presence of bacteria and lactic acid may cause the starches and sugars to ferment and produce a condition of distress. It is for this reason that people who are suffering from mental strain are likely to be the victims of indigestion. As the food reaches the proper stage of digestion peristaltic action gradually forces it through the sphincter-controlled pylorus into the small intestine.

The contribution of the small intestine to the food-preparing service. The small intestine is a convoluted tube about twenty-two or twenty-three feet long and about one and one-half inches in diameter. This section of the alimentary canal is the most active in food digestion. Here the food is mixed with the pancreatic juice from the pancreas, the bile from the liver, and the intestinal juice from the walls of the intestine. The pancreatic juice contains three very potent enzymes. Of these the amylase, very similar in its action to ptyalin, attacks the starch, which escapes the digestive processes of the mouth and decomposes it into semi-digested sugars, especially maltose, or malt sugar. The lipase breaks down the fats, forming glycerin and fatty acids. The trypsin (believed by some to be a complex ferment) may be said, for our purpose, to attack both the undigested proteins that may have passed the stomach and the peptones and proteoses produced by the action of the gastric juice. Under the influence of this enzyme the proteins are almost completely reduced to soluble amino-acids which are the fully digested end products.

Under the stimulus of food in the duodenum, or upper part of the small intestine, bile in relatively large quantities is poured in from the liver. It enters through the lower portion of the same duct that carries the pancreatic juice. When the duodenum is empty the sphincter muscle closing the bile duct contracts, and the bile itself backs up into the gall bladder, where it accumulates until it is needed. The bile, like the pancreatic juice, is alkaline and assists materially in the digestion of fat by saponification of the fatty acids. The addition of bile to a mixture of olive oil and pancreatic juice increases five to ten times the amount of oil digested. The bile also serves as an excretory medium in which toxins, metals, and products of cellular disintegration are believed to be eliminated from the system. The human bile is of a clear, watery consistency and is yellow, brown, or greenish in color. In the adult the daily secretion may reach five hundred cubic centimeters or more, and it is this constituent that gives the feces their characteristic brown or yellowish color.

The intestinal juice also contains several enzymes which carry the process of digestion through to completion. Among these are erepsin, which transforms into amino-acids any partly digested proteins that may have escaped the action of the trypsin. Several sugar enzymes are also present which break the more complex sugars down into simple glucose or similar soluble and diffusible forms of sugar. Among these are invertase, which acts on cane sugar; maltase, which attacks the malt sugars; and lactase, which digests the milk sugars.

Many factors affect the rate of digestion in the small intestine. Investigations indicate that under favorable conditions food may begin to enter the large intestine from two hours after eating to a little over five hours and that normally from nine to ten hours are required for it all to pass from the small intestine.

The function of the large intestine in the food-preparing service. The large intestine consists essentially of three parts: the ascending colon, the transverse colon, and the descending colon, which terminates in the rectum. It is normally between four and five feet long and in its largest parts is approximately two and a half inches in diameter. Throughout most of its length it is full of folds and pockets, in this respect being quite different from the small intestine.

A further addition of intestinal juice to the food mass occurs in the large intestine, but the secretion contains no digestive enzymes. Here further decomposition is carried on by the enzymes previously added to the food and by bacterial action. Extremely large numbers of bacteria of different species are found in the contents of the large intestine. The slow movement of the food is especially conducive to their multiplication. Certain tests have shown that from a fourth to a third of the water-free material evacuated is composed of bacteria. The final products of food decomposition found here are many and varied. Moreover, organic material added through the bile, the various secretions, and from the walls of the alimentary canal itself add to the complexity of the mass.

Considerable absorption of water takes place in the large intestine, as is shown by the way in which the feces are dried down. Under normal conditions this absorption is desirable. As to the degree of auto-intoxication that occurs in the average individual from toxic products there is a very wide divergence of opinion. Some extremists believe that deleterious toxic absorption here can be avoided only by the most vigorous colon management; others hold that the degree of danger to health from this source is but slight if indeed there is any at all. As is usually the case the safe ground probably lies somewhere in between. If pronounced constipation occurs, there is doubtless some undesirable absorption; toxic products, such as indican, recovered from the urine under such conditions, furnish concrete evidence of the fact. Some of the conditions of distress, such as headache and other unpleasant symptoms, no doubt are the result of an overloaded condition of the large intestine which causes pressure on the nerve endings in the colon. Nevertheless the average person who has one good bowel evacuation each day is not likely to suffer auto-intoxication from this source. However, this daily defecation is a very important matter, and its time can usually be fixed fairly definitely by habit. People who experience difficulty in getting proper elimination can usually correct the condition by (1) responding promptly to the least desire to evacuate; (2) eating a reasonable amount of bulky, indigestible material such as is

TABLE I. SUMMARY OF THE PROCESS OF DIGESTION

Section of alimentary canal	Digestive juice secreted	Source of digestive juice	Enzymes contained	Kinds of food acted upon	Intermediate products of digestion	End products of digestion — the traveling forms
Mouth	Saliva	Salivary glands	Ptyalin	Starch	Malt sugars	
Stomach[1]	Gastric	Walls of stomach	Pepsin Rennin	Proteins	Peptones and proteoses	
Small intestine	Pancreatic	Pancreas	Amylase Lipase	Starches Fats	Malt sugars	Glycerin and fatty acids
			Trypsin	Proteins Peptones Proteoses		Amino-acids
	Intestinal juice	Intestinal wall	Erepsin	Proteins Peptones Proteoses		Amino-acids
			Invertase Lactase Maltase	Cane sugar Milk sugar Malt sugars		Simple sugars, such as glucose
	Bile	Liver	None	Fats		Assists in digesting fats

[1] The stomach also secretes hydrochloric acid, which is not an enzyme but which stimulates action of gastric enzymes and partly sterilizes the food.

found in leafy vegetables and bran products; (3) eating fruit to stimulate visceral action; (4) drinking six or eight glasses of water daily; and (5) tak-

ing regular exercise, such as gymnasium work, calisthenics, golf, walking, or other exercises that will involve the abdominal muscles.

Digested food must be absorbed. After food in the alimentary canal is prepared for organic use by being liq-

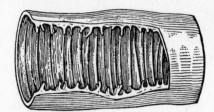

FIG. 35. Folds of the Living Membrane of the Small Intestine[1]

uefied, it must next be taken up by the walls of the digestive tract in order that it may pass into the blood stream and be distributed throughout the body. This passage of liquid food through the walls of the alimentary canal into the blood stream is called *absorption*. Until absorption does take place the food is, in reality, still outside the body.

It is probable that some food is removed all along the alimentary canal from the stomach to the rectum. But the intestines are the main organs of food absorption as well as digestion, and in absorption, as in digestion, the small intestine is much the more active of the two. For this there are three reasons. The small intestine is several times longer than any other section of the alimentary tract. Its inner surface is also greatly increased by little projections called villi and by an elaborate system of folds. Some authorities state that the actual absorbing area is about one hundred square feet, or equal to the floor space in a room ten feet square. The small intestine also receives, as has already been pointed out, the most potent digestive juices secreted by the body; so by the time the food has passed the upper part of the small intestine it is highly liquefied and ready to be removed.

The digested food reaches the blood stream through two paths. One is through the lacteals and lymph vessels; the other is through the blood capillaries and veins. We shall

[1] From Kimber and Gray's *Textbook of Anatomy and Physiology*. By permission of The Macmillan Company, publishers.

study the lymphatic system more extensively later, so it will suffice to say here that the lymph vessels are tubes of various

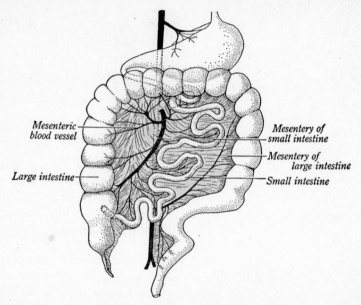

Mesenteric
blood vessel

Mesentery of
small intestine

Mesentery of
large intestine

Large intestine

Small intestine

FIG. 36. The Intestinal Mesentery

sizes which carry liquid food, waste products, dissolved oxygen, and other dissolved substances from place to place through the body. The termini of these tubes in the intestinal walls are called lacteals. The lacteals absorb practically all the products of the digestion of fat, such as glycerin and fatty acids after they have been recombined into fat droplets in the epithelial cells of the villi. The blood capillaries, under normal conditions, take up the amino-acids and the simple sugars.

A pathway through which the blood and the lymphatic vessels pass as they carry these food products from the intestines to other parts of the body is afforded by a thin membrane called the mesentery. This membrane suspends the intestines and holds them in a more or less neat and orderly mass, as may be observed by examining the viscera of a slaughtered animal.

The lacteals, like tiny rivulets at the headwaters of a stream, flow together to form larger lymphatic ducts and finally empty their contents into a large lymph vessel called the thoracic duct. The thoracic duct, taking its origin near the region of the kidneys, extends upward along the backbone just inside the body cavity through the thorax, and pours its flow of lymph, including the fats, into the blood stream under the left shoulder. In the carcasses of dressed beeves or hogs, or even when dressing a chicken, the thoracic duct may be seen easily.

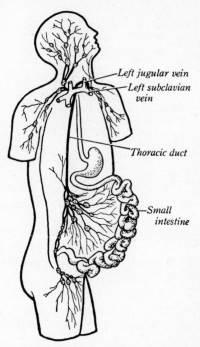

Left jugular vein

Left subclavian vein

Thoracic duct

Small intestine

Fig. 37. The Principal Lymphatic Areas[1]

The lacteals are the lymphatic capillaries in the intestine

The tiny blood capillaries converge to form larger vessels, and in the end their contents are carried to the liver through the portal vein. Here some proteins may be stored, although most of them are carried directly to the tissues of the body. The sugars, on the other hand, may be stored temporarily in the liver cells as glycogen or as a fat. Some sugar may be stored as glycogen in the muscles also. The liver is the largest gland in the body. It weighs about three pounds and forms an admirable storehouse for the large amount of sugar thrown into the blood stream after the digestion of a good meal. In the liver cells the stored products are gradually transformed into sugar again, which is slowly but constantly fed out to the copious blood stream that passes

[1] Redrawn from Kimber and Gray's *Textbook of Anatomy and Physiology*. By permission of The Macmillan Company, publishers.

through this organ. By way of the blood stream, then, energy-supplying sugar ultimately reaches all parts of the body.

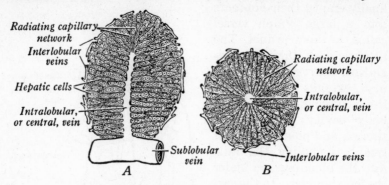

FIG. 38. Lobules of the Liver

A, longitudinal section of the lobule; *B*, cross section of the lobule [1]

The transportation service is performed by the blood and lymph circulations. To function properly every living cell in the body must constantly be supplied with certain necessary materials. From every cell also the waste products of metabolism must be continually removed to prevent clogging with waste and poisoning from self-generated toxins. To carry these necessary substances to the cells and to transport the undesirable ones away from the cells requires an efficient transportation line. In reality there are two transportation systems that aid each other in performing this important service. They are the blood circulation and the lymphatic circulation.

The blood as a transporting medium. The circulatory system in the higher animals is composed of blood inclosed in a system of tubes of varying size. This fluid is kept flowing every minute from the day of birth to the hour of death by the action of a strong muscular pump called the heart. In lower animals, such as the insects, spiders, mollusks, and worms, the blood may be almost wholly transparent and may be propelled about through the body either by a simpler form of heart or by contractile blood vessels.

[1] From Kimber and Gray's *Textbook of Anatomy and Physiology*. By permission, The Macmillan Company, publishers.

In man we find the blood to be a very interesting and complex liquid tissue. In the main it consists of a more fluid part called the plasma, in which cells specialized for certain definite purposes are carried about freely. This plasma is approximately 90 per cent water, in which are dissolved various food and waste products, together with substances designed to aid the body in combating disease germs and certain hormones manufactured to coördinate the physiological processes.

Exceedingly large numbers of red corpuscles are found floating in the plasma. In healthy, normal blood the number ranges from about 4,500,000 to 5,000,000 per cubic millimeter and varies with altitude, state of health, and other conditions. The color of these red corpuscles is due to an iron-containing substance called hæmoglobin, which enables them to accept oxygen in loose chemical composition and to carry it to all parts of the body where it is needed. The oxygen compound formed with the hæmoglobin is oxyhæmoglobin. When the red corpuscles reach points in the tissues where the oxygen supply is low, the oxygen in some way is disengaged from the hæmoglobin and again becomes a gas in solution available for use. Blood well supplied with oxygen has a bright-red color; but when the content of this gas is reduced, the red corpuscles take on a dark-red or crimson hue. It is from its characteristic bright scarlet color, as we shall see later, that arterial blood can be distinguished from venous blood.

The white blood corpuscles, or leucocytes, are not nearly so abundant as the red ones and have a wide variety of size, form, structure, and function. In reality they are gray, not white, and are found in the plasma at the rate of from five thousand to seven thousand per cubic millimeter. The number is increased rapidly by infections such as pneumonia, appendicitis, or abscesses. This increase in white-corpuscle count is a valuable aid to the physician in diagnosing certain diseases and bodily infections. In other diseases, such as typhoid fever and malaria, the number of leucocytes is characteristically reduced. The white blood corpuscles are very active in the healing of wounds. When an injury of any kind occurs, huge numbers of those leucocytes are carried by the blood stream

2

to the site of the trouble, where they migrate through the capillary walls and assist in healing the wound. Subsequently they become active in repairing and regenerating the tissues to close up the wound. The white blood corpuscles are also very active in combating disease, as will be made clear in a later chapter. These cells are amœboid in character, and the fact that they can squeeze through tiny apertures in the walls of the capillaries and go exploring about among the cells that compose the tissues is another characteristic which greatly facilitates their work. There are other minute disks, called platelets, in the blood that aggregate at the ends of severed vessels and help to form clots.

The systemic circulation supplies blood to all parts of the body except the lungs. Leading out from the heart is a huge artery called the aorta, which divides and subdivides until its branches reach and supply all parts of the body except the lungs. The walls of these arteries are, relatively speaking, thick and muscular and expand and contract with each heartbeat. At the end of the minute arterial divisions the inner lining of the tiny arteries extends on in the form of very slender and extremely thin-walled blood vessels called capillaries (Latin *capillaris*, "hair"). The capillaries traverse every part of the body from head to toe except the cartilages, hair, nails, cuticle, and cornea of the eye. Without special knowledge of them it is difficult for one to appreciate just how thoroughly the tissues are supplied with capillaries. Dr. Krogh of Denmark, who received the Nobel prize for his work on the human circulation, tells us that if all the capillaries in the average human body could be opened up and spread out flat, their total area would just about cover a city block; that if they could be placed end to end along with the other blood vessels, the whole would reach around the earth two and one-half times. He completes the picture by stating that the cross section of an active muscle strand having the diameter of a knitting needle would present about eight hundred capillaries. The bore of these tiny blood vessels is so minute that the larger blood corpuscles have to stretch out and elongate greatly to slip through them. The thin walls, as has already been mentioned, appear to have microscopic apertures through which the amœboid white blood corpuscles can leave the blood

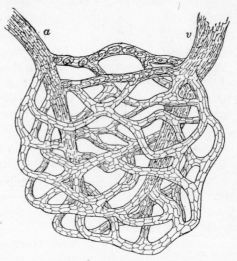

stream and squeeze through into the lymph spaces surrounding the cells of the tissues. The blood while passing through the capillaries gives up its transported food materials and oxygen to the tissues and receives in turn its cargo of waste products to be carried out of the body.

The capillaries lead into very fine veinlets which unite confluently to form larger veins. Finally, a single large vein from the head and arms and another from the lower part of the body pour their contents

FIG. 39. The Relation of the Capillaries to the Vein and the Artery

a, artery; *v*, vein [1]

into one of the heart chambers. Many of the veins of the body have flaplike valves that permit the blood to flow toward the heart but prevent its flow in the opposite direction. These circulatory channels, which lead from the heart to all parts of the body and then return again, form what is known as the *systemic circulation*.

The pulmonary circulation carries blood to and from the lungs. Contrasted with the systemic circulation is the *pulmónary circulation*, which carries the blood from the heart out through the lungs and then returns it to the heart. The artery leading to the lungs is called the pulmonary artery and is the only artery in the body that carries impure, venous blood. After passing through the dense meshwork of the pulmonary capil-

[1] From Kirkpatrick and Huettner's *Fundamentals of Health*. Ginn and Company, publishers.

laries, the return flow of the blood is made through the pulmonary veins. These veins, as contrasted with the pulmonary artery, are the only veins that carry fully oxygenated blood.

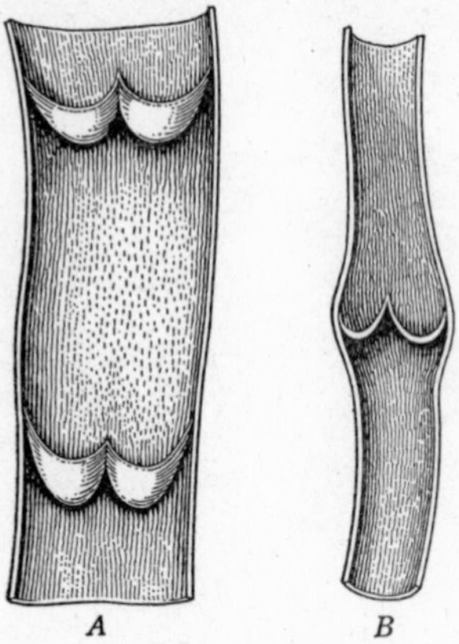

FIG. 40. The Character of Valves in Veins

A, a vein laid open; *B*, longitudinal section of a vein showing the valve closed [1]

The heart forces the flow of blood through both circulations. The heart serves as both pump and driving motor for the circulation. It is a large muscular organ which occupies a position slightly to the left side of the thorax and is composed of four chambers, separated from one another by a middle partition and flaplike valves. The two chambers at the large end are called *auricles*, and the heavier-walled, larger chambers at the pointed end are called *ventricles*. From the two large incoming veins, the *superior vena cava* and the *inferior vena cava*, the venous blood from the systemic circulation flows into the right auricle. From there it is forced by contraction of the auricular wall into the right ventricle. About a tenth of a second later the right ventricle contracts to force the blood out through the relatively short pulmonary circulation. Valves close the path behind to prevent the blood from flowing backward into the auricle. The left auricle receives the blood from the pulmonary veins and, by its contraction, forces the blood through the valves in its floor down into the left ventricle. The walls that form the left ventricle are much thicker and stronger than the walls of the other chambers because it has the hardest work to do. By the powerful contraction of the left ventricle the blood is forced all over the body through the systemic circulation. The moment the left ventricle contracts, the muscular arteries expand to permit the blood's immediate expulsion from the heart. Then, as the ventricle expands and

[1] From Kimber and Gray's *Textbook of Anatomy and Physiology*. By permission, The Macmillan Company, publishers.

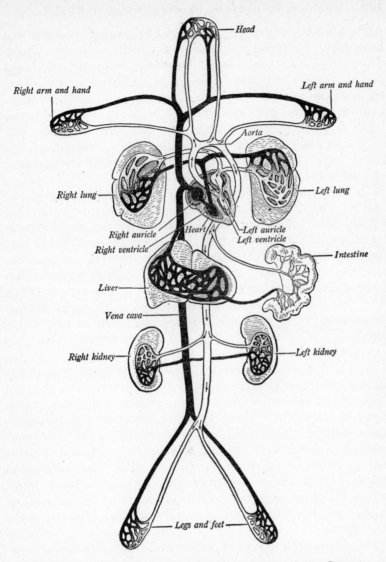

Right arm and hand

Left arm and hand

Head

Aorta

Right lung

Left lung

Right auricle

Left auricle
Left ventricle

Heart

Right ventricle

Intestine

Liver

Vena cava

Right kidney

Left kidney

Legs and feet

FIG. 41. Diagrammatic Representation of the Circulatory System

The clear lines are arteries; the dark lines are veins. Note that the pulmonary system and the systemic system are connected only through the heart [1]

[1] From Kirkpatrick and Huettner's *Fundamentals of Health.* Ginn and Company, publishers.

fills again with blood, the arterial walls gradually contract to keep the blood flowing steadily through the vascular system. The rate of heartbeat varies, depending upon the individual and his degree of activity. The average pulsation is from about 66 to 72 per minute. Violent exercise speeds heart action up in order to carry more food and oxygen to the tissues. During sleep or other periods of inactivity the pulse rate may fall as low as from 55 to 60 per minute. The movement of the blood back through the veins is also assisted by pressure exerted during muscular contraction.

Part of the transportation service is performed by the lymphatic system. Practically all cells of the body are bathed by a thin, watery fluid which is called tissue fluid by some and lymph by others. This fluid occupies tiny slitlike spaces between the capillary walls and the cells that comprise the tissues. So numerous are the slits that almost every cell throughout the body abuts on one or more of them. For this reason these tiny fluid-filled spaces are often referred to as water fronts. As a result of this provision it would not be far wrong to say that every cell of the body lives in a watery medium, much as a protozoan or a fish does.

The lymph acts as a retailer between the blood as it flows through the capillaries and the cells that form the tissues. Liquid food, plasma, oxygen, chemical salts, and white blood corpuscles leave the blood stream by way of the capillary walls and enter the water front. From there the dissolved substances diffuse into the cells to supply their needs. Waste matter escapes back through the cell walls into the lymph spaces. From the water front this metabolic waste may diffuse into the capillaries to be eliminated at special points in the body, or it may move away through the lymph spaces themselves.

Projecting into the fluid which fills the tissue spaces are the closed ends of millions of tiny tubes that converge to form larger tubes. These tubes are called lymphatics. They are filled with lymph, plasma, waste products, food, and other substances received from the water front. The lymphatic tubes continue to unite until finally they pour their contents into the blood stream under the shoulder. The largest of these tubes, the thoracic duct, carries not only lymph but also the

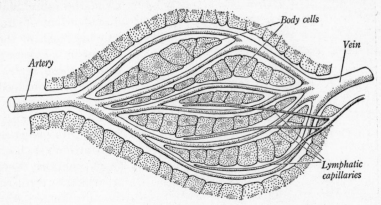

FIG. 42. The Relation of Tissue Cells to the Capillaries

emulsified fats absorbed by the lacteals, as we have already found (Fig. 37). The other main lymph duct leads in from the upper right side of the body. The larger lymphatics have valves much as the veins do, to insure a flow in but one direction. Interrupting the tubes at various points are lymph nodes. These structures are believed by many to perform the double function of straining out and destroying bacteria that may escape from infected areas into the lymph stream and of producing leucocytes that replenish the white corpuscle supply in the blood.

Unlike the blood the lymphatic circulation has no pump to force it along. On the contrary, the lymph flows slowly, impelled along by an inequality of pressure within the lymphatics themselves and by the alternate squeezing and relaxing of muscular action. One reason that exercise should be taken daily is that it promotes a proper flow within the lymphatic system.

Through the action of the blood stream and the water front, then, supplemented by the flow of lymph within the lymphatics, there is provided an efficient and constantly functioning transportation service. In this manner necessary products are carried to the tissues and undesirable substances are conveyed away. The speed of these transportation lines is

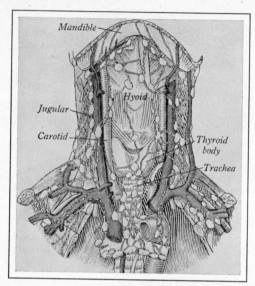

FIG. 43. The Lymph Nodes of the Neck and Upper Part of the Thorax[1]

also admirably regulated to meet the varying needs of the organism. Exertion throws more carbon dioxide into the blood stream from the muscles. As we have learned, its presence there stimulates both heart action and breathing through the autonomic nervous system. These responses accelerate the flow of blood through the circulatory system, and the lymph is pushed along faster both by the increased pressure from direct muscular action and by the more frequent expansion and contraction of the walls of the thoracic duct.

The oxygen supply is furnished by the lungs. In higher animals, if oxygen for respiration is to be obtained from the air, there must be a supply station — a place where the molecules of this gas can be delivered to the red blood corpuscles. This is afforded by the lungs, whose structure equips them splendidly for this service. One-celled animals take in the required oxygen directly through the cell walls; many lower forms have gills of different kinds; insects are fitted with an intricate system of internal tubes that communicate with the air through a row of orifices called spiracles, arranged along either side of the body; frogs have lungs, but they are little more

[1] From Gerrish's *Text-Book of Anatomy.* By permission of Lea and Febiger, publishers.

than hollow bags, for oxygen is also absorbed through the skin. In man, as well as in other mammals, the lungs are spongelike structures with myriads of tiny, thin-walled air vesicles. These moist, delicate walls are thoroughly permeated by pulmonary capillaries. The lungs communicate with the outer air through the noncollapsible trachea and nasal passages. Air is forced into the lungs by atmospheric pressure when the action of the diaphragm and intercostal muscles expands the

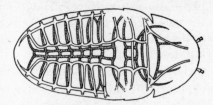

FIG. 44. The Tracheal System of a Cockroach

The cross-lined tubes distribute oxygen through the body of the insect [1]

thorax. The diaphragm is a big sheetlike muscle that separates the thoracic cavity from the abdomen. The intercostal mus-

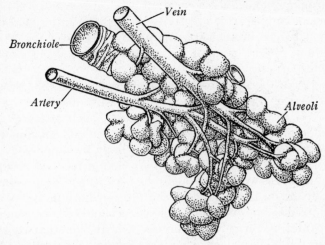

FIG. 45. Air Sacs of the Lung, showing their Relation to the Arteries and Veins

cles are those that to some degree move the ribs. The reverse action of these thoracic muscles expels the air. In this way inspiration and expiration follow each other in continuous order.

[1] From Burlingame and others' *General Biology*. By permission of Henry Holt and Company, publishers.

The oxygen of the inspired air comes in contact with the thin, moist walls of the air sacs. These sacs expose a surprisingly large surface through which oxygen may diffuse into the capillaries. In man the inner surface of the lungs measures something like two thousand square feet, which is about one hundred and twenty-five times the area of the external body surface. Once inside the capillaries, the oxygen unites with the hæmoglobin to form oxyhæmoglobin. In this form it is transported to the points where it is to be liberated again and used.

The food and oxygen are used through assimilation. Living animals make two demands upon the food within the cells. In the first place, it must furnish the materials for growth and repair; in the second place, energy must be liberated from the food to maintain the normal body temperature and to supply the power required by the tissues and organs to do their work. The process of rebuilding the worn-out protoplasm of the tissues from amino-acids is a constant one. It is necessary during periods of comparative inactivity as well as during periods of exertion. Even hibernating animals, whose activity is at a minimum, require a small but continuous supply of proteins for this purpose.

Energy is liberated from the liquid food in the cells through the process of chemical change. This change results in the formation of certain intermediate products, such as lactic acid. By oxidation of a portion of these intermediate substances or by oxidation of carbohydrates energy may be supplied for the reconversion of at least a portion of the intermediate products into substances which may again be used as energy-producers. If reconversion is not needed, the intermediate products may be oxidized to end products such as carbon dioxide and water. This oxidation furnishes an excess of heat over that required to keep the body at its normal temperature. The excess heat escapes through radiation at the surfaces.

The transference of this required oxygen from the air into the tissues of the body and the removal of the carbon dioxide from the tissues to the air make up the process called respiration.

Warm-blooded animals require large quantities of heat to keep the body temperature approximately constant. In man the average normal body temperature is 98.6 F. A change of a few degrees either above or below the normal is accompanied by grave consequences. Yét, despite this rather exacting demand as to temperature relations, the body is not an economical producer of heat, as we have found. It has been estimated that 73 per cent of all the heat produced is lost through radiation from the skin and other body surfaces such as the lungs.

As was previously stated, we found that glycogen was abundantly stored in the liver to be gradually fed out into the blood stream in the form of glucose as conditions required. Two rabbits were fed a heavy meal of clover, and after a sufficient time had elapsed for digestion and absorption a post mortem on one showed the liver cells to be well filled with glycogen. The other rabbit was chased about the yard for some time, then it was killed and its liver examined. The cells were found to be almost depleted of their glycogen. When frogs go into hibernation they are found to have livers well supplied with glycogen, but when they emerge in the spring the liver cells contain scarcely any of it.

In man, as in other warm-blooded animals, the circulation is very rapid. One author states that all blood in the body makes its complete circuit in about two minutes. For every thousand ounces of blood that passes through the liver about an ounce and a half of glycogen is removed and carried out. When the body, because of exertion, requires more sugar, the circulation is speeded up by increased heart action. A group of basketball players after seven minutes of playing were found to have raised their pulse rate from an average of 72 to 120.

When an excess of food is eaten, the body disposes of it in different ways. A large part of the carbohydrates and fats is converted into animal fats and stored as adipose tissue to be drawn on as a reserve supply in leaner days. It is for this reason that heavily fed animals and men with inordinate appetites become fat. When an excess of protein food is eaten, however, the body seems unable to store much of it. After the bodily needs for this class of foods have been met, the oversupply, for the most part, is deaminized and the carbohydrate component is utilized in the same manner as the other energy-producing foods. The nitrogenous part of the molecule is reduced to urea. Most of this is probably eliminated as waste. In this way those who eat too heavily of meats and other protein foods are likely to throw a heavy burden upon the kidneys and other excretory organs.

When fuel is consumed in a furnace many waste products are formed. Carbon dioxide and carbon monoxide are generated, water vapor is produced, and sulfur compounds are liberated. In addition considerable incombustible mineral matter accumulates in the form of ash. In a similar manner, as a result of metabolism many waste products are formed in the animal body. Carbon dioxide and water vapor are produced, and both nitrogenous and mineral waste tend to accumulate. To remove these waste substances promptly, a special service for their removal is provided.

Waste is removed from the human body by the excretory system. The waste products from the human body, as well as from the body of every animal, must be removed constantly and continuously. In man this is accomplished by several organs, each of which merits our attention.

The alimentary canal. All indigestible and unabsorbed food residues, together with numberless intestinal bacteria, are voided through the alimentary tract. These products themselves, to be sure, are not, strictly speaking, metabolic wastes. However, they must be removed and, along with these cast-off materials, much real body waste is eliminated. Worn-out and broken-down cells are sloughed off from the inner wall throughout the digestive tract. Bile flows into the intestine and carries with it toxins, metallic salts, and the end products of cellular disintegration.

The skin. The chief secretion from the skin is sweat. Most of the perspiration dries from the surface of the body without the individual's becoming conscious of its presence. But the total amount of water exuded is considerable. The average amount in an adult for a twenty-four-hour period is from five hundred to six hundred cubic centimeters; but the quantity varies widely and, under some conditions, the amount given here may be liberated in an hour. Although the principal function to be served in perspiration is the regulation of body temperature, much water is secreted in this way, and small quantities of chemical salts are removed. The chief salt removed is sodium chloride, but traces of urea, uric acid, and other products may also be found. In profuse sweating considerable quantities of urea may be eliminated. It is for this reason that in cases of Bright's disease or other afflictions of the kidneys sweat baths are often prescribed by the physician. The artificially produced excessive perspiration eliminates much nitrogenous waste that would otherwise clog and poison the system.

The lungs. The lungs eliminate large quantities of water vapor and carbon dioxide. The condensation produced when one breathes on a cold window pane is evidence of the former, and the latter may be vividly demonstrated by exhaling the breath through a pipette into a lime-water or barium-water solution. Under free and easy breathing the average male adult inspires and expires approximately five hundred cubic centimeters of air, or about a pint, with each breath. During the few seconds that the air is in the lungs the percentage of oxygen it contains is reduced and the carbon dioxide is increased, as the following table[1] will show:

	Percentage of oxygen	Percentage of carbon dioxide	Percentage of nitrogen
Inspired air	20.96	0.04	79
Expired air	16.02	4.38	79
	4.94 loss	4.34 gain	00

The total elimination of carbon dioxide by the average adult for a twenty-four-hour period has been estimated at one hundred and forty gallons at normal atmospheric pressure; and the water exhaled during the same period would amount to about a pint. The water excreted through the lungs could be thrown off by the body in some other way should breathing stop, but not the carbon dioxide. It must be eliminated

[1] Diana C. Kimber and Carolyn E. Gray, *Textbook of Anatomy and Physiology*, p. 370. By permission of The Macmillan Company, publishers.

through the lungs. Should this function cease, death from suffocation would promptly follow. A person could live but a very short time if all breathing were cut off.

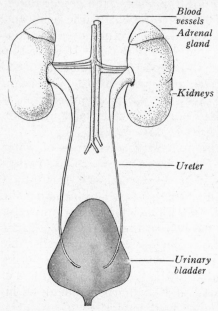

The kidneys. The kidneys are very important excretory organs. Principally they eliminate water and nitrogenous materials. These organs are two bean-shaped structures a little over four inches long that lie embedded in fatty capsules on either side of the spinal column. Anatomically they consist of a central cavity surrounded by a heavy wall of more or less spongy tissue. The lobed, central cavity connects with the ureter through a set of branched openings, each of which is called a calyx. The ureter leads to the bladder. The

Labels on figure: Blood vessels — Adrenal gland — Kidneys — Ureter — Urinary bladder

Fig. 46. The Urinary System, showing the Relation of the Kidneys to the Circulatory System

blood flow to the kidneys is heavy, and certain fine ramifications of the renal arteries form globular masses of finely twisted capillaries called glomeruli. Fitted around each glomerulus is a spherical, cuplike structure which receives the water and its dissolved nitrogenous waste as they are copiously extruded through the thin walls of the arterial capillaries. After receiving this urine, as the combination of water and bodily waste is called, the cup conducts it through long, tortuous, confluent tubes which finally pour their contents into the ureter. The ureter carries the urine to the bladder, where it is temporarily stored. By this arrangement, water passes from the alimentary canal into the blood stream and, after passing round through the organism, is again strained out in the kidneys. In this way the major part of the metabolic waste is removed. In reality it is a flushing system to keep the body washed out and in a healthful condition.

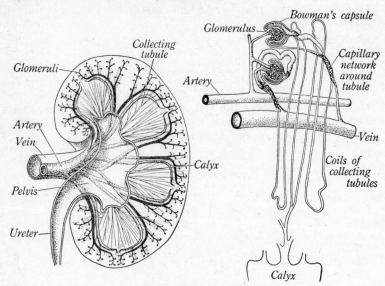

FIG. 47. The Finer Anatomy of the Kidney

On the right, tubules much enlarged. Note the arterial capillary **not** in connection with the kidney cells

QUESTIONS FOR STUDY

1. Why is it essential for a person to eat good, wholesome food?

2. Give the functions of each part of the alimentary canal.

3. What are the principal digestive enzymes used by the body and what purpose does each serve?

4. Trace the digestion of fats and tell how they enter the circulatory system.

5. Of what does the circulatory system consist and what are its chief functions?

6. Describe the exchange of gases in the lungs.

7. Describe the action of the heart.

8. Make a list of the component parts of the blood and give the general functions of each.

9. By what means is food passed along the alimentary canal?

10. What did Dr. Beaumont learn from the study of Alexis St. Martin's stomach?

11. What are the principal factors responsible for the secretion of digestive juices?

12. Why is it best not to scold or punish a child either immediately before or after eating?

13. Why does the body need oxygen?

14. Why is it necessary to have a vigorous circulation?

15. What provision has the body made for the storage and future utilization of excess food?

16. What are some of the effects of exercise upon the action of the body?

REFERENCES

BURLINGAME, L. L., HEATH, H., MARTIN, E. G., and PEIRCE, G. J. General Biology (second edition), pp. 86–110.

CANNON, W. B. The Wisdom of the Body, pp. 98–137.

CLENDENING, LOGAN. The Human Body (second edition revised), pp. 71–202.

HARTRIDGE, H. Bainbridge and Menzies' Essentials of Physiology (seventh edition).

KIMBER, DIANA C., and GRAY, CAROLYN E. Textbook of Anatomy and Physiology (eighth edition), pp. 219–328, 351–515.

KIRKPATRICK, T. B., and HUETTNER, A. F. Fundamentals of Health, pp. 239–340.

STILES, P. G. Human Physiology (sixth edition), pp. 186–384.

WILLIAMS, J. F. Textbook of Anatomy and Physiology (third edition revised), pp. 284–458.

Chapter IV · Animals make Adaptive Responses to their Environment

Introduction. Thus far it has been seen that man's body, and the bodies of all other higher animals as well, are in a sense a splendidly balanced organization of machine-like levers and an elaborately constructed system of motor-like muscles to move these levers. Moreover, we have found that provision has been made for a supply of energy and the removal of waste in order that the body may be kept in a working condition. Only one feature is lacking, then, to make the machine automatic, to make it respond to outside influences and to do of its own accord those things necessary to preserve itself and to continue its existence. That feature is a directing agent — one that will order the motors when to act, what to do, and that will give them the necessary impulse to accomplish this action. This necessary agent of control is found in the nervous system.

The nervous system is the most bafflingly complex and intricately functioning portion of the human body. To liken it to a telephone system, with its simple direct connections between parties and with its hook-ups through other centrals to complete many diverse and long-distance circuits, is to use a poor figure indeed; yet this is perhaps the best simile that can be employed. The nervous system may receive a stimulus from outside the body and, in response, may initiate a direct, simple reflex action, and that be the end of it; the stimulus may, on the contrary, be carried simultaneously to other nerve centers, and a reflex follow that involves action in several parts of the body; or the stimulus may be referred from its original center to the brain, and there set up a whole train of correlated physical movements, mental reactions, and emotional responses that affect the behavior of the individual in a most profound and far-reaching manner.

2   119

Because of its delicacy and complexity very much yet remains to be found out about the nervous system. In our dis-

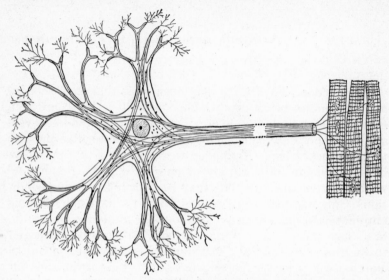

FIG. 48. A Motor Nerve Cell highly Magnified

The fiber between the body of the cell and the muscle is sometimes
several feet long

cussion, then, we shall confine ourselves largely to a simple study of the nervous system as a directing organ of control.

The nerve cell is the unit of nervous correlation. The fundamental unit of the nervous system is the nerve cell, or neuron, as it is often called. Its unique features consist of a high degree of specialization in the direction of irritability and conductivity. Neurons vary somewhat in character, but in general the nerve cell consists of a cell body and certain processes, or branches, that extend out from this central body. Within the cell body a definite nucleus is found, together with a network of fine fibrils too delicate to be seen unless highly magnified. The processes consist of a group of relatively short, rapidly tapering, highly branched, unsheathed extensions of the cell proper called the *dendrites*, and usually one much longer sheathed branch, of almost uniform thickness, called the *axon.* The term *dendrite* (Greek *dendron*, "a tree") is applied because of the highly divided condition of these cell extensions. The axon, like an insulated electric wire, is covered with a layer of fatty substance, and the whole is generally incased in a sheath. This fatty layer gives the axon a white appearance and is believed to be of the nature of an insulator to keep the nerve impulses from "spilling off" as they course along the nerve tract. The axons of nerve cells vary greatly in length, depending upon their location in the body. When they take their departure from an inner nerve center, such as the spinal cord, axons may extend unbroken to the peripheral part that they serve. For this reason some axons are extremely long. The longest extend from the spinal cord, down the legs to the feet, and are sometimes half the length of the man's body. We must not confuse an axon with the comparatively cable-like nerves which extend through the body and which are large enough to be easily dissected out. These nerves are often made up of thousands of slender axons bound together by connective tissue. In this respect they are more like the lead-covered cables of a telephone system containing the many insulated strands within than they are like a single telephone wire. Judging from Dr. Spiedel's recent work on tadpoles the axons appear to grow out over this whole distance from their respective nerve cells. How thousands of them can push out through the same trunk line and yet each reach its own particular muscle or gland far out at the periphery of the body is one of the unexplained mysteries of life. At present there is not the slightest knowledge to indicate how nature accomplishes this feat. If a single axon be cut the outer portion dies, and under favorable conditions the lost segment may be regenerated along the same path, and the appropriate connections again be made. This regeneration proceeds slowly, and it is for this reason that a wounded region of the body often feels numb and peculiar for some time after the injury has apparently healed. A peculiarity of the dendrites and the axons is that they are one-way tracks. Nerve impulses coming into the nerve cell always travel over the dendrites; those going out are conducted along the axon.

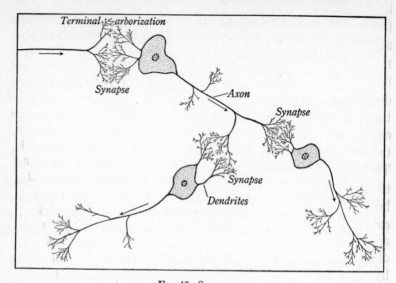

FIG. 49. Synapses

The direction of travel of the nerve impulse is shown by the arrows

Within the nerve cell are peculiar granular bodies called Nissl's bodies, which are supposed to represent a store of nervous energy. Their quantity is variable, depending upon the degree of cell fatigue. Rested cells show a relatively larger amount of Nissl's bodies. Their quantity, however, decreases with fatigue, fever, or asphyxia.

Separate nerve cells are believed in most cases to have the termini of their processes in communication with the terminal processes of other nerve cells. The axon, for instance, ends in a highly branched brushlike structure. The outer ends of this brush of fine branches lie close to the fine terminal endings of the dendrites belonging to the adjacent nerve cells. These microscopic gaps, believed to exist between the fine endings of the axons and dendrites, are called synapses. The ease with which a nerve impulse can be conducted across the synapse is variable, depending upon conditions. When the connection is good, the impulse flows freely; when it is poor, the passage is slower and is accomplished with difficulty.

Under certain conditions the synapse, like an open switch in an electric line, is believed to block the nerve current altogether. Thus anæsthetics and narcotics perhaps produce their effect largely at the synapses.

The exact nature of a nerve impulse is not known. In some respects it resembles an electric current; in others it is quite unlike it. For one thing its passage is much slower. In warm-blooded animals, such as man, it has been estimated that the nerve impulse travels about three hundred and sixty feet a second; if it traveled directly from the toe to the brain of a six-foot man, it would require about one sixtieth of a second. Experiments also show that when a nerve impulse passes, a slight amount of carbon dioxide is generated. This indicates that the impulse is some sort of metabolic action involving chemical reaction.

The nerve cell is the most remarkable cell in either the human body or the body of any other animal. In addition to the functions already enumerated it has other peculiar powers. For instance, those in the higher brain centers, at least, have the power both to step-up and to step-down nervous impulses. Anyone who has seen a man lose his temper and fly into a rage in response to an offending word, or who has observed with what composure the self-possessed man can hear the most false and vituperative attacks upon himself, has seen these properties of nerve cells manifested in human conduct. Just how these reactions are initiated in the nervous organization or the extent to which nerve tissue and other organs are involved is frequently unknown. But that what the individual does is a response to nerve stimuli cannot be denied. Nerve cells also have the power to interpret stimuli. For example, a stimulus to the optic nerve is interpreted as light or color; one to the auditory nerve as sound; one to the olfactory nerve as smell; and so on for the whole gamut of external forces that affect the body and, in a similar manner, for many internal ones. Moreover, these interpretations for any particular nerve cell are largely specific. By this we mean that if any sense organ, such as the eye, is stimulated by a force other than that to which it normally responds, it still persists in interpreting the stimulus as light or color. It is for this reason that when one receives a blow on the head, and the optic nerve is severely jarred, he "sees stars" rather than experiences a sense of touch through that channel. Finally, many nerve cells have the power to store up the effects of incoming stimuli and in some way discharge them again at a later time. The lapse of time between reception and discharge may be but a few seconds or it may be many years. This ability of nerve cells to store stimuli, however, is the very basis of memory and will be more adequately dealt with later.

The correlating mechanism consists of receptors, adjustors, and effectors. It is obvious that for an organism to respond to an outside influence appropriately and thereby successfully adjust itself to its environment, it must have some mechanism by which this can be done. In other words, it must have some specialized part of the nervous system to receive the stimulus; it must have another part to work the incoming stimulus over and to evolve and send out a motor stimulus; and, finally, it must have a muscle or muscles to receive the motor impulse and, by the responding action, make possible the bodily movements necessary to the adjustment. The three aspects of this correlating mechanism are called the receptors, the adjustors, and the effectors. In the lower animals these functional parts are very simple. As animals grew larger and more complicated the parts of this adjusting mechanism became more intricate. Ultimately it emerged in man as organs of correlation so sensitive and complex, and capable of such a wide variety of response, that it is impossible for even our most brilliant physiologists and neurologists to trace fully their correlated action. Let us trace briefly its development.

This mechanism of correlation is present in the amœba in a very primitive form. Although it consists of a single cell this simple, microscopic animal is, under ordinary conditions, capable of successfully adjusting itself. If in its flowing movement across a microscope slide, let us say, its naked mass of protoplasm is made to come in contact with a needle point that is uncomfortably warm, this impulse is immediately conducted across the cytoplasm, and a pseudopodium is thrust out on the opposite side by means of which the animal "backs away" from the object of irritation. The adjustment has been made, and in this case the protoplasm of the single cell acts as both receptor and effector.

The hydra is a small, saclike aquatic animal, perhaps a quarter of an inch in length when fully extended. It is found living in the fresh waters of almost any locality in the United States. It is attached at one end, and on the other end it has a mouth surrounded by food-gathering tentacles. Through this one mouth it takes food and also eliminates solid waste. Its body wall is composed of two layers of cells: an outer and an inner. Interspersed among the cells of the outer layer are found certain slender cells terminating in a short, fine, hairlike projection at the outer end and extending at the other end into a flat, expanded, contractile

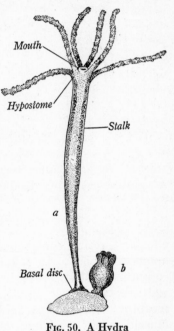

Fig. 50. A Hydra

a, greatly extended; *b*, contracted

portion lying at right angles to the main axis. When the outer end of this cell is excited, a stimulus is received that is conveyed to the flat inner portion of the cell, which then contracts, pulling the animal or a part of it into a new position. Here we have a single cell performing both the function of receptor and that of effector.

The sea anemone is an animal a little higher in the scale of life than the hydra. In it is found stages of further development in the nervous correlating mechanism. Here slender receptor nerve cells may be found interspersed among the cells that form the outer layer of the animal's body. These receptors terminate at the inner end in a fine ramification of nerve-cell processes that form a junction with certain contractile cells lying at right angles to them. When the receptors are stimulated, a nerve impulse is conducted within to the

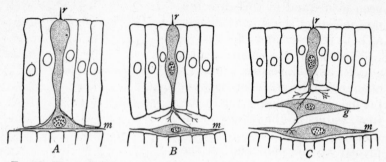

FIG. 51. The Coördinating Mechanism of the Hydra and the Sea Anemone

A, a single cell functions in the hydra as both receptor and effector; B, a receptor cell (sea anemone) in communication with a contractile cell; C, a similar arrangement (sea anemone) with an intervening ganglion cell; r, the sensitive part of the receptor cells; m, effector; g, ganglion cell

effectors, which contract and move that portion of the sea anemone's body. In this case the correlating mechanism is composed really of receptors and effectors.

In certain other portions of the sea anemone's body the receptor cells are separated from the effector cells, or simple muscle cells, by an intervening cell. This intervening nerve cell receives the sensory stimulus from the receptor, and in response sends out a motor nerve impulse that causes the effector to contract and produce movement. Here we have the complete correlating mechanism in its simplest form; namely, a sensory nerve cell, a motor nerve cell, and an effector muscle. The nervous correlating mechanism, from the sea anemone on through the animal kingdom up to and including man, was retained with these three essential parts. As progress was made in the worms, the mollusks, the crawfish, and vertebrates lower than man, each part — the receptors, the adjustors, and the effectors — was developed in extent, complexity, and specialization to meet the demands of the organism. When we reach man we find a correlating mechanism, as has already been indicated, so extensive, so highly specialized, so delicate in its response, and capable of such a wide variety of bodily adjustments that it places man in a class by himself. Physically he is farther separated from the lower animals in respect to his nervous system than in relation to any other part of his body. Man's nervous correlating mechanism may be said to be the wonder of the organic world. Not only does it make possible the highly coördinated muscular movements of his body, but, far more remarkable, it forms the very seat of all his intellectual processes and his emotions. It is impossible to understand man at all, either as a biological organism or as a spiritual and social being, without some knowledge of this extraordinary correlating mechanism. For our purpose we must describe and study it in a very elementary manner. At many points too we shall approach the very limits of human knowledge in this field and shall have to withdraw and pass on, leaving the picture hazy and incomplete. But what is known makes a marvelous and interesting story.

The correlating mechanism is highly specialized in man. Man's nervous system forms the greater part of his correlating mechanism. Hormones (secretions from the ductless glands) secreted at one point in the body may be carried to another by the blood and there arouse action without the intervention of a nervous impulse. But for the present the correlating aspects of the hormones will be disregarded. The nervous system consists of a highly specialized set of receptors and an extremely intricate and extensive system of adjustors. The adjustors may be taken roughly to include the brain and spinal cord, together with all sensory (or afferent) nerves that bring impulses in from the sense organs and all motor (or efferent) nerves that carry impulses out to the muscles or glands. There are some other nerve centers, or ganglia (ganglion, an aggregation of nerve cells outside the brain and spinal cord), which will be noted in due time but need not be stressed here.

Receptors. We usually think of the receptors as falling into five classes: those of sight, hearing, smell, taste, and the tactile senses. The sight receptors are located in the eye, this whole organ being designed to gather a proper amount of light and focus it on the retina, at the rear. In this respect the eye is much like a photographer's camera. The retina contains the receptor cells in the form of rods and cones. The receptor impulses are conveyed as stimuli from the retina to the brain through the optic nerve, where they are interpreted as light and color.

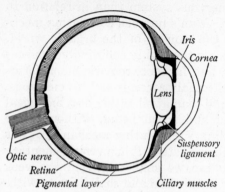

Fig. 52. A Horizontal Cross Section of the Eyeball

The illustration shows the principal structures of the eye

The nerve cells receptive to sound are located in the ear, which is a most complicated organ. The ear functions as a device to gather up the sound waves, amplify them through the agency of the eardrum, and then convey them in the form of vibrations into the *spirally coiled* part of the ear, where the receptor cells are located. This spiral compartment, called the cochlea, is filled with a fluid which vibrates in response to the eardrum and in so doing agitates the hairlike termini of the auditory receptors that line certain portions of its interior. The impulses so generated pass into the brain over the auditory nerve, where the sensation of sound is interpreted.

The nerve receptors sensitive to odors are located in the nasal passages. Small particles of matter wafted about by the air dissolve on the moist membranes of the nose, and the solution so formed stimulates the olfactory receptors to send an impulse to the brain over the olfactory nerve. These receptors are often much more acute in certain other animals, such as dogs, than they are in man. They fatigue easily, and in a short time men become insensitive to odors that at first were extremely pronounced.

The receptors of taste are located mainly on the upper surface of the tongue and along the sides. Because of their fancied resemblance to flower buds they are often called "taste buds." These buds are arranged along the lateral sur-

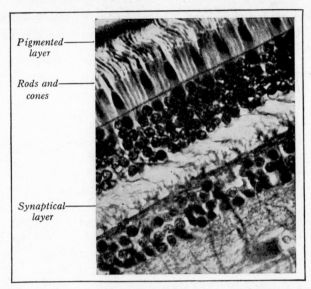

Pigmented
layer

Rods and
cones

Synaptical
layer

FIG. 53. The Rods and Cones and their Relation to the Pigmented Layer
of the Eye

An enlargement (about eight hundred times) of a photomicrograph made at the
University of Colorado. It was prepared from an H. E. section of the retina in the
laboratory of the late Dr. H. J. Prentiss, Professor of Anatomy, Histology, and
Embryology at the University of Iowa. (Courtesy of James M. Shields, M.D.)

faces of chinks, or crevices, in the tongue. Dissolved substances
irritate the fibrous endings of the taste receptors extending
beyond the surface and set up nerve impulses which we interpret
as taste. People with little knowledge of the taste receptors
believe they are capable of tasting a great variety of flavors.
Such does not seem to be the case. The taste buds seem sensitive to but four principal kinds of stimuli. These stimuli produce the sensations of sweet, sour, bitter, and salty. All the
other flavors which add so much to the pleasure of eating are
derived chiefly from the olfactory nerves. We confuse their
messages with those of the gustatory sense and interpret
them as taste.

The tactile sense, though a very common one, is not a simple
matter. It includes responses to a whole group of stimuli quite
varied in character which are thrown together and roughly

classified under this head. In addition to the sensation of touch, it includes those of cold, heat, pain, muscle sense, and

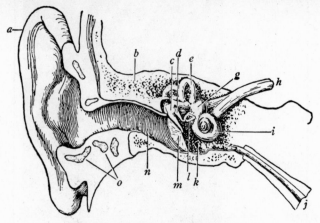

FIG. 54. A Section of the Human Ear

a, pinna; *b*, bone of skull; *c, d, l*, bones of the middle ear; *e*, semicircular canal; *g*, vestibule; *h*, auditory nerve; *i*, cochlea; *j*, Eustachian tube; *k*, middle ear; *m*, tympanic membrane; *n*, outer ear; *o*, cartilages supporting pinna [1]

perhaps others. These receptors include different types of end cells located among the skin cells and at other points in the body, such as the muscles and joints. Our sensations as to the position of a limb, muscular fatigue, and the weight of a lifted object come from the latter sources.

The value of these special senses to man and other organisms is at once evident. They form the very threshold between the body and the outer world of objects. They are the only channels through which stimuli from the whole world of our environment may enter in to make us aware that such a thing exists. Without these organs or something else to take their place man would be as dead and insensible to other objects in his universe as a post apparently is; but, with these sense doors open, whole floods of nervous stimuli pour into his central nervous system continuously. From these raw sensations he distills ideas and chains of thought, he builds up

[1] From Woodruff's *Foundations of Biology*. By permission of The Macmillan Company, publishers.

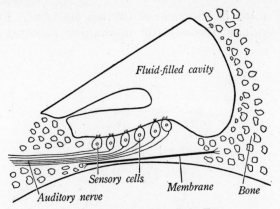

FIG. 55. Receptor Cells of the Ear

Note the relation of these cells to the nerve fibers and the tough membrane

convictions and attitudes, and from them, as original sources, he derives all the pleasures of life as well as all the sorrows.

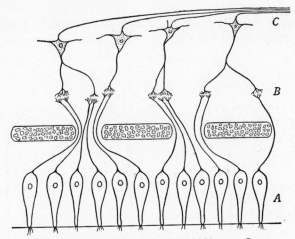

FIG. 56. Nerve Mechanism of the Olfactory Sense

A, receptor cells; *B*, synaptical connections; *C*, olfactory nerve fibers

Adjustors. Adjustors, as we have found, were first intro duced into the animal world in the form of a single cell inter. jected between the receptor cells and the contractile cells.

This we noted in the case of the sea anemone. From this simple beginning the adjustor mechanism enlarged its func-

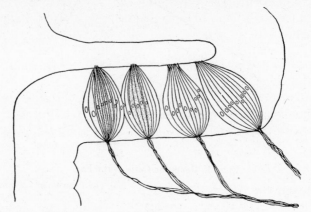

FIG. 57. Receptor Cells of the Taste Buds and their Relation to the Nerve Fibers

tional scope rapidly. As animals were forced to meet more trying conditions of life the situation demanded more efficient

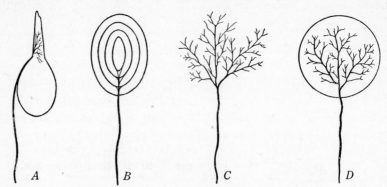

FIG. 58. Receptor Cells

A, touch; B, pressure; C, pain; D, temperature

and complex adjustors. The movement of a single part of the body, at first sufficient, had to be superseded by intricate movements involving many parts of the body. These movements had to be correlated. Motor impulses had to be generated and simultaneously sent to widely separated muscles. Finally, along with the expanding adjustors, a sort of intelligence appeared. This intelligence center enlarged its sphere of influence rapidly and kept its base of operation within the adjustor mechanism. In the end this part of the nervous system came to be the seat of man's psychic powers, the base of his feelings, his memory, his will; in fact, it came to be the physical expression of about all we mean when we say *man* as distinguished from other animals.

The adjustor mechanism in man, as it may be seen, comprises the whole of his nervous system lying between the receptor cells and the contractile muscles. It includes all sensory and motor nerves that find their ramifications in every part of the body; it takes in the spinal cord and the brain; and, to complete the picture, the ganglia that comprise the autonomic nervous organization must be added.

To examine this adjustor part of the nervous system more closely we find the spinal cord, which is composed of both gray and white matter, to be on the average about eighteen inches long. The gray matter throughout the nervous system is made up of aggregates of nerve-cell bodies; the white matter consists of bundles of sheathed fibers. At the upper end of the spinal cord is the brain. In reality the brain is nothing more than the greatly expanded and highly differentiated front end of the cord itself. Along the spinal cord are found thirty-one pairs of spinal nerves which supply all parts of the trunk, arms, and legs. Each spinal nerve is joined to the cord by a dorsal root which is a sensory nerve and by a ventral root which is a motor nerve. The dorsal root of each nerve brings in nerve impulses from all parts of the body to the central ganglia. For this reason this root is often called an afferent (carrying-in) nerve. Likewise the ventral root, since it sends out motor impulses to the muscles and the secretory glands, is often referred to as an efferent (carrying-out) nerve.

The brain is by far the largest and most imposing part of man's nervous system. Dr. Keith calls it "nature's master

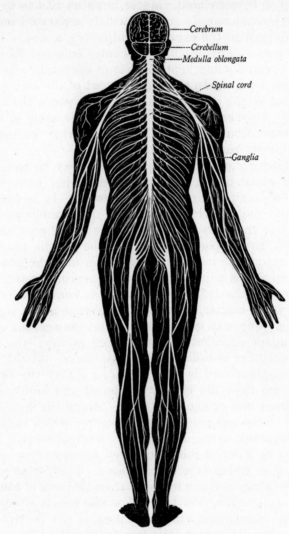

FIG. 59. Man's Adjustor Mechanism

contrivance." Many animals had lived and become extinct before nature devised a brain. When it finally did come in, it developed gradually, like most per-manent living organs, as animals evolved. In Amphioxus, a little ani-mal which seems to have been the forerunner of the vertebrates, its central nervous system is all cord and no brain. The rise of the true brain began in the fishes, but the spinal cord here is heavier than the brain. In reptiles the brain takes the lead in weight, and by the time man is reached the brain is about fifty times heavier than the spinal cord. The average male Caucasian has a brain that weighs approxi-mately forty-eight ounces, or just three pounds. This central organ in man actually weighs more than the brain of any other animal save that of the elephant and the whale.

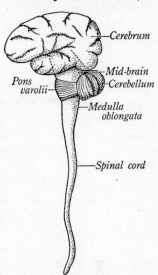

Fig. 60. Principal Parts of the Central Nervous System

For our purpose the brain of man may be roughly divided into three parts: the cerebrum, the cerebellum, and the medulla oblongata. A deep cleavage furrow divides the cere-brum into two hemispheres.

The medulla oblongata is in reality the more or less swollen top of the spinal cord. Altogether there are twelve or thirteen pairs of cranial nerves that take their origin from the brain somewhat after the fashion that the spinal nerves take their origin from the spinal cord. These cranial nerves supply the head, face, and neck. The auditory, gustatory, and olfactory nerves are in this group. One pair, called the vagus (wanderer) nerves, extend far down into the trunk, and, in addition to branches sent to the throat and neck, it sends other branches to the esophagus, stomach, heart, lungs, pancreas, and both the small and the large intestine. Several pairs of these cranial nerves, including the vagus, arise from the medulla.

2

Another peculiar feature is found in the medulla. The ganglia, hundreds of them, all up and down the spinal cord, are intricately connected by thousands of ascending and descending fibers. Many of these connecting fibers extend up through the medulla into the cerebellum and the cerebrum. The peculiar feature referred to is that the fibers ascending into the higher parts of the brain cross over from one side to the other in the medulla; that is, the fibers extending from the right hemispheres of the cerebellum and cerebrum, respectively, connect with ganglia in the cord that supply the left side of the body. For this reason an injury, or lesion, on the right side of the brain may paralyze parts of the left side of the body. The necessity for this crossing over of the nerve fibers from one side to the other is not apparent.

The cerebellum seems to be connected principally with the function of muscular coördination, especially with those movements that have to do with locomotion. A dog from which the cerebellum has been removed, completely fails at first to coördinate its locomotor movements. It rolls about on the ground as if convulsed with pain; yet it probably is not suffering except from bewilderment and mystification. Ultimately it regains a good deal of its muscular control. But its muscular coördination without the cerebellum is never fully restored. Its gait is uncertain and wabbly. The fact that the harmonious coöperation of the muscles concerned in walking can be even partly regained, however, indicates that other coördinating organs are at work. It is believed that the eye, the semicircular canals of the ear, the muscles, the tendons, and the joints are also involved in locomotion and body equilibrium.

The cerebrum is the most conspicuous part of the brain. Its large, convoluted hemispheres not only occupy the whole space in the fore part of the cranium but extend back over and to a large extent cover up the other parts of the brain. In the rabbit the cerebrum makes up somewhat more than half the bulk of the entire brain. In man it exceeds four fifths of the total weight. The outstanding characteristic of the cerebrum is its cortex, or outer part, which is composed largely of nerve-cell bodies and their short, highly ramifying processes that make possible a connection with many other neighboring cells. The two halves have an extensive network of white fibers below, which not only connect the halves and different parts of the same half but also send thousands of other branches downward — some to the cerebellum and many to communicate with ganglia at all levels of the spinal cord.

Not only is the cerebrum the most conspicuous part of the brain, but it is believed to be the source of consciousness and all higher intellectual processes. As Walter says:

> With an increase of cerebral function the instinctive reflexes take more and more to the background, and therein is the great distinction between "lower" animals, which are largely at the mercy of their environment and heredity, and the "higher" animals, which to an increasing degree have risen above the environing conditions and have become more and more "the captains of their souls." The most prized possession of mankind is the "capacity for individuality"; yet even what passes "for free will" has its basis in the neurones and reflexes built up in the brain, that after all must be regarded as the mechanism *through which* consciousness, memory, imagination, and will are effected.[1]

The cerebrum plays such an important part in the life of man that it is almost certain he could not live without it. But by three successive skillful operations, with considerable time intervening between each two for recovery, a dog was entirely deprived of its cerebral hemispheres. The animal could walk but was idiotic. It ranged in a bewildered manner about the laboratory where it was kept, making detours around objects in its path. It took the same pains to avoid spots of sunlight on the floor that it did to avoid material obstacles. It seemed unable to learn anything after the operation.

The central nervous system is the basis of man's behavior. In general, man's correlating mechanism may be said to result in action on three levels. These levels are not sharply defined, however, and action on one, two, or even all three of them may take place simultaneously. To illustrate: Let us suppose that

[1] H. E. Walter, *Biology of Vertebrates*. By permission of The Macmillan Company, publishers.

it is the morning of the Fourth of July and that a barefooted boy is giving expression to his patriotism by shooting fire-crackers on the lawn.

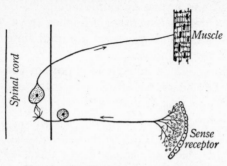

FIG. 61. The Mechanism of a Simple
Reflex Action

Let us further assume that as he lights the fuse of a cracker, he steps on the hot ashes of the remains of another cracker that had been exploded some time before. As soon as the heat receptors in the bottom of his foot are stimulated, a sensory impulse shoots in over the afferent nerve of the adjustor mechanism to the ganglia in the spinal cord. Here a motor impulse is immediately generated and sent rushing back over the efferent nerve or nerves to the muscles of the leg, which instantly contract to lift the foot off the uncomfortably hot object. Perhaps the boy is so intent on getting the firecracker whose fuse he has just lighted safely away from his body that he does nothing more than lift his foot. He may not even step, but stand momentarily on one foot. This represents the simplest type of bodily adjustment. Such action is called a *simple reflex* because the impulse on the sensory tract travels directly to the spinal ganglia and is reflected back immediately in the form of a motor impulse to move the affected member out of danger. The whole correlation takes place before thought could possibly intervene. Its value to the organism lies in the speed with which the imperiled bodily member may be withdrawn from the source of possible injury. Many movements of the body are of this sort. Closing the eyes in the presence of flying glass, jerking the hand away when suddenly pricked with a needle, and dodging when a menacing blow is directed toward the face are all examples of simple reflexes. All higher animals and most of the lower ones are provided with these simple reflex mechanisms to serve in emergencies requiring immediate action.

Action on the second level takes place when the sensory impulses are sent along the spinal cord to centers higher up and even to the cerebellum itself. Suppose the boy in question had stepped on a very hot, actively burning firecracker stub. The acute sensory impulse would have overflowed the first center and stimulated the centers all up the spinal cord and also the cerebellum to send out motor impulses to other parts of the body. The boy probably would have hopped about, grabbed his burned foot in his hand, made facial grimaces, and the character of his breathing would have been altered — all because of these motor impulses that poured out from the higher ganglia as they were stimulated by the ascending sensory impulse.

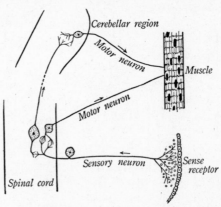

FIG. 62. **Mechanisms of Nerve Responses on the First and Second Levels**

Note the nerve leading to the cerebellar region and the additional one to the muscles

Now suppose the strong sensory impulse had passed beyond even the cerebellar region and had gone on up to the great maze of ganglia in the cerebrum. In addition to complex physical movements, thought processes too might have been induced. The boy probably would have applied pain-relieving agents to the burn, he might have put on his shoes, or, had he been old enough, he might even have reflected upon the personal safety and economic wisdom of using fireworks as a means of patriotic expression.

These illustrations give some idea of how the stimulating of a single receptor may cause nerve impulses, by reason of man's extensive adjustor mechanism, to be sent to all parts of the body and how they may finally emerge in the form of the most intricately and highly coördinated bodily movements and responses. Even long trains of ideas which form the materials of connected thought may be unleashed in this manner.

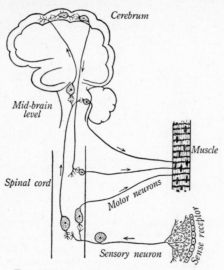

FIG. 63. Nerve Mechanism on the First, Second, and Third Levels

Note the additional nerves, one to the brain and another to the muscle from the brain

The autonomic nervous system supplements the central nervous system. In addition to man's central nervous system, which is usually said to be voluntary because for the most part he can exercise control over it, there is a second system, the autonomic (or involuntary) nervous system. It is not a separate system in that its action is distinct from the central nervous system. It would perhaps be more proper to designate it as an adjunct to the voluntary system. As such it is intimately connected with the central system, and its influence is closely correlated with motor impulses sent out by the spinal and cranial nerves. By reason of its synaptic connections with these nerves, it sends impulses to the eye, the ear, the salivary glands, the esophagus, stomach, intestines, pancreas, lungs, heart, kidneys, rectum, bladder, and sex organs, to the walls of the blood vessels, and to the skin. The action of all smooth muscles, as well as of the heart muscles, is under its control. The extent of its autonomic nerve ramifications and its great importance in the body may be seen from its control over the organs enumerated above. Some organs, such as those involved in urination, defecation, and breathing, are subject to the control of both the autonomic and the central nerves. It is for this reason that a child may learn to exercise a large measure of control over the first two as to time and place and to attain some degree of control over the last. But for the most part all the visceral portions of the body are completely under the sway of the involuntary nerves, and the voluntary nerves are freed from the endless routine task of innervating and controlling them. The autonomic system, as Walters says, "by taking over most of the routine drudgery of living . . . releases the central nervous system for higher evolutionary adventures."

In reality the autonomic system falls into two groups of nerves. One consists of motor cranial nerves rising from the medulla and others from near the base of the spinal cord. These two sets of nerve centers connect synaptically with ganglia lying either in the walls or in the regions of all the organs named above and perhaps others. When these nerves connect with ganglia outside the walls of the organs, relay fibers extend from the ganglia into the organs themselves. The vagus nerve, noted before, is one of the cranial autonomic group. The necessity of making all these autonomic connections is doubtless why the vagus nerves wander about so widely through the viscera. This part of the system is sometimes called the *parasympathetic system.*

The other group of nerves of the autonomic system is often called the *sympathetic system.* In man it consists of two rows of ganglia lying outside the spinal column on the ventral side and extending within the body cavity from the base of the skull to the coccyx. The ganglia of each row are connected up and down with extensive fiber systems, giving them something of the appearance of beads loosely strung. Motor branches of the spinal nerves relay through these ganglia on to the same organs reached by the parasympathetic nerves. This means that practically all organs in the whole body acting in an involuntary manner are reached and influenced by nerve branches from both the parasympathetic and the sympathetic portions of the autonomic system. The effect of this double innervation is most interesting. One set of nerves in each organ opposes the action of the other set. One group tends to stimulate action; the other, to retard it. For example, the salivary glands have branches from both sets of

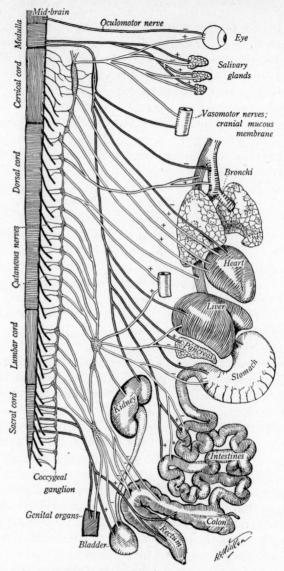

Fɪɢ. 64. Diagram of Autonomic Nervous System

Dark lines, parasympathetic nerves; light lines, sympathetic nerves

nerves. In the case of one of these glands the parasympathetic impulses rise in the medulla and are relayed in through ganglia near the gland itself.
If this nerve from the medulla be stimulated by an electric battery, a copious flow of saliva follows. On the other hand, if the nerve coming from the sympathetic ganglion be similarly stimulated, a reverse action takes place and the flow of saliva is almost stopped. The nerve fibers from the two systems balance each other, the action of the gland being in the direction of the most effective impulse. The character of the impulse is in turn determined by the needs and condition of the body.

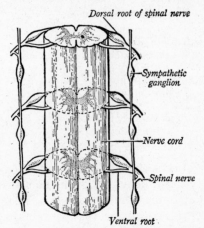

Dorsal root of spinal nerve

Sympathetic ganglion

Nerve cord

Spinal nerve

Ventral root

FIG. 65. Simplified Diagram showing the Connection of the Dorsal and Ventral Nerve Roots with the Sympathetic Chain[1]

We have found that the heart muscle has power within itself to initiate contraction. But the *rate* of contraction in the heart is balanced and controlled by the opposing divisions of the autonomic system. If, through violent exercise, carbon dioxide accumulates in the blood, it is believed to act as a stimulus to respiratory centers in the medulla. More effective impulses are sent out through the sympathetic centers by fibers leading through the cord to the appropriate spinal nerves, and the heart beats faster. As the carbon dioxide in the blood is reduced, less effective nerve currents pass down from the medulla, and the quieting effect of the balancing parasympathetic ganglia has a chance to exert its influence. As a result heart action slows down again.

The following contrasted bodily responses to the opposing effects of the two divisions of the autonomic system show

[1] From Kirkpatrick and Huettner's *Fundamentals of Health*. Ginn and Company, publishers.

vividly how they balance each other. The list is not complete, but serves as an illustration.[1]

Parasympathetic	Sympathetic
1. Contracts pupils of the eye.	1. Dilates pupils.
2. Contracts ciliary eye muscles so that eyes are accommodated to see objects near at hand.	2. Lessens tone of ciliary muscles, so that eyes are accommodated to see objects at a distance.
3. Contracts bronchial tubes.	3. Dilates bronchial tubes.
4. Slows and weakens heart action.	4. Quickens and strengthens heart action.
5. Increases contractions of the gastro-intestinal tract.	5. Decreases contractions of the gastro-intestinal tract.
6. Increases secretions of glands.	6. Decreases secretions of glands.

It is through their influence on the autonomic system that certain drugs produce their effect on the body. For instance, atropine (the drug that oculists use to dilate the pupils of the eye) paralyzes the parasympathetic system and thus gives the sympathetic system free sway. When it is used, the pupils dilate, the salivary glands dry up, the heart beats faster, and the intestines puff up. But if the vagus nerve be cut before the atropine is administered, no such effects result, which shows that the drug exerts its effect through the autonomic system. Nicotine, which has a tendency to speed up the heart and increase blood pressure, is believed to exert its effect through the involuntary nervous system. The action of the autonomic nervous system is also closely related to emotional states of the body, as we shall see later.

The most important sense organs are located in the anterior end of the body. It is usually the anterior end of the body which first comes in contact with the environment. For this reason nature has centralized the most acute and highly specialized sense organs in the head. Located where they are, they come in contact with the forces of the environment first and bring quick, continuous messages to the brain of what they have experienced. This is probably why the spinal cord became greatly swollen with ganglia at the anterior end, and it is the chief reason why the brain developed there.

[1] Adapted from Kimber and Gray's *Textbook of Physiology*. By permission of The Macmillan Company, publishers.

Three of these receptors, the eye, the ear, and the nose, have become distance receptors; that is, they not only carry impulses into the brain as to what has been seen, heard, and smelled, but they also enable one to form a judgment as to how far away the stimulating object is. This ability to judge distance is of the utmost value to the organism. If the stimulus comes from immediate contact with the environment, suitable response and adjustment of the body to the situation are likely to be made through a direct, simple, and rapid reflex action; but if the object from which the stimulus comes is some distance away, then the response necessary to adjust oneself successfully need not be made so quickly. This gives one time for choice of action. He has time, too, to bring to bear upon that choice all the valuable experience of the past. In other words, the individual, if he be a person, can make that choice in the light of all that he has learned in the past as to what is the proper thing to do to make his adjustment a successful one. Suppose a man engrossed in thought is crossing the street. Suddenly he hears an automobile horn and, glancing up, sees a car bearing rapidly down upon him. From its distance away he judges that he has time to avoid it by proceeding in the direction in which he is walking. He continues to walk, then suddenly hears a horn from the opposite side. Glancing in that direction, he sees a second car rushing down and cutting off his projected path of escape. He is in a dilemma. He might try to escape by dodging in either direction. But experience has taught that such a course is likely to be disastrous. So, glancing both ways almost simultaneously, he notes the distance between the lines of progress of the two cars and decides that the space between the two will permit him to save himself by standing perfectly still as the cars pass him on either side. Moreover, from past observation he is led to believe that the drivers will probably favor him by shifting their courses slightly to give him the advantage of more space between the cars as they go by. With this ability to judge distance, and with past experience to help him to choose wisely, he comes unharmed out of the traffic danger in which he was caught. It is doubtless this same judgment of distance and also the memory of past experience that lead a cat to climb a tree rather than trust its fate to flight when it sees a bulldog a few paces away charging down upon it.

Memory enables one to use past experience. We have found that receptor impulses are usually immediately transformed by the adjustor ganglia into motor stimuli that result in prompt muscular action. One exception to the immediate reflex motor impulse which ordinarily follows an incoming sensory stimulus may be found in the gray tissue of the cerebral hemispheres. In some way certain cells within the cortex have the power to receive stimuli relayed to them from lower centers and to hold this stimulus stored up within the cell for a time. Later the impulse is released from these cells, is relayed out through the motor centers in the spinal cord, and results in action. The city council may pass an ordinance that all automobile drivers who hear the siren of an approaching fire truck must pull up to the curb immediately and wait until the truck is past before proceeding on their way. The published ordinance may be read by a citizen, and if the city be a small one, a month may elapse before an occasion to obey the order arises. Then, on the afternoon of this later date, the citizen, engrossed with other affairs, is driving down one of the main thoroughfares. Suddenly he hears a warning blast of the fire siren behind him. Now the impulses that were stored in his cerebral cells a month earlier are promptly released. They pour out to the appropriate muscles in the form of motor stimuli, and the car is turned in to the curb while the fire truck unimpeded rushes past. Just how this stimulus gained from the printed page could be stored in the brain cells, and its effect be promptly released at some future time, is a profound mystery. No one has been able to explain it satisfactorily. But to the fact that it is done our whole experience bears evidence.

At this point the reader will realize that we have been discussing the biological basis of *memory*. It is in some such way that we store memories of past events. The value of this is obvious; for it is only by reason of this ability that we are able to profit by past experience. The more useful the memories the individual has stored in his brain cells, the better is he equipped to meet the demands of life successfully. Another peculiarity about these stored impulses is that they may subsequently be released many times. After reading the ordinance just referred to, the citizen would not only pull up to the curb once when he heard the fire siren but would continue to do so as long as he could drive an automobile.

Stimuli stored in the brain may be associated. Not only is man's brain capable of storing nervous impulses coming in from the receptors, but it is also capable of establishing some sort of connection among them. For example, suppose we have a child who has reached the age where he is capable of gathering sense impressions from the outside world but who has never yet had an orange presented to him. Then an orange is placed before him. Immediately he begins to receive sense impressions from it. He sees the yellow color of the orange and its size and shape; he feels its rough surface and smells its inviting odor; later he tastes it and gains another impulse through his gustatory sense. Each of these receptor reports registers itself in his brain cells as a stored-up impulse — a physical memory. If the first stimulus in each case is not sufficient to make a permanent impression on the cerebral cells, the respective stimuli will be repeated and reënforced each time that the child is introduced to an orange. In this way, by repetition, receptor impulses become thoroughly impressed upon the brain cells, and permanent memory follows. But this is not all. These individual stored stimuli gained from the sight, touch, smell, and taste receptors do not remain separate and distinct within the brain cells. In the language of the psychologist they become associated. When this result has been accomplished in the brain, a *single* stimulus coming from an orange, such as its taste, its smell, its appearance, or even its feel, will touch off some sort of reaction in the brain cells that enables the individual not only to recall that particular quality of the orange but to recall all the other qualities too. This recalled, associated memory is psychologically known as an idea or a concept. The act of so recalling is called associative memory, and is certainly one of man's most valuable psychic powers. It means the ability to build up within the brain cells associative connections between a group of related receptor impulses. Then, later, when one stimulus is released, the whole chain of memories in their proper sequential relations comes trooping through the mind in the form of an associated memory. Ideas, as well as sense perceptions, may be stored, associated, and recalled in this way. It is for this reason that a traveler can sit down and relate a whole train of incidents and experiences gained through his journey. It is this association too that often enables one to recite a whole poem when he recalls the first line. After one has had a number of experiences — has seen, read, and heard many things — the ability to associate these experiences within the brain cells in an efficient and effective manner is perhaps the next most valuable capacity in the development of a strong, intelligent mind. In its more intricate aspects it is the recall of these related ideas that enables one to make comparisons, to reason, to reach conclusions, and to make decisions.

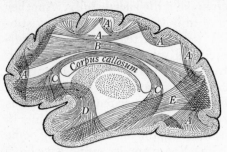

FIG. 66. Longitudinal Section of Cerebral Hemisphere of Man showing Association Fibers

A, between adjacent convolutions; B, between frontal and occipital areas; C and D, between frontal and temporal areas; E, between occipital and temporal areas [1]

This power of association can be wonderfully strengthened and developed by proper practice and study habits. It should be the ambition of every young student who wishes to possess a strong and effective mind to develop his associational centers to the utmost.

Just how the brain cells function to perfect these associational hook-ups is not known. It is believed, however, to be

[1] From Williams's *Textbook of Anatomy and Physiology*. By permission of W. B. Saunders and Company, publishers.

accomplished through extensive association fibers and their synaptic connections. The accompanying diagram presents a view of what are conceived to be association fibers between various parts of the brain.

Man exhibits many types of nervous reaction. The excitation of the receptor nerves throws the nervous mechanism into gear and results in some kind of action, either expressed or repressed. Just what response would be made in each case will vary somewhat with the individual. It will doubtless be modified by his age, experience, and heredity. Nevertheless, disregarding the matter of individual differences, we find that all men behave in much the same manner. It is with these common elements of behavior that we are concerned here: with what man does because he is a man — a representative of the highest type of animal.

Reflex action. Reflex action has previously been discussed. It is exemplified by a quick movement of the foot whenever the bare sole is tickled, or, more complexly, by jerking the hand away, making a face, and perhaps jumping about should the hand accidentally come in contact with a very hot object. Such responses result from nerve impulses that do not reach the cerebrum before the action is made. It is true that the same nervous impulses that initiate these movements may rise to the level of the cerebral cortex a moment later. There they may in turn set off all sorts of mental and emotional reactions. But the reflex response itself has usually been well begun (if not completed) before the higher brain centers function. Such behavior is nature's safeguard against sudden and unexpected dangers before which the more slowly reacting thought processes would be helpless. For this reason reflexes are highly important both to man and to other animals.

Volitional action. Volitional action is action controlled and directed by the will. It is what one does as the result of choice or compulsion. In this sense it is quite different from a reflex action and doubtless is intimately connected with one's associational powers. A certain stimulus is received, ideational processes take place in the brain, and action follows. For instance, a man with $5000 to invest hears over the radio that a favorite stock has reached the price at which he has decided to buy. This auditory stimulus sets off a chain of associational impulses that lead to a decision. In a moment he walks to the telephone, calls up his broker, and instructs him to purchase the stock. This is action by free choice.

Sometimes volitional acts emerge without any apparent stimulus to initiate them. Ideas seem to flow without any impulse to set them going, and action of an appropriate kind follows. It is probable, however, that such is not the case. There may be no evident external stimulus, and doubtless often there is none. But in such cases it is likely that the whole train of nervous reactions necessary to result in an appropriate response was initiated by an internal stimulus. It may have come from the muscles, the joints, the continuously active organs of the body, or it may have been set going by the stimulus of a certain associated brain cell not otherwise necessarily connected with the particular action in mind. We have all sat and daydreamed. Whole trains of inconsequential, apparently unrelated ideas, one after another, have drifted through our minds. In such cases it was doubtless the associational stimulus from a brain cell involved in the end of one idea that touched off a brain cell involved in the initial stages of the following idea.

Habitual action. One of the interesting and valuable aspects of volitional action is that it may become habitual. At first an act may be initiated by a stimulus which requires elaborate, deliberate, and laboriously formed associations before it can emerge in the form of action; but once the whole train of procedure from stimulus through to muscular execution has been accomplished, the act is believed to be slightly easier to do a second time. Especially is this true if the repetition follows reasonably soon. When a child first attempts to write, stimuli excite his receptors. A paper may be laid before him, a pencil may be placed in position in his hand, the copy of a word that he is to try to write may be written on the blackboard, and, finally, he may be encouraged by the kindly voice of his teacher to make the attempt. The stimuli from all these sources reach the brain, and a whole train of associational connections is inaugurated. These connections at first are made hesitatingly, falteringly, and with great uncertainty. Motor impulses subsequently are sent out that lead to writing movements. These also are slow at the outset, poorly coördinated, and badly executed; but once the feat of writing the word is achieved, the task is slightly easier to do a second time. Continued repetition makes the entire process from beginning to end easier to do and more accurate. Finally, such proficiency is reached that the initial stimulus is all that is needed to set the whole process of writing going. Words of all kinds flow from the pencil with ease and skill. In short, the mechanics of writing has become wholly automatic. This automatic result is habit. Just how the feat of writing, so painfully difficult at first, is made so easy as to become automatic no one knows. We express the idea in various ways. We say that "brain paths" are established, that appropriate synaptic connections are made, or that "brain grooves" are worn; but these expressions are but clumsy attempts to express facts which man has not yet been able to explain.

Regardless of just how they are formed, habits are of inestimable value to man. They liberate him from the necessity of giving time, effort, and attention to the purely routine and mechanical in life. They liberate his mind so that it may be free for the higher mental pursuits. If man always had to center his attention on the pure mechanics of writing, as the child did, he would have little opportunity to think what he was going to write. It is only by proper habit formation that man's highest powers and abilities are freed to express themselves.

It is precisely for this reason that good habits form such a large part of the intelligent, productive person's life. The more he can delegate the purely physical and mechanical in life to the realm of habit, the freer he is for higher pursuits and responsibilities. Practically nothing is more important than proper and extensive habit formation in the program of any young person who wishes to reach his highest mental efficiency.

It is for the reasons just explained that bad habits are so hard to displace with good ones. Before a bad habit can be broken off, the conceived neural connections between receptor,

2

stimulus, and the muscular effector have to be broken down.
Often this can be accomplished only by careful attention,
perseverance, and determination. Once the stimulus has been
inhibited from producing its accustomed action, however, the
neural path becomes weaker. Each repression of the act
weakens it still more. Finally, all connection is lost, and the
habit is broken. The person seeking to accomplish this result should never let a lapse occur. If he does lapse he
restrengthens the old, undesirable bonds by just so much. Of
great assistance in breaking a bad habit, too, is the substitution of a desirable one in its place. In this way one's energy
and attention are given over to the process of building up new
bonds, and the old ones are discontinued through disuse.

Instinctive action. Instinctive action is action in conformity
to an inherited pattern of response. Whenever the organism
is subjected to a certain stimulus, a chain of nervous impulses
is set up which emerges in action of a specific kind. No voluntary thought processes are involved, and the whole behavior
reaction is almost as independent of the will as is coughing
when the throat is tickled. Action among the lower animals
is mostly instinctive. Especially is this true among the insects;
practically the whole scope of their life behavior is instinct-
bound. They have to react to their environment as they do,
even though the conditions of that environment may have so
changed as to render their particular type of response perilous.
For instance, if in certain cocoon-spinning caterpillars the
larva is interrupted in its task when the cocoon is half spun,
and the completed part of the case is removed, the caterpillar
pillar promptly resumes its spinning. Instead of beginning
the whole cocoon again, however, it starts at the point
where its labors were arrested, and completes the second
part of the cocoon. This done, the helpless animal passes
into the chrysalis state and becomes quiescent, even though
such action means certain death. Fabre gives us another
picture of the insistence of instinctive action to follow its
course. A certain Sphex wasp stings crickets and drags them
into its burrow to become food for the newly hatched wasp
larvæ. When a mother wasp has stung her victim and dragged

it to the entrance of her burrow, she leaves it there momentarily and goes inside. This is done apparently to look about and see that all is well. She then promptly emerges and drags the cricket in. Fabre experimented. While the wasp was on the inside reconnoitering, he removed the cricket a few inches away from the threshold. The wasp came out, looked about, and, spying her victim, again dragged it up to the door of her burrow. She again left it while she went in to explore a second time, although she had just taken this precaution a few seconds before. Fabre removed the cricket forty times with the same result. The wasp was so instinct-bound that she never caught the idea — made another association — of dragging the cricket directly in. As Huxley says, "Drag cricket to the threshold — pop in — pop out — pull cricket in" was all she knew.

The value of instinct, of course, is to insure action of an appropriate kind in the presence of a certain need. It works well as long as the conditions remain unchanged; but it permits of no freedom of choice, no variation of behavior. All lower animals possess well-established instincts, but the compelling force of instinct was modified more and more as higher types evolved. The child has some instincts, such as to cry, to suckle when the mother's breast is presented, and others. But, of all animals, man is the least bound by this blind force. It is for this reason that he is capable of building up new associations in the brain and of making different reactions and combinations of reactions to varying conditions. It is this characteristic more than any other that separates man psychically from the beast and makes him an independent, self-directing being.

Emotional response. This term *emotional response* is used to express action and behavior in response to feeling. Many things that both people and animals do appear to be done in a rather matter-of-fact and machine-like manner. On the other hand, some actions are accompanied by strong emotions that affect the whole character of their expression profoundly. These states of feeling among both man and animals are of great importance. In man they are the seat of his highest joys and his deepest sorrows. They color the whole personality and are highly potent in determining whether the individual is alert, active, and dynamic or whether he is slow-going, plodding, and phlegmatic.

Emotions have a physical basis and are touched off, so to speak, by some receptor excitation, either external or internal. These stimuli promptly reach the autonomic nerve centers, and from them a complex of motor impulses is diffused through the body, especially to the unstriated muscles. The smooth muscular tensions and reactions which follow, together with the accompanying nervous impulses involved, rise to the level of consciousness, and the individual experiences a certain feeling. This feeling is the emotion. In many cases (perhaps we shall find in all cases when our knowledge is complete) the autonomic stimuli are greatly amplified and modified by the effect of certain hormones. These in some way are released into the blood stream and are carried to the nerve centers, where they do their work.

For our purpose the emotions may be said to fall into two natural groups: the *agreeable* and the *disagreeable*. The first group produce feelings that are associated with conditions conducive to bodily well-being; the second group, on the other hand, produce feelings that are associated with conditions inimical to bodily well-being. These emotions, when well defined, are almost invariably accompanied by some type of bodily expression. For instance, the amused person smiles, the grief-stricken one weeps, and the angry one fights. It is these bodily reactions that give a clue to the biological significance of the emotions and indicate their value to the organism. To illustrate: The bodily changes that are likely to accompany the disagreeable emotions are as follows:

1. The hair is raised so as to stand on end.
2. The pupils of the eyes are dilated.
3. The face takes on a blanched appearance.
4. The mouth becomes dry owing to the suspension of the salivary-gland secretion.
5. The heartbeat is greatly accelerated.
6. The digestive function is completely suspended, both the secretion of the digestive juices and the muscular movements of the alimentary canal being suppressed.

These bodily changes will readily be associated with the needs of the organism in times of emergency. Some of them serve man by enabling him to put up a better "bluff"; others render him more efficient. The elevation of the hair is of little value to man, since his capillary equipment is so scanty. But we have all seen a cat or a dog, which just a moment before appeared to be nothing more than a docile house pet, suddenly transformed into a beast of apparently great ferocity. The standing hair did not make the animal physically one whit more dangerous, but it did add profoundly to its ability as a bluffer. Perhaps growling, barking, snarling, and showing the teeth belong to the same category. The dilated pupils permit of a wider range of vision, especially from the corners of the eyes. The pale face, dry mouth, accelerated heart, and suspended digestion are all indicative of marked metabolic changes that greatly increase the power of the animal should combat become necessary.

The flow of digestive juices stops and the peristaltic movements cease, so that energy which is being expended in digestion may be temporarily diverted to the skeletal muscles. A cat that had been worried and angered showed stagnation of food in the stomach from three to six hours on different occasions. An angered dog, instead of producing the usual sixty-five or seventy cubic centimeters of gastric juice, secreted but nine cubic centimeters of very poor quality. The quickened heartbeat also hastens the flow of blood to the muscles and brain, thereby increasing their sugar and oxygen supply. The result is a greater release of energy. In helping to produce these bodily conditions the hormone adrenaline, of which we shall learn more in a later chapter, is a potent factor. It is for these reasons that an angered man or animal is actually stronger than an unangered one. It is now evident why the emotion of anger may be of great value to an organism, especially in times of emergency, when all the power and strength that can be mustered are needed for self-preservation. Agreeable emotions are produced by other stimuli in a manner similar to that of anger. They contribute to the individual's joy of living and also to his environmental and social adjustments.

Plants and animals are similar in some respects, but in others they are very dissimilar. In the past three chapters an attempt has been made to give the reader an idea of the plant and the animal as living, working organisms. The difference between the lowest plants and the simplest animals is not always evident; in fact, among the one-celled plants and animals there are certain individuals which are in some ways so much like a plant and in others so much like an animal that with a great deal of justice they might be claimed by either the botanists or the zoölogists. From that point on, however, plants and animals in respect to similarity diverge farther and farther until when flowering plants and mammals are reached we scarcely think of them as being alike. It is true that in certain fundamental characteristics, such as the structure of protoplasm, its activities and functions, there is still considerable likeness, but these basic similarities are so covered up and obscured by outward marks of dissimilarity that we scarcely ever think of the two as being related.

To contrast higher plants and animals briefly, green plants make their own food; animals cannot make food, but must secure it either directly or indirectly from plants. Higher plants are stationary and spend their whole life rooted to the soil; higher animals move about with great facility and often, indeed, with surprising speed. Plants have strong, rigid bodies; animals have skeletons, but the parts are so articulated as to bend with great ease. Higher animals have developed highly specialized sense organs through which they may become acutely aware of many elements of the environment; plants have no such organs, and to many forces of the environment they appear to be entirely mute and unresponsive. Moreover, animals are possessed of an intricate nervous system and a highly centralized brain which, coupled with an efficiently responding muscular mechanism, enables them to respond to the environment in a multitude of ways impossible to the plant.

QUESTIONS FOR STUDY

1. Contrast the action of the sympathetic and parasympathetic nervous systems.

2. Discuss the theory as to the biological basis of memory.

3. How does man differ from other animals with respect to his nervous system?

4. What are the principal functions of the nervous system in man?

5. What is the value to man of habitual action?

6. Explain how the nerve cell is believed to function.

7. Describe the correlating mechanism of the body.

8. Trace the development of the nervous system from that of lower animals to man.

9. When one is involved in an accident, what changes may occur in the nerve impulses that influence behavior.

10. What is involved in the sensation of sight?

11. Of what importance is the front end of the body?

12. How does one hear?

13. What nervous mechanisms are used when one forms associated memories?

14. Describe the difference in the nervous mechanisms involved in action on the different levels.

15. How does a child learn to walk?

16. How can a man drive an automobile and at the same time carry on a conversation?

17. What is believed to be the biological basis of an emotion?

18. Of what value to the organism are the physiological accompaniments of disagreeable emotions?

REFERENCES

BEST, C. H., and TAYLOR, N. B. The Human Body and Its Functions, pp. 221–333.

BURLINGAME, L. L., HEATH, H., MARTIN, E. G., and PEIRCE, G. J. General Biology (second edition), pp. 123–171.

CLENDENING, LOGAN. The Human Body (second edition), pp. 53–68, 203–206, 223–265.

CLEVELAND, F. A., and others. Modern Scientific Knowledge of Nature, Man, and Society, pp. 358–378.

GATES, A. J. Elementary Psychology, pp. 33–103, 183–213.

KIMBER, DIANA C., and GRAY, CAROLYN E. Textbook of Anatomy and Physiology (eighth edition), pp. 41–206.

KIRKPATRICK, T. B., and HUETTNER, A. F. Fundamentals of Health, pp. 180–238.

WALTER, H. E. Biology of Vertebrates, pp. 587–739.

WELLS, H. G., HUXLEY, J. S., and WELLS, G. P. The Science of Life, Vol. I, pp. 97–139; Vol. IV, pp. 1147–1390.

WILLIAMS, J. F. Textbook of Anatomy and Physiology, pp. 56–283.

UNIT II

Synthesis and Decomposition form a Cycle in Nature

THE phrase "dust thou art and unto dust shalt thou return" briefly portrays the cyclic history of living bodies. It expresses a profound biological truth as well as a philosophical estimate as to the end of man.

The bodies of all living organisms are built up from the inorganic elements of air and soil. In a sense they are for the most part the products of the synthesizing activities of green plants. All elements and compounds held in this manner are securely locked up by the powerful forces of chemical affinity. Since living organisms are constantly growing and multiplying, if there were no return of these organic elements back to their original source, the natural supply would soon be exhausted. As a consequence life in all its forms would soon pass from the earth.

Fortunately there is another group of organisms which constitute the wrecking crew of nature. They are the agents of decay—the bacteria and fungi that, for the most part, decompose dead bodies. They tear down and resolve these carcasses, great and small, back again into the simple compounds of air and soil. In this rôle the microörganisms of decay become nature's benefactors. They carry on one set of activities that make the great cyclic processes of nature possible.

Chapter V · The Cycle of Metabolism in Living Things makes the Continuance of Life on the Earth Possible

Introduction. The synthesis of carbohydrates, fats, and proteins by plants, and the metabolic use of these foods by both plants and animals, result in many organic products that are insoluble. Leaves, twigs, fruits, and flower parts drop from plants; hair, feathers, and skin cells are constantly being sloughed off by animals. Moreover, when an individual organism dies, regardless of whether it is an amœba or an elephant, a protococcus or a redwood, practically all substances built into their tissues are in the insoluble state. If no natural processes provided for the return of these elements to the soil in a soluble form, the result would be tragic. The earth would become cluttered up and filled with dead organic remains. All inorganic food elements used in the production of the tissues comprising these organisms would be securely locked up. In this state the chemical substances involved would be wholly unavailable. They would be as effectively removed from organic circulation as money locked in a bank vault is removed from commercial exchange. Under these conditions soil would become unproductive, and in the end life would be cut off from the earth.

Fortunately, however, nature has provided a great group of organisms whose chief business seems to be to tear complex substances down. In a measure higher plants and animals do this through digestion. In the past few chapters we have discussed largely the synthetic processes that have made the life of higher plants and animals possible. In this chapter we shall turn to the activities of the decompositional and allied agents in nature. We shall consider those agents that carry organic products of all kinds back to "dust" again; in a

161

word, those forces that contribute to soil fertility. In all these processes, both building up and tearing down, chemical changes are concerned and accompanying shifts of energy follow. It is through these wholesale and varied exchanges of energy that life, of whatsoever form, is maintained.

There are many kinds of decomposition. The decompositional changes in nature are many and varied. The higher organic compounds are extremely complex; consequently when they are decomposed the process is not a simple one. Not only may many intermediate products of decay be formed before the end products are arrived at, but a whole series of organisms may be involved as causative agents. Carbohydrates may be fermented, yielding a long list of products and by-products. In the absence of sufficient oxygen proteins may putrefy, producing toxins and foul-smelling gases. Metabolic processes normally carried on by all respiring organisms are decompositional in their nature. To some extent also chemical changes produced by the purely inanimate forces of weathering may promote decomposition. From this it may be seen that decay in all its aspects is a highly involved matter. But its importance as a natural process will warrant a brief study of the kinds of decay, its agents, and the products formed. At every step it will be seen that enzymes also are involved. Decomposition, as well as synthetic processes, has to be speeded up.

Some items in the following list, perhaps, may never have been thought of as processes of decay; but they result in tearing complex bodies down into simpler ones and, for this reason, should be placed in this class.

Digestion. Digestion as carried on by most plants and animals is, in reality, a kind of decay. In their passage through the alimentary canal of man, for instance, starches are split into malt sugars and these, in turn, into the simple sugars; fats are hydrolized to form glycerin and fatty acids; and proteins pass through a series of changes before they emerge as simple amino-acids. The whole movement, then, is a tearing-down one, and in this sense man, the horse, and the dog must be considered to be active agents of decay.

Autolysis. All cells at death contain enzymes capable of producing digestive changes. These are said to be autolytic (Greek *auto* + *lusis*, "self-dissolving") enzymes. During life they appear to be held in check and are confined in their activities to the solution of such substances as will promote the welfare of the cell. It is generally believed that the only reason why the stomach does not digest itself is because of certain inhibitors produced by the cells which hold the pepsin in check and permit it to work on the food proteins only. But in death all restraint seems to be off, and these chemical accelerators turn upon the contents of the cell itself and begin to decompose them. This condition holds for plants as well as for animals. So it is probable that the initial stages of decomposition are self-inflicted, and then that these are later supplemented by the work of outside organisms. The presence of these autolytic enzymes is one reason why meat should be thoroughly chilled immediately after an animal is killed, and then stored at a low temperature. If it is not, undesirable enzymatic action is greatly speeded up by the normal temperatures.

Decomposition through metabolism. Whenever energy is released from the food, to be used by the body in functional processes, the food is changed chemically by enzymatic action. This means that the food is torn down, and this, in a sense, is decay. For example, the carbohydrates are broken down through respiration into water and carbon dioxide, with a certain release of energy. Fats liberate an even larger measure of energy; their chemical components too come out in the end as water and carbon dioxide. Proteins also are decomposed through respiration. When oxidized they yield various by-products, such as water, carbon dioxide, uric acid, and urea. In animal bodies the uric acid and the urea must be eliminated mostly by the kidneys. Plants, on the other hand, appear to be able to recombine these nitrogenous by-products of respiration into proteins again by synthesizing them anew with other carbohydrate molecules. This nitrogen "frugality" on the part of plants is of immense economic importance. It makes possible the production of much larger crops than could otherwise be grown on poor land.

Carbohydrate fermentation. All through nature carbohydrate products have a tendency to decay through fermentation (Latin *fermentare,* to "boil"). The type with which the ancients were familiar was the active one occurring when fruit juices sour. This term was applied because the rapid ebullition of carbon dioxide at the surface of the fermenting juice resembled the effect produced by boiling. Starch, cellulose, wood, sugar, and similar substances are attacked by fermentative action. The by-products may be alcohols, acids, and other products. If the decompositional process is carried through to the end, carbon dioxide and water emerge as the final products, just as in complete respiration. In fermentations, so abundant in nature, organisms are always involved. Enzymes are secreted that produce a shifting and recombination of atoms and molecules. In these rearrangements energy is liberated to supply vital force to the fermentative agents.

Putrefactions. Putrefactions, as already noted, are decompositions of protein matter in which putrid, offensive odors are generated. If carried to completion, protein decomposition results in the formation of water, carbon dioxide, ammonia, sulfur, and phosphorous compounds, besides others; but because this type of decay involves a long series of changes, many intermediate products arise. Among them may be marsh gas, hydrogen sulfide, mercaptans, skatole, and indole. Hydrogen sulfide is the offensive gas liberated by rotten eggs; skatole and indole are prominent constituents of human feces. Because of their high nitrogenous content, meat, milk, eggs, cheese, beans, and peas are prolific sources of putrefaction. Everyone has experienced the foul odors emitted by a decomposing animal carcass. Here the putrefactive action is at its height.

Agents active in decomposition. The organic agents that carry on decay may be grouped under three classes: animals, bacteria, and fungi.

Animals. We rather recoil from thinking of the normal digestive processes that take place in the human body and in those of other animals as being decay. But they are, nevertheless, as we have found, and must be so considered. What takes place when a fallen tree in the forest is slowly resolved into brown, powdery, rotten wood dust by decay is not essentially different from the digestion of food in the animal body.

Bacteria. Bacteria are minute unicellular plants found practically everywhere in nature. These tiny plants, together with other one-celled plants and animals, are often called microbes. Bacteria themselves are perhaps the simplest form of life known. Dr. Greaves refers to them as being "at the bottom of the ladder." They were first recognized and accurately described by Anton van Leeuwenhoek, a Dutch cheese merchant and janitor of the town hall, who found more pleasure in grinding microscope lenses and examining drops of stagnant water than he did in his commercial activities.

Structurally, bacteria are composed of a diminutive bit of protoplasm surrounded by a cell wall. They have nuclear matter, but are so low in the scale of life that the chromatin is scattered throughout the cell instead of being aggregated into a definite cell body as in higher organisms. They were formerly thought to keep a permanent shape, size, and form. Recently, however, accumulative evidence seems to indicate that many bacterial forms are capable of a sort of Dr. Jekyll and Mr. Hyde existence; that is, they may appear as definite cells at one time and as more or less disorganized protoplasm at another time.

Bacillus   Coccus   Spirillum

Fig. 67. The Morphological Shapes of Bacteria

When manifesting the cellular form, bacteria may take one of three general shapes: spherical, rod, or spiral. The spherical forms are known as *cocci*, the rod-shaped ones as *bacilli*, and the spiral ones as *spirilla*. The term *bacteria* was derived from the rod-shaped cells, which were among the first forms studied. It comes from the Greek *baktērion*, meaning "a stick."

In size bacteria are all very small. The largest forms are invisible to the naked eye, and the smallest ones are believed to be below the range of vision through the most powerful microscope. The organism causing relapsing fever is one of the largest known, yet it would take fifteen hundred of them placed end to end to reach an inch. One of the smallest forms, the Pfeiffer bacillus, formerly believed to be associated with influenza, is so tiny that it would require one hundred and twenty-five thousand of them placed end to end to cover a line an inch long.

Bacteria normally reproduce by simple cell fission, and under favorable conditions the rate at which they multiply is prodigious. Fission may be completed in twenty minutes. Disregarding the retarding influences of toxic accumulation and limited food supply, it has been estimated that with optimum conditions a single bacterium might result in 281,500,000,000 offspring in two days. In three days the total weight might reach 148,356,000 pounds, and in five days the organisms would fill to a depth of one mile all space occupied by the oceans.

Although not in reality a form of reproduction, many bacteria have the ability to form resistant spores. Spores are an adaptation to tide the organisms over a particularly unfavorable period of existence. In spore formation the protoplasm dries down, rounds up, and the cell puts on an especially tough, resistant wall. In some species these spores are the most resistant living cells known to man. Some have been known to survive a quarter of a century, and in one case viable spores were found in air-dried soil that had been kept in jars standing in the laboratory for a period of fifty years. Dr. Charles B. Lipman of the University of California recently reported that by the use of special methods he had recovered bacteria from rocks which had been deposited over a million years ago. When planted on agar cultures the spores of these microörganisms became active. The spores of the dread anthrax bacillus can survive boiling for several hours. It is because of spore production that so much precaution must be taken in canning and preserving foods. The ordinary vegetative bacteria are easily killed by boiling; but if the food products happen to contain spores, methods of canning which will

kill spores must be adopted, else fermentations may follow that will develop dangerous toxins. The spores of bacteria are characteristic, and for those species that produce them they offer a valuable means of identification.

Fungi. Fungi have life habits very similar to those of bacteria, but they differ from bacteria in that they are higher forms of life. They include such plants as the yeast cells, bread mold, mildews, toadstools, mushrooms, and the grain rusts. They are either unicellular or filamentous and, more perhaps than anything else, are characterized by a complete absence of chlorophyll. This means, of course, that they cannot make their own carbohydrates, but are dependent upon other organisms for their supply of this kind of food.

FIG. 68. Spores of Bacteria

The hairlike filaments usually aggregate themselves into a more or less tangled mass called the *mycelium*. The mycelial threads may thoroughly penetrate the organic material upon which the fungus is feeding, or they may form largely on the surface, sending tiny rootlike threads down into the substrate to absorb nourishment. Bread mold produces its mycelium largely

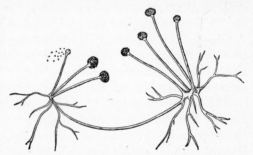

FIG. 69. Mycelium of Bread Mold

on the surface of the bread or other material upon which it is growing. It is this external growth that is recognized as the mold itself. The tough, skinlike covering often found on unsealed preserves and jellies [is nothing but a mass of interwoven fungal filaments living on the material beneath. Active yeast cells live disseminated through the sugary solutions upon which they feed. In the case of many fungi

2

the filaments thoroughly penetrate the living or dead tissues upon which they subsist. Rotting wood or rust-infected wheat stems are thoroughly permeated with the threadlike growths of these dependent plants. Regardless of the general form which the fungus body takes, the method of getting food is similar in all cases. Enzymes are secreted which break down the organic matter, reduce it to a solution, and make it available to the multiplying cells or ramifying fungus filaments. It is for this reason that mildews on damp clothes or leather are so destructive to these materials, and it is for the same reason that fruits with broken skins or potatoes with damaged epidermis are so susceptible to decay. Rapidly decaying oranges and lemons are having their tissues completely consumed by fungal filaments.

In some fungi the filaments become so closely impacted into a more or less spongy structure as to lose their characteristic filamentous appearance. This is especially true in the fruiting bodies of toadstools, mushrooms, and puffballs. The umbrella-shaped mushrooms bought on the market are not the real plant itself, any more than an apple is the apple tree. On the contrary, they are the fruits of the mushroom fungus. A microscopic examination of the humus soil or rotting leaves or straw from which the mushroom fruits push up will show the whole substrate to be thoroughly impregnated with the true vegetative filaments of the fungus plants.

The widespread and at times vindictively rapid growth of fungus infestations is due to their method of reproduction. They reproduce both asexually and sexually by means of spores. Since these plants live either on dead and decaying hosts or on living hosts, they do not have to produce highly differentiated and specialized structures such as roots, stems, and leaves. About all they need is food-absorbing filaments, and the rest of their activity can be given over to spore production. For this reason the number of spores produced is sometimes prodigious. Perhaps the reader has noted the tiny black specklike bodies thickly disseminated through the mycelium of bread mold. Every one of these is a minute spore case containing thousands of spores, each of which is potentially

capable of producing a new bread-mold growth.   In some cases
chains of successive spores are formed and pinched off at the
ends of the ordinary vegetative filaments.
In this way the whole powdery surface of
the greenish fungus that infects oranges
and lemons becomes a source of citrus-fruit
infection.   In the earlier stages of grain-
rust infections the enormous numbers of
spores formed in tiny bunches at the surface of the leaf and stems give the plant
a reddish-brown appearance.   Sometimes
these spores are so abundant at harvest
time as to cover the horses, parts of the
harvester, and the clothing of the men who
operate it with a reddish-brown dust.   It
is for this reason that the plant produc-

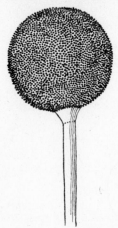

FIG. 70. A Spore Case
of Bread Mold

Note the immense num-
ber of spores it contains

ing these spores is called grain "rust."
By means of these red spores, at times especially favorable for their growth, the
infection spreads over a large field at an
almost incredible speed.   The farmer expresses the situation by saying that his grain has been "struck" with rust.

Mushrooms and toadstool fruits
bear gills whose surfaces are literally covered with spores that
when ripe are forcibly hurled off and
are caught by the wind and wafted
about. It is astonishing how many
spores a single fruit will produce.
Dr. Buller, a famous Canadian student of fungi, estimates that a single
fruit of the common meadow mushroom may produce 1,800,000,000
spores. The spores liberated by a
single shaggy-mane fruit were estimated at over 5,000,000,000, and a
big shelf fungus may produce as
many as 100,000,000,000 spores per

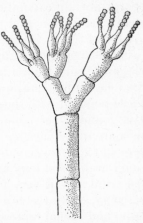

FIG. 71. The Conidiophores
of Penicillium

year. The meadow and the shaggy-mane mushrooms are among our most abundant and highly prized edible fungi.

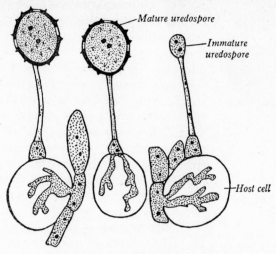

Fig. 72. Uredospores of Grain Rust

Bacterial and fungal spores are to be found almost everywhere. Because of their tiny size and abundance, bacteria and fungal spores are almost universal in their distribution. The food we eat, the water and milk we drink, and the air we breathe are full of them. Regardless of how carefully food is prepared, it usually contains many of these microörganisms. Most well water contains from a few bacteria and spores to several hundred per cubic centimeter. Streams usually have more, and those polluted with the sewage from towns and cities may contain millions per cubic centimeter. Productive agricultural soils are quite variable in their bacterial and fungal content, depending upon conditions of moisture and temperature. Good arable soil may have a hundred thousand organisms to the gram and, if very favorable conditions prevail, the number may easily reach a million. Deeper in the soil the bacterial and fungal representatives drop off rapidly, but in some of the looser soils of western North America numerous microörganisms are found even at a depth of ten feet.

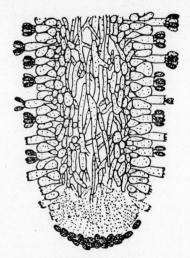

FIG. 73. Diagram of the Basal Cells in the Gill of the Mushroom, showing the Spores [1]

Milk as secreted by the cow is normally free from bacteria; but milk drawn from the udder always contains them. The number varies with different cows and with the care and cleanliness exercised in the dairy barn. Under extremely favorable conditions fresh milk as taken from the cow may contain as few as twenty bacteria per cubic centimeter; if the milking is done under unsanitary conditions, there may be thousands of organisms in each cubic centimeter; if the cow is diseased, the number may run into millions.

Except immediately after a rain or snow, there are myriads of dust particles floating about in the air. Everyone of these is a potential dirigible to carry microörganisms about. Buller has found that the air is often full of the spores of certain fungi. He says that if their diameters could be increased a hundred times without increasing their rate of fall we could see them with the naked eye and that they would often darken the sky. The United States Department of Agriculture has studied grain-rust migrations. By placing spore traps on the

FIG. 74. Dissemination of Mushroom Spores [2]

[1] By Buller, in *Jahrbücher für Wissenschaftliche Botanik*, Band LVI, 1915.
[2] From A. H. R. Buller's *Researches in Fungi*, Vol. I. Longmans, Green & Co., publishers.

wing struts of airplanes, investigators have found spores and pollen grains numerous up to a height of two miles.

In these ways decomposing organisms of all kinds are continuously sown about thickly over the earth. Not even a tiny pinhole can be made in a can of fruit and then resealed without introducing fermenting yeasts and bacteria. Wherever they find lodgment and begin their work, organic compounds of all classes are slowly torn down, step by step, and carried back to form a part of mother earth again. In this state the end products of decay are again ready to be built into organic food and thus be started anew on their cyclic round of life and decay.

Some organisms are parasites. In the foregoing pages we have considered agents of decay only — those microörganisms that get their food and energy by decomposing the dead bodies and remains of other organisms. There is another set of bacteria, fungi, and protozoans that do not wait for their victims to die but attack them while still living. This is the great group of parasites. They include some higher plants and animals also. They attack the tissues of living things and, by the secretion of potent enzymes, dissolve the cells while the individual is still alive. These comprise our great group of pathogenic (or disease-producing) organisms. Their object — food and energy — is the same as that of their less rapacious relatives, but along with their decomposition activities they produce pain, suffering, and often death.

Many products of fermentation and decomposition are of great value to man. We have found that the end products of decay are of great importance in nature. Many of the intermediate products also possess great value, as we shall see.

Alcoholic fermentation. Despite their frequent misuse by man the products of alcoholic fermentation take an important place in modern industry. In this process sugars are broken into alcohol and carbon dioxide by the action of yeast cells. A few bacteria and other fungi may produce fermentation also, but the products are seldom the same. For instance, the poisonous fusel oils, often found in "bootleg" liquor, are probably the products of molds and bacteria that infest the mash.

It was the work of Pasteur that proved fermentation to be the work of yeast plants. Even the souring of grapes and other fruits whose skin had been broken while still on the plant was demonstrated to be due to wild yeasts blown about in the air. In this process the simple sugar, glucose, is decomposed into two molecules of alcohol and two molecules of carbon dioxide. The enzyme is zymase secreted by the yeast plant. Thus,

$$C_6H_{12}O_6 + \text{zymase} \longrightarrow 2\ C_2H_5OH + 2\ CO_2 + \text{energy}$$

The strictly fermented liquors cannot reach an alcoholic content of more than from 12 to 15 per cent. Alcohol itself is toxic to the yeast cells, and they perish when this alcoholic strength is reached in the fermenting liquor. Whiskies, brandies, and other high-grade liquors of alcoholic content are made possible by fractional distillation. The fact that alcohol boils at about 78° C., whereas water requires 100° C. at sea level, makes the separation of alcohol from the accompanying watery products possible.

In breweries and distilleries the sugar is first obtained from starch by a process called malting. This consists of soaking the grain — barley, rye, or corn — and then placing it in moist chambers to sprout. Under these conditions the embryo secretes both amylase and maltase, which, respectively, hydrolyze the starch into malt sugar and glucose. When the proper stage of starch conversion is reached, the grain is then dried to arrest germination and conserve the sugar. The grain is subsequently ground and soaked in water to form the dilute sugary solution in which the yeast inoculation multiplies and thrives.

Commercial yeasts, such as compressed and dried yeasts, are made from very rich yeast cultures prepared by special techniques. The yeast cells are usually mixed with starch, flour, or meal and pressed into cakes. If the specially prepared cakes are kept moist and cool and sold while still fresh, the product is called compressed yeast; if the yeast is mixed with meal and dried before being packed, the product is dry yeast. In dry yeast the plants gradually lose their vitality. When used, the viable yeast cells still remaining must have time to multiply before satisfactory leavening results can be obtained.

In panary (Latin *panis*, "bread") fermentations the yeast cells are employed for the carbon dioxide they generate rather than for the alcohol. Only flours containing gluten, which forms a stiff, elastic mass when mixed with water, can be used. As the carbon dioxide is liberated it bubbles up through the dough, becomes enmeshed, and causes the dough to puff up and become light. The alcohol formed is mostly driven off during the subsequent baking process.

Lactic-acid fermentations. Cow's milk, as well as that of goats and other animals, contains a considerable quantity of lactose, or milk sugar. This substance is readily attacked by lactic-acid bacteria and other organisms which ferment it. Under favorable conditions the acid soon accumulates to the point of coagulating the protein content of the milk, and the curd so produced forms the basis for many kinds of soft cheese, such as the common cottage, or Dutch, cheese. In the harder cheese the milk proteins are coagulated by rennin, an extract from the lining of the calf's stomach, and, after having the watery whey pressed out, are permitted to ripen to the desired state. In the ripening of special kinds of cheese, temperature, humidity, and other aërial conditions are important. The process, chemically speaking, is a complicated one and in most cases is imperfectly understood. Protein enzymes are known to be involved and oftentimes specific types of fungi. In the well-known Roquefort cheese, whose cut surface presents a greenish, mottled appearance, the particular fungus is a species of the green mold Penicillium. The *decompositional* products formed by this fungus and perhaps other organisms give the characteristic Roquefort flavor. The mottled effect is produced by the green mycelial threads matted together throughout the holes and crevices of the cheese. Camembert cheese is similarly flavored by another species of Penicillium. Limburger cheese owes its pronounced odor and flavor to a continuance of the so-called ripening process until bacterial *digestion* and *decomposition* are well advanced.

In modern creamery practice butter is made from cream that is first sterilized to kill undesirable, wild, and possibly pathogenic bacteria. It is then inoculated with pure bacterial cultures. Then, by permitting the controlled fermentation to proceed just so far, the desired flavors and degree of acidity necessary to produce good butter are obtained.

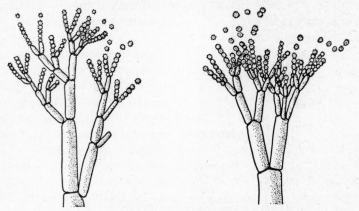

FIG. 75. Left, Penicillium Camemberti; right, Penicillium Roqueforti

These two fungi give flavor to Camembert cheese and Roquefort cheese

Sauerkraut is made from clean shredded cabbage leaves mixed with about 2 per cent of salt and usually packed in an earthen or wooden container. The brine draws the water and part of the soluble organic materials out of the leaves by osmosis, so that all interstices become filled with the liquid. The rapid respiration of the injured leaf sections continues for a time and rapidly reduces the oxygen content. In this condition the whole mass ferments. The process is probably initiated by yeast cells and then carried forward by bacteria of different kinds assisted, perhaps, by leaf enzymes. The original organisms to provide the culture probably came from the surface of the cabbage leaves themselves. Usually about 1 per cent of lactic acid is formed, and this provides the characteristic sour taste. Other desirable flavor components are doubtless added by the action of other species of bacteria.

Acetic-acid fermentation. Alcohol is oxidized by acetic-acid bacteria into acetic acid, or vinegar. Thus,

$$\underset{\text{(alcohol)}}{C_2H_5OH} + O_2 \longrightarrow \underset{\text{(acetic acid)}}{CH_3COOH} + H_2O$$

Some brands of vinegar are produced by fermentation of fruit juices to produce the alcohol. The alcohol is subsequently

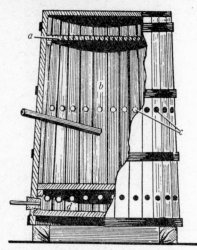

FIG. 76. Cask used for Making Vinegar Commercially[1]

oxidized to form the cider vinegars. Other fermented juices and even distilled alcohol in solution may be used.

In the case of home-manufactured vinegars the "hard" cider is usually permitted to stand in partly filled barrels or casks until spontaneous fermentation occurs. The acetification may be initiated by the addition of some "mother of vinegar" taken from another active culture. This slimy, ropy "mother of vinegar" is a mass of acetic bacteria that cling together by the slimy, gelatinous cell walls which they develop. Commercial cider vinegar is made by allowing the fruit alcohols to trickle slowly down through well-aërated tall wooden cylinders filled with beechwood shavings inoculated with acetic bacteria. The shavings provide a large surface for the growth of the organisms, whose great numbers carry on acetification rapidly, and the acid drips through holes provided in the bottom of the container. Beech shavings are used because this wood imparts no undesirable flavors to the vinegar.

Retting. The principle of decomposition is employed in the preparation of certain fiber crops for commercial use. The long, slender, tough fibers of such plants as flax and hemp are found in the outer portion of the stems. The cellulose fibers are grouped into bundles and are held together within the bundle by a pectic substance which structurally is known as the middle lamella. In reality the middle lamella constituted

[1] From E. D. and R. E. Buchanan's *Bacteriology.* By permission of The Macmillan Company, publishers.

Fig. 77. Retting Flax[1]

the original thin cell walls of the growing tissue. These cells were later converted into tough, thick-walled fibers by the addition of successive layers of cellulose to the middle lamella. In order to get the long fibers separated so that they may be spun into thread, this middle lamella must be dissolved. The process by which this is done is called *retting*; it depends upon the decomposing action of bacteria. The flax or hemp to be retted (rotted) is either placed under water or left exposed to the dew and rain. Bacteria related to those that cause butter to become rancid attack the middle lamella under these conditions and decompose it, letting the fibers fall apart. The fibers after being cleaned and dried are spun into thread or twine. Linen cloth may then be woven from the flax thread, or burlap and rope may be manufactured from the hemp twine.

Food-poisoning may result from the products of decomposition. There are many forms of food-poisoning due to the decompo-

[1] From Frye and Atwood's *New Geography, Book II*. Ginn and Company, publishers.

sitional action of bacteria. For our purpose but two of the most common ones will be noted.

Botulism. The first attack of botulism to be successfully investigated occurred among the members of a church choir in the little town Ellezelles, in Belgium. After rehearsal the members sat down to refreshments. The chief dish was some pork sausage which unwittingly had been improperly cured. Several of the victims died. From that incident and perhaps from others of a similar nature the disease took the name botulism (Latin *botulus*, "sausage"). Within recent years there have been several more or less isolated outbreaks in America. The poison is a deadly toxin secreted by the bacteria and is so potent that a single taste of the vitiated food may be fatal, even though it is not swallowed. In white mice experimental doses so dilute as to contain theoretically but a few molecules have killed the animal.

It has been found in this country, mostly in connection with canned goods such as beans, peas, asparagus, ripe olives, and meats. Any food with the slightest odor or evidence of being unwholesome should be discarded without risking even a taste of it. This bacterium is a spore-forming one and escapes the otherwise sterilizing temperature of canning while in this condition. Fifteen minutes of hard boiling decomposes the toxin and renders the food harmless if eaten immediately.

Ptomaines. Ptomaines, unlike botulisms, are not toxins, but are the poisonous products of protein decomposition called amines. There are several kinds of amines, and they vary widely in their potency. They are not nearly so dangerous as botulism, but are sometimes fatal. They are sufficiently poisonous, however, to make one very cautious about eating any kind of stale or "overripe" proteinaceous foods.

Certain phases of the carbon and nitrogen cycles are dependent upon decomposition. The process of decomposition is important in so far as it carries any elements required by plants from their organic condition back to the constituency of the soil. Sulfur, phosphorus, calcium, magnesium, and other elements are again made available in this way. But the two elements required in relatively larger quantities by plants are carbon and nitrogen. In providing an adequate supply of these elements decay is invaluable.

The carbon cycle. The carbon used by plants is taken from the air in the form of carbon dioxide. It comprises only about .03 per cent of the atmosphere, but in this diluted form it is indispensable if the usual cycles of plant and animal life are to continue to appear on the earth. Carbon is built into every cell wall of higher plants, forms an important constituent of carbohydrates, fats, and proteins, and is a necessary component of the protoplasm itself. This element is equally indispensable to animals. The average human body contains about twenty-five pounds of carbon. Yet despite its wide use by all organisms, the small percentage of carbon in the air makes a constant replenishment imperative if the processes of nature are to proceed uninterruptedly. The late Dr. Cyril G. Hopkins, of the University of Illinois, estimated that two luxuriant corn crops, such as grow in parts of that productive state, would practically exhaust the carbon dioxide in the air above the field if there were no replenishment from the outside. The weathering of rocks also removes large quantities of this gas. But fortunately the forces of nature provide for the constant restoration of the normal amount of this oxidized carbon.

Some of the supply comes from volcanoes and some from artificial combustion, but by far the greater amount is derived through organic decomposition. It has been estimated that the people of the world alone liberate about fifty million tons of carbon a year as a component of the by-product of respiration. But the chief source of carbon dioxide in the atmosphere is decay. The earth is covered with the débris of plants and animals. The roadsides, street corners, vacant lots, pastures, grassland, and forest are constantly giving off a volume of this important gas. The forest floor may be covered with a layer of leaves and humus several inches deep. To this are added the massive trunks of fallen trees and oftentimes a heavy deposit of detached twigs. In cultivated fields the earth is fully impregnated with the dead roots of former crops. Moreover, in regions where agricultural lands are fertilized with barn-yard refuse great quantities of carbon dioxide are liberated from this decaying mass. It is probable that one of the fertilizing effects is the increase of carbon dioxide. So, in all these ways, decomposition provides for a replenishment of carbon in the atmosphere. This in turn makes the continuance of life on the earth possible.

The nitrogen cycle. Nitrogen, as we have found, is an indispensable constituent of the protoplasm and is therefore necessary to life. The body of the average man contains something like twelve pounds of nitrogen. One would think that with the almost limitless supply of this gas in the air the nitrogen problem for plants would be a simple one. Such, however, is not the case. The condition of the ordinary green plant in this respect is much like that of the fabled king Tantalus. Tantalus had been a very tyrannical, cruel, and oppressive king while on the earth. After death, when he appeared before the judge to receive his reward, his record was found to be a very black one. Accordingly, as a fitting punishment, he was condemned to stand stationary near the bank in a river of clear, cool, sparkling water that came up to his waist. Over his head swayed the boughs of fruit trees laden with all sorts of the most attractive, fragrant, and luscious fruit. Under these conditions Tantalus must endure the pangs of eternal hunger and thirst. Whenever he reached up to get the fruit the boughs swayed gently out of his reach. Whenever he stooped to slake his thirst the water receded before his chin. It was from this myth that we got our word *tantalize.* So it is with green plants. Their aërial parts extend up into an atmosphere 78 per cent of which is nitrogen. They are literally bathed with it. Yet, for all practical purposes, they cannot appropriate a bit of it. What they use must be taken from the soil in the form of a soluble chemical salt. How to manage the soil so as to secure at all times an adequate supply of soluble nitrogen for growing crops is one of the major problems of the agriculturist.

Here, again, decay enters the picture through a process called nitrification. The chief source of nitrogen supply is decaying organisms. When the proteins and protoplasm break down, the nitrogen is finally liberated in the form of a soluble nitrate which plants can absorb and use. In reaching this end product of decay three sets of organisms are involved. The nitrogenous organic compounds are first acted upon by a group of bacteria and fungi that produce nitrogen in the form of ammonia as one of the products. This was discovered in 1893 by Müntz and Condon. Previously it had been known that fertile soil contains some ammonia. But that year these two men demonstrated that ammonia does not develop in inclosed soil which has first been heated to kill the soil microörganisms. This particular decompositional process is known as ammonification, and by it the microörganisms get the necessary energy to live and function.

A second set of bacteria attack the ammonia compounds and extract energy by oxidizing this nitrogenous ammonia into nitrites. These nitrites, which cannot be successfully used by higher plants, are in turn acted upon by a third set of bacteria and further oxidized into nitrates. These nitrates, then, in the form of chemical salts, such as $Ca(NO_3)_2$ or KNO_3, form the most desirable source of nitrogen for higher plants. The first experimental evidence that ammonia may be oxidized into nitrates was obtained by Schlösing and Müntz in 1877. Sewage containing ammonia was made to trickle through glass tubes a meter long filled with clean sand and limestone. Nitrates were produced and could be extracted from the filtrate that issued from the tubes. However, as soon as the microörganisms in the sewage were killed by treating the filters with chloroform the ammonia trickled through unchanged. Organisms have been reported recently that oxidize ammonia directly into nitrates without the intermediate nitrite step.

Nitrogen fixation. Although the fixation of raw nitrogen is not a decompositional process, it is so closely connected with the nitrate supply of green plants through nitrification that it should be considered here. Direct nitrogen fixation is also the product of bacteria. Of these there are two groups: the *symbiotic* bacteria and the *free-living nitrogen-fixing* bacteria.

The symbiotic (Greek *symbiosis*, "a living together") bacteria live within the roots of higher plants, where they produce tumor-like growths called nodules or tubercules. The legumes, such as peas, beans, the clovers, alfalfa, and a few other plants, are especially susceptible to infection from this source. These bacteria are found in the soil and enter the host plant through the root hairs. Once inside the root, the bacteria become located in the cortex of the root, where they produce the characteristic tubercles. Here they find life most favorable. Being unable to make carbohydrates themselves, they steal their supply from the host plant. The water of the soil, also absorbed by the roots of the host plant, always contains some free nitrogen in the form of a solution. This nitrogen was dissolved out of the air as the water came in contact with it when it fell as rain or, later, in the soil. This dilute nitrogen solution ordinarily passes entirely through the green plant, and, in transpiration, the gas emerges from the leaves as free nitrogen. But here is where the symbiotic bacteria come in. Living in these nodules and stealing their sugar, they also steal water containing the dissolved nitrogen. But, unlike the green plant in whose roots they live, the bacteria have the power to combine the carbohydrates and the nitrogen into proteins to build up their own bacterial bodies. Under these conditions life for the bacteria is extremely easy. They multiply in the nodules, appearing by the millions. Moreover, they suffer the consequences of their inordinate appetites and soft living. They develop very large misshapen bodies called bacteroids. These may take the form of an X, a Y, a T, or

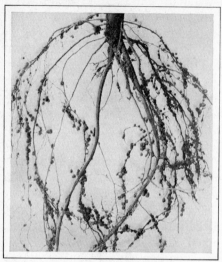

U. S. Dept. of Agriculture

FIG. 78. Bacteria in Nodules on Bean Roots collect Nitrogen from the Air

other shape. In this stage (which is more likely to occur during the early half of the growing season) the nodules are plump, juicy, and have a fresh, active appearance.

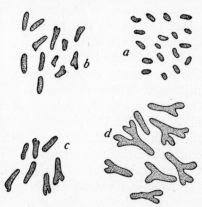

But in the case of these nitrogen-fixing bacteria, as in the case of most robbers, retribution always overtakes them sooner or later. In some way, not as yet understood, the host plant now turns on the invaders, so to speak, and appropriates the contents of their bacterial bodies for its own benefit. Nodules examined toward the close of the season are often found to be more or less hollow and of a brownish appearance, show-

FIG. 79. **Development of Root Bacteria**

a, root bacteroid; *b, c, d*, forms in nodules of *Vicia sativa* (Beijerinck) [1]

ing that the bacteria have for the most part been consumed.

In this way the legumes themselves greatly supplement the limited supply of nitrates normally obtained from the soil. Consequently these plants are very rich in nitrogenous matter. They make excellent forage for growing animals and milk cows. Alfalfa, clover, or pea hay is indispensable in first-class dairies. The seeds of the legumes, such as peas, beans, and lentils, are also the richest protein foods produced by crop plants, and it is partly for this reason that these seeds are valued so highly as human foods.

Legumes also form one of the plants used in any modern crop-rotation plan. It has been found that to maintain the fertility of any continuously cropped agricultural land, systematic rotation is necessary. The legumes are employed principally to restore the nitrogen depleted by other field plants. If the legumes are plowed under, — roots, stems, and leaves, — as is commonly done, the total annual addition of nitrogen

[1] From E. O. Jordan's *General Bacteriology*. By permission of W. B. Saunders Company, publishers.

2

to the soil may amount to from one hundred to two hundred pounds an acre. As long as this nitrogen is a component part of the legumes it is securely locked up and unavailable for other plants. It can be unlocked only through nitrification, as already explained.

In addition to the symbiotic nitrogen-fixing bacteria, there are many other kinds that live free in the soil and fix nitrogen. These organisms get their carbon and hydrogen from some source, such as decaying humus or perhaps soil algæ, of which there are large numbers, and combine the free nitrogen of the atmosphere with these to build up the protein content of their own bodies. Then, when these bacteria die, the inevitable process of decay breaks down the proteins and, through nitrification, resolves them into nitrates to be used by higher plants. The amount of nitrogen that a single bacterium would contribute through decay is almost negligible; but when millions of bacteria add their individual mites, the total becomes very important. Hall, an investigator at Rothamsted, England, which is the oldest and most famous experiment station in the world, noted an annual increase of one hundred pounds of nitrogen per acre. He believed most of this to be due to these nonsymbiotic nitrogen-fixing organisms.

QUESTIONS FOR STUDY

1. What are some of the chief kinds of decomposition and how do they affect man?

2. How does autolysis contribute to decomposition?

3. What are the agents of decomposition?

4. Describe and name the morphological forms of bacteria.

5. Explain the differences between asexual reproduction and sexual reproduction.

6. Outline the life history of bacteria.

7. By what means do fungi multiply rapidly?

8. What is the chief difference between ptomaine-poisoning and botulism and how is each caused?

9. What precautions must be taken in canning vegetables? Why?

10. What part of the mushroom fungus is used for food?

11. What are parasites? Give some important examples of each.

12. Of what value to man are the various products of fermentation?

13. What is pasteurization and why is it practiced?

14. What is the process of retting?

15. Make an outline of the carbon and nitrogen cycles and explain their value to man.

16. What is nitrogen fixation and of what economic value is it?

REFERENCES

BUCHANAN, E. D. and R. E. Bacteriology.

BURLINGAME, L. L., HEATH, H., MARTIN, E. G., PEIRCE, G. J. General Biology (second edition), pp. 265–284.

GREAVES, J. E. and E. O. Elementary Bacteriology.

HILLIARD, C. M. A Textbook of Bacteriology and Its Applications.

KENDALL, A. I. Civilization and the Microbe.

UNIT III

Living Organisms have evolved in Response to a Changing Environment

HOW long ago life first appeared on the earth can only be conjectured. Some place the date as possibly a billion years ago; others, at a time less remote. The exact time of its appearance, however, is not so important as the changes through which it is believed to have passed since it first appeared.

Such evidence as we have indicates that the first forms of life were very simple protoplasmic structures. They may have even been undifferentiated as to their plant or animal nature. Since plants are the chief food-makers, it may be that primitive protoplasm was more plantlike than animal, and that animals later became differentiated from the primeval plants. Here, again, scientists can do little better than to form hypotheses.

But one thing seems certain. When life did appear, it did not remain simple in form. Geology shows that the environment of plants and animals has constantly changed and is doing so now. These changes made it imperative that plants and animals change too in order to adapt themselves successfully to new conditions. If they could not do this they must succumb, as they seem often to have done.

The evidence of this age-long change in plants and animals, which we call organic evolution, and a consideration of the forces that lay back of it make one of the most interesting chapters of biological science.

Chapter VI · The Principle of Change seems to be a Fundamental Law of the Universe

Introduction. A few years ago the president of the American Association for the Advancement of Science, in his inaugural address, made substantially this statement: "The greatest problem confronting botanists is how the living forms of plants came to be." This same problem comes to us today with equal force. In fact, in the light of recent developments, if one expanded that statement to say that the greatest question confronting all scientists is how the present forms of material things both living and nonliving came to be, he would not be far from right.

Two theories have been proposed to explain the origin of both animate and inanimate things. To this problem of origin there have been two general answers: one is *special creation* and the other is *evolution*. In the main the special creationists hold to the belief that everything was brought into existence by divine command; that in this way the stars, the sun, the planets, and all other heavenly bodies were spoken into existence. Moreover, that in six days (of twenty-four hours each or more) plants, animals, and, as a crowning achievement, man himself were brought into being. At the end of this period the books of creation, so to speak, were closed and living things, especially, have remained much as they were originally made. Certain slight variations might be admitted, but certainly no variations wide enough to make possible the advent of a new, derived species.

This, it will at once be recognized, is a static theory. Things were made and, aside from superficial changes, have remained as they were; in other words, they have stood still.

Opposed to this static view is the belief in evolution. The evolutionary theory holds that nothing has been stationary, but that all things, both animate and inanimate, have changed. Things as we find them, be they stars, planets, butterflies, or elephants, have come to be what they are through a slow, gradual process of transition from preëxisting ancestral forms. These evolving forms are better prepared to survive and are therefore more successfully adapted to the environment than the ancestral lines from which they came. Moreover, that process of change is not completed. It may be going on today; indeed, in many cases it is going on, perhaps just as fast as it ever has in ages past. From this point of view the evolutionary concept is a dynamic one. Things are not made but are still in the process of making.

It is perfectly evident that any persistently inquiring mind, especially in regard to the question of life on the earth, may let his imagination run backward from form to form and from species to species until he comes to the first organism — the original type of life. At this point he puts the perplexing question "Where did *it* come from?" As was stated in Chapter I, the honest scientist must say "I do not know." He may have his hypothesis as to how life came into existence, but when it comes to considering the question in the light of a fact he has no established conclusion. Science at present cannot answer that problem. Any conviction one has is a matter of intellectual faith and not a conclusion arrived at on the basis of evidence. Although it does not clarify the mystery much, perhaps as good an answer as has ever been given is embodied in the statement "In the beginning, God."

But with this problem of primeval matter and life the evolutionist is not greatly concerned. He accepts the fact of their original existence in some form and busies himself with what has happened to them — what changes they have passed through since they came to be.

Both animate and inanimate things are believed to have evolved from preëxisting forms. It is perfectly obvious from the foregoing that if evolution be a fact it, then, has affected both living and nonliving things. On the one hand, it has had to do with the natural changes connected with inanimate matter; on the other, it has had to do with the changing transformations effected by nature among living things. The former is *inorganic evolution*; the latter is *organic evolution*. Since considerable space has been given to the changes and transformations occurring in matter, we shall not give much time to that phase of the problem now. We shall recapitulate only the main ideas of inorganic evolution and then pass to the question of organic evolution which is to occupy the center of the stage in this chapter.

Cosmic evolution. Both within our galaxy and without, clouds of nebulous matter are known to exist. As to what this matter is or where it came from originally, conjecture alone can be made. But, to trace the history of a star within our own universe, it is believed to be formed from a detached cloud of nebular material that in some way establishes a center of condensation and begins to rotate. At first this nebular mass is extremely rarefied and of very low density. Gradually, under the influence of gravity through a period of millions, if not billions of years, the particles settle together to produce a young red star such as Betelgeuse. Eons of time pass; and the heavenly body ages, passing through such temperature gradations as to present yellow, white, and finally a blue-white middle-aged star. These terrific temperatures arise partly from intermolecular contractions, but perhaps mostly from subatomic changes. After a period of what may be termed stellar adulthood the star begins to wane, passing through the stages of white, yellow, and red again in the inverse order. Ultimately it may become a dark derelict drifting about in the heavens. In this condition, or even before it loses its luminosity (as in the case of the companion of the dog star, Sirius), the density of the star may be so high that a piece the size of a baseball would weigh several tons.

Not only may stars pass through transitional changes, but even groups of stars, or island universes, are believed to do so too. Giant nebular masses, including enough matter to form a galaxy, are thought to establish a center of rotation. Scattered throughout the huge cloud are individual centers of condensation which form the stellar nuclei referred to in the preceding paragraph. Through an, at present, indeterminable but enormous period of time, the whole giant nebula is believed to evolve into an island universe that goes whirling along through space. Whither it is going, or what its purpose is, remains a profound mystery. It is probable that our galaxy itself forms one of these colossal aggregations of stars. At present perhaps a million or more of these island universes have been sighted with the aid of the big telescopes. It is probable that this number would hardly constitute a beginning on the total number existing in space.

We have also found that the solar system itself is believed to have been formed by an evolutionary process. According to the planetesimal hypothesis the original material of our solar system constituted a stellar body of some kind, perhaps an old, defunct star. Another heavenly visitor in its progress through space passed close enough to the solar mass to form huge tides set up by the gravitational pull. Finally the stress became so great that immense arms hundreds of millions of miles long were extruded with explosive force from opposite sides of the central body which subsequently was to form our system. The visitor in its passage also imparted a pull to our sun-to-be in such a manner as to give it rotational motion, which wound up the arms in spiral fashion. Within these arms were denser nuclei around which the planetary bodies finally condensed.

Geological evolution. One of these centers of condensation found in the spiral arms formed the beginning of our earth, which at first was relatively small. Planetesimals pulled in by gravity caused it to grow to its present proportions. Whether the earth body itself was built originally of solid matter or whether it passed through the successive gaseous, molten, and solid states is a controverted question. Nevertheless, regardless of which view is correct, the earth ultimately approached its present form and condition. During the transition volcanic action has extruded from the interior enormous masses of lava and volcanic ash; terrific earthquakes have torn and rent its crust; great mountain ranges have been thrust up and compensatory sea beds have been formed by isostatic readjustments. The forces of weathering, erosion, transport, and deposit have also added their assistance to carve and sculpture the face of the earth and bring it to its present condition.

Atomic evolution. Before Henri Becquerel discovered the radioactive disintegration of uranium salts in 1896, atoms were thought to be unyielding, indivisible objects. Ancient alchemists had deceptively claimed the transmutation of elements and were on the lookout for the "philosopher's stone" to turn dross substances into gold. Scholarly men, however, gave small credence to these extravagant claims. They preferred to believe with Newton that atoms "were so very hard as never to wear or break in pieces." Becquerel's experience put a new face on things. Reputable physicists soon were able to show that uranium was actually decomposing of its own accord, and that in so doing it was emitting alpha, beta, and gamma rays. The beta rays were soon found to be electrons; the alpha rays, to be stripped helium atoms. Moreover, the other products of uranium disintegration formed a whole series of less complex substances emerging finally into lead as the lowest one. Whether lead can be further reduced radioactively into elements of lower station is not positively known. Theoretically, it seems quite probable that it can, although this goal has not yet been reached experimentally.

Later Rutherford and Chadwick by bombarding nitrogen gas with alpha particles succeeded in tearing its molecules to pieces and producing a simpler element, carbon, with hydrogen as a by-product. In this way aluminum was also made to yield magnesium and hydrogen; phosphorus produced silicon and hydrogen; and so on for a whole series. Here was certain evidence that more complex elements might be transformed into simpler ones. But could the transitions go in the other direction? Could two simpler elements be combined in such a way as to produce a more complex one? Frankly this result has been achieved only on a very small scale in the laboratory; but that it is possible and may actually occur in nature seems wholly reasonable. The cosmic rays, as Millikan has suggested, may be emitted by the "packing effect" when hydrogen atoms are combined to form helium. More recently two other discoveries important in this connection have been reported. One is to the effect that a particle called the neutron has been found. This particle, as its name indicates, is electrically neutral, and it has been suggested that it may prove to be the fundamental element upon which all other atoms are produced. The other discovery is that of an isotope of hydrogen which has an atomic weight of 2 instead of 1. This newly discovered hydrogen may be found to be, as has been proposed, an intermediate form indicating the steps by which helium may be built up from ordinary hydrogen.

From this short summary it may be seen that the concrete evidence supporting the conception of inorganic evolution is but sketchy and incomplete; still it is increasing from year to year, and the idea itself may be said to be tentatively established. However, synthetically speaking, the forces involved and the course which the process follows must be said, as yet, to be almost wholly hypothetical. The unquestioned proof of the theory must await the results of much careful experimental investigation.

The idea of organic evolution was first advanced by the early Greeks. To turn now to organic evolution, this idea also is believed by many to be a comparatively new one. Most people doubtless think of Charles Darwin as being the originator and promulgator of the theory. Such, however, is not the case. Though the popular notion of evolution is probably of more recent origin, the idea of progressive change among living things is an old one. Indeed, throughout all ages of history there have been in most eras a few bold, original minds to whom the doctrine of static creation seemed wholly inadequate. One can, in truth, trace the idea of origin through change far back to the early Greeks.

Empedocles (495–(?) B.C.) may be said to have been the father of the evolutionary idea. He believed all elements of nature to be played upon by two ultimate forces: one a combining force of love, the other a separating force of hate. All organisms first appeared not as complete individuals but as parts of disorganized individuals — arms without shoulders, eyes without sockets, and heads without necks. As love triumphed over hate these parts united fortuitously. All kinds of hideous combinations occurred — beings with men's heads and animals' bodies; double-headed monsters; even creatures with two chests and one head. But of these unnatural products, only those adapted to live successfully survived; the rest perished.

Aristotle (384–322 B.C.) assumed the existence of an intelligent design in nature which acted as a "perfecting principle." The action of this perfecting principle resulted in a complete gradation in nature. He considered that inanimate material formed the lowest types of objects. From this nonliving matter plants sprang. Plants in turn gave rise to simple animals, such as the sponges and others found in marine waters. From these lower animals rose the higher animals with feelings, sensibilities, and powers of locomotion. Finally came man as the highest point in a long and continuous series of transformations. Aristotle considered the whole process to be presided over by the "Efficient Cause, the Prime Mover, or God."

Gregory of Nyssa, who lived in the fourth century A.D., a bishop and early theologian, attempted to harmonize the teachings of the Greek philosophers and the scriptural account of creation. He taught that creation was potential. He believed that God imparted to matter its fundamental properties and laws. At first this matter was chaotic and without form. Subsequently, from this disorganized material the objects and completed forms of the universe gradually developed under Divine direction.

Augustine (353–430 A.D.) was a church father and bold defender of that institution. He wrote *The City of God*. This book was a courageous defense of the church against the claims of the dissolute Roman rulers, who attributed the decline of the Empire to the effeminizing influences of Christianity. To write such a book at that time greatly endangered Augustine's life.

Augustine rejected the idea of special creation in favor of a developmental concept. In attempting to explain the scriptural passage "In the beginning God created the heaven and the earth" he says:

> In the beginning God made the heaven and the earth, as if this were the *seed* of the heaven and earth, although as yet all the matter of heaven and of earth was in confusion; but because it was certain that from this the heaven and the earth would be, therefore the material itself is called by that name.

Here is a distinct statement of the idea of change as involved in the formation of completed objects and creatures. Moreover, Augustine says that we ought not to think of the six days of creation as being equivalent to our solar days, but as perhaps much longer periods of time.

All through the Dark Ages there seems to have been little controversy about the origin of things; but with the advent of the thirteenth century we find another important churchman, Thomas Aquinas (1225–1274 A.D.), definitely expressing his views. He embraced Augustine's ideas on this matter and confidently expounded them. He even went so far as to say that the belief which holds that "on the third day plants were actually produced, each in his own kind" is a view favored by the superficial reading of Scripture only.

From Aquinas on, many writers have expressed their ideas regarding the genesis of things and how they came to be. Most of these opinions, however, we shall disregard, since they add nothing of a material nature to our historical sketch. Buffon, a French naturalist, believed in the transmutation of species, as did Erasmus Darwin, the grandfather of Charles Darwin. But we encounter the first of what we may term "unqualified evolutionists" in Jean Baptiste Lamarck (1744–1829). This man had such a checkered career, and his name is linked so indelibly with evolutionary history, that it will be of interest to examine his record briefly.

Lamarck came from a military family, but being the eleventh child his father decided to diversify the family vocations somewhat by placing Jean, against his will, in a Jesuit college at Amiens to study for the clergy. To this move Lamarck never became reconciled. Consequently after his father's death the son immediately forsook his clerical training. He succeeded in buying a broken-down horse and hastened to join the French army, then campaigning against the Germans. In the first battle he distinguished himself for bravery. His company being subject to the direct fire of the enemy, the officers were killed one by one until Lamarck, who was a sergeant by order of succession, came to the command. Although the French army retreated, Lamarck refused to move with his squad until directed from headquarters to do so.

Fig. 80. Lamarck, the Proponent of Use and Disuse[1]

Later an injury to his neck, the result of a practical joke by one of his comrades, unfitted him for further military life. He returned to Paris and took up the study of medicine. While in training he acquired a consuming interest in the natural sciences and later devoted his whole time to the study of botany.

After about nine years' work he published his *Flora of France* in 1778. It won high recognition. He took an active part in the reorganization of a famous natural history institution which later came to be known as the *Jardin des Plantes* (Botanical Garden), and joined Saint-Hilaire in inviting Cuvier to join the staff. Cuvier was "later to be advanced above him in the Jardin and in public favor, and . . . was to break friendship with Lamarck and become the opponent of his views."[2]

At fifty years of age Lamarck forsook the study of botany for that of invertebrate zoölogy. It was his work in this field that brought him into conflict with Cuvier, who had developed into a dynamic personality, a brilliant anatomist, and an inspiring teacher. As a result of his study and dissection of animals, Lamarck completely abandoned his former ideas

[1] From Thornton's *British Plants*.
[2] W. A. Locy, *Biology and Its Makers*. By permission, Henry Holt and Company, publishers.

as to the fixity of species. He came to believe positively in a transmutation of species, or evolution, and, after expressing this conviction in 1800, held to it unwaveringly to the end of his life.

FIG. 81. Cuvier

Cuvier was a believer in the fixity of species and a dynamic opponent of evolution

Lamarck explained evolution on the basis of *use* and *disuse* supplemented by the *principle of heredity*, by which he believed acquired characters could be passed on from parent to offspring. Thus the long legs of wading birds had been acquired, as he explained it, by a constant attempt of their ancestors to wade out into deep water. As the length of leg accumulated through use, this character became fixed in the species and was passed on to succeeding generations by heredity. The final result was long-legged birds, such as the crane and the heron. In the same way Lamarck explained the development of webs between the toes of swimming birds, the length of the giraffe's neck, and many other organic structures.

It was Lamarck's evolutionary views that clashed with the belief in the fixity of species as held by the dominant Cuvier. But for this unfortunate contingency perhaps Lamarck, instead of Darwin, might have been recognized as the founder of the theory of evolution. As it was, the slower, more methodical Lamarck was completely eclipsed by the versatile, intrepid, and almost scintillating personality of Cuvier. The latter was not only a great anatomist with a large following of students but also an admired friend of Napoleon. For this reason every "new folly" of Lamarck's, as Cuvier styled it, was laughed out of court and practically silenced.

The theory of evolution must await a new champion — a man of almost world-wide observation, of prodigious labors, of profound conviction, and, withal, a man with a sterner jaw and a more determined purpose.

Charles Darwin was born on February 12, 1809. He at first gave little promise of those characteristics that were to distinguish his after life. When he was sixteen he left a certain Dr. Butler's school. In a short autobiography written for his own children as he approached the close of his life, he says of himself at that time:

When I left the school I was for my age neither high nor low in it; and I believe that I was considered by all my masters and by my father as a very ordinary boy, rather below the common standard in intellect. To my deep mortification my father once said to me, "You care for nothing but shooting, dogs, and rat-catching, and you will be a disgrace to yourself and all your family." But my father, who was the kindest man I ever knew and whose memory I love with all my heart, must have been angry and somewhat unjust when he used such words.[1]

Even at this period, however, Darwin had begun to manifest some of those interests that were to mark him as a scientist. He was an ardent collector of minerals. He tells us that he would also have collected insects, but after consulting his sister he decided that it was not right to kill insects just for the sake of making a collection. He loved to fish, and after being told that he could kill his fishworms with salt and water he relates that "from that day I never spitted a living worm, though at the expense probably of some loss of success."[2]

After leaving Butler's school he was sent to Edinburgh University to begin his medical studies. He found the lectures at Edinburgh "intolerably dull," except those in chemistry, and writes that "Dr. Duncan's lectures on Materia Medica at 8 o'clock on a winter's morning are something fearful to remember."[3] He seems to have drifted along for two years with an interest chiefly in the personalities that he met and in the meetings of different societies which he attended. He disliked the operating clinic thoroughly. Those days were before the discovery of anæsthetics, and he shrank

[1] Francis Darwin, *Life and Letters of Charles Darwin*, Vol. I, p. 30. By permission, D. Appleton and Company, publishers.

[2] Ibid. p. 28.   [3] Ibid. p. 29.

2

from the human agony and suffering that he witnessed in the clinic. Finally, on two occasions he attended the operating theater in the hospital "and saw two very bad operations, one on a child, but I rushed away before they were completed. Nor did I ever attend again, for hardly any inducement would have been strong enough to make me do so; this being long before the blessed days of chloroform. The two cases fairly haunted me for many a long year." [1] Nevertheless, this aversion to suffering did not prevent him from becoming a dead-shot and a zealous hunter of birds.

After two years Darwin's father, perceiving that he did not wish to be a physician, proposed that he become a clergyman. After reading several books to decide whether he could subscribe to the creed of the Church of England or not, Darwin finally concluded that he could and was sent to Cambridge to take up his studies. Here, however, he seems to have exhibited little more interest in the salvation of men's souls than he did in the healing of their physical bodies; for he says, "During the three years which I spent at Cambridge my time was wasted, as far as the academical studies were concerned, as completely as at Edinburgh and at school." [2]

At Cambridge, though, Darwin met two personalities whose influence was to change the whole current of his life. These men were Henslow and Sedgwick. Henslow was a botanist; Sedgwick was a geologist. These two men were ardent students of nature at first-hand and used to take Darwin with them on their field excursions. In fact, he accompanied Henslow so faithfully whenever opportunity offered that he came to be known by some of the college men as "the man who walks with Henslow." But despite his dislike for his studies in general, no one should make the mistake of picturing Darwin as an inactive, mentally indisposed, sluggish-minded fellow. He was far from it. While here he developed, largely through association, a fine appreciation for good pictures and for first-class music. He tells us, "I acquired a strong taste for music,

---

[1] Francis Darwin, *Life and Letters of Charles Darwin*, Vol. I, p. 33. By permission, D. Appleton and Company, publishers.
[2] Ibid. p. 40.

and used very often to time my walks so as to hear on week days the anthem in King's College Chapel." [1] He seems also to have recovered from some of his childhood aversions; for "no pursuit at Cambridge was followed with nearly so much eagerness or gave me so much pleasure as collecting beetles." [2]

It was at this time that England had formed the policy which contributed much toward making her the first maritime power in the world. It was the practice to send out exploratory vessels in order to map and chart the seas and to gather such information as would be of value in building up over-sea commerce. It was the custom to send along with each of these vessels a promising young naturalist to report on the animal and plant life and other natural products of the regions visited. In 1831 it was proposed to send the ship *Beagle*, under command of Captain Fitz-Roy, to explore the southern coasts and lands of South America. Henslow suggested Darwin's name to the government as a naturalist to accompany the expedition, and he received the appointment.

Darwin's father, however, objected strongly to his going, presumably because he had not distinguished himself as a student. But he said to Charles, "If you can find any man of common sense who advises you to go I will give my consent." Luckily for Charles his uncle, Josiah Wedgwood of pottery fame, thought the idea of Darwin's going a good one and indorsed the plan. Since the elder Darwin had always "maintained that he [Wedgwood] was one of the most sensible men in the world," he at once consented in a kindly manner to Charles's taking the appointment.

This voyage, Darwin said, was by far the most important event of his life and determined the whole course of his career. When he departed from England he carried with him a volume of Lyell's *Principles of Geology*. Henslow, a confirmed special creationist, had recommended it to him as a book to read, but admonished him "on no account to accept the views therein advocated." Lyell, an original investigator of the

---

[1] Francis Darwin, *Life and Letters of Charles Darwin*, Vol. I, p. 42. By permission, D. Appleton and Company, publishers.

[3] Ibid. p. 43.

formations of the earth, took the view that the present is the child of the past; that is, that the strata as they are found at any time have evolved from past conditions due to the constant interplay of geological forces. As Darwin read this book the suggestion was a seed sown in fertile soil. The *Beagle* was out five years, and during that time Darwin changed from a rather indifferent youth to a deeply interested, careful, painstaking student of the natural sciences. He collected plants, birds and other animals, marine specimens, and samples of minerals. He made original discoveries and kept a careful journal of all his findings and observations. He noted the living armadillos and cumbersome sloths of Brazil. When he studied the fossil bones of the huge extinct Glyptodon and Megatherium unearthed in the Argentine he was struck by the resemblance of these two skeletons, respectively, to the two strange animals he had seen in Brazil. We are told that one day while he was sitting out on the pampas reflecting, the thought occurred to him that perhaps these living and fossil animals were related by descent and change. The principle of origin by change, that Lyell had proposed for the rocks, might also be true for living things.

When the *Beagle* returned to England in 1836 Darwin wrote a journal of his travels that is still a classic in travel literature. He prepared a monograph of the great variety of barnacles he had collected and upon which he worked for eight years. He wrote many other scientific papers that attracted favorable attention and won for him wide recognition as a careful and painstaking investigator.

The problem preëminent in his mind, however, after his return was the *species question*, or how the then existing forms of plants and animals came to be. He got no light upon this perplexing question for almost a year and a half. He himself had come to believe thoroughly in the principle of origin by descent, but *how* did this descent come about? He did not care to present his convictions on this matter to the world and get laughed out of court as Lamarck had done before him. He wanted to determine both the forces back of these changes and the evidences that proved both their existence and the evolutionary effects that they produced. Finally he chanced to read the essay *On Population* by Thomas Malthus, an English clergyman and political economist. In this book Malthus portrayed the geometrical increase of population, but held that the food supply can be enlarged only arithmetically. As a result he foresaw in the future an acute competition between men for the bare necessities of life — a competition that would tend to limit population and hold it down commensurate with the supply of food available. Here was a fundamental idea that was eagerly grasped by Darwin's alert mind. It "served as a spark falling on a long prepared train of thought." Perhaps this very principle, reflected Darwin, was acting among all living things in nature and constantly serving as a sorting process to preserve and perpetuate the best-fitted individuals. It seems almost the irony of fate, too, that Darwin should have received his key idea from a profession that later was to furnish his most bitter and irreconcilable critics — those that were both to impeach his motives and to denounce his conclusions.

But for Darwin's thorough, painstaking nature the central idea was only the starting point. He worked four years longer before he set forth in writing even his first expressions on the matter. He worked twenty-one years before he presented his theory in its final form as embodied in his *Origin of Species*, published in 1859. Even this work he says was incomplete. These twenty-one years were years of surprisingly thorough and prodigious labors. He studied, traveled, wrote, and conversed with eminent scientists everywhere. History offers no better example of the tremendous labor and painstaking effort that it takes to produce a really great piece of original scientific work. Subsequent investigation has proved that at minor points Darwin was wrong in his conclusions. But the central idea of evolution has never been shaken. It stands as impregnable today as it did then. Moreover, it is probable that in Christendom no other book ever written, save the Bible alone, has effected the change in men's philosophies, beliefs, and creeds as has the *Origin of Species*.

Darwin conceived certain factors to underlie evolution. The *Origin of Species* is a book of more than five hundred pages; hence in this section we must condense heavily and, at the same time, disregard a great deal of related material. But it will perhaps not be doing violence to Darwin's conception to say that in the main he conceived five factors to lie back of evolution. These, he believed, working together, produced the permanent changes which he felt certain had taken place among plants and animals. He made no attempt to explain scientifically how life first appeared on this planet. But, once it was here, he felt convinced that changes had occurred. He regarded these transformations as the structural and functional responses which living things had made to geological and climatic changes as well as to the interactions of plants and animals themselves. More than this, he believed that these slow changes were going on in his day and would continue to do so in the future.

FIG. 82. Charles Darwin

Darwin was the promulgator of evolution by natural selection. He was a careful observer and a deep thinker

Overproduction. Plants and animals everywhere reproduce with great prodigality. Linnæus, the great Swedish botanist, estimated that if a single plant produced just two seeds a year, and its offspring continued at the same rate, in twenty years the number of individuals would amount to one million. Certainly no ordinary plant is such a slow breeder. Most plants produce seeds by the thousands. A single elm tree has been observed to develop a bushel or more of so-called seeds. The total elm trees possible from this yield would mount into the hundreds of thousands. Schmucker tells us that one full-sized mullein plant has been estimated to produce six hundred and thirty thousand seeds in one season.

Animals also reproduce on a grand scale. A few birds, such as the Fulmar petrel, lay but one egg and the condor produces two. But most birds lay from four to a much larger number. Even the ostrich may deposit a score or more of eggs in a single hatch. One female codfish is believed to spawn as many as two million eggs at a time; while the prolific oyster, on the basis of observation, has been reckoned to lay from nine million to eleven million eggs during a single season.

Mammals reproduce more slowly than fish and many birds, but even here the rate is fast enough for the progeny to increase surprisingly fast. Under favorable conditions, slowly reproducing man has been known to double his population in twenty-five years. Were this rate to continue unabated, in less than a thousand years there would literally not be standing room on the earth for his progeny. The elephant is believed to be the slowest breeder of all known animals. Breeding probably begins when the elephant has reached about fifteen years of age. Assuming that reproduction will continue at the rate of about one calf every two and one-half years until the parents have reached ninety years, the total offspring from a single pair of elephants would be thirty young. If this rate were to continue for approximately seven hundred and fifty years, something like 1,358,000,000 elephants would have descended from the original two progenitors.

Darwin believed prodigality of reproduction to have been of great importance because it leads to a crowded condition and hence to a sharp competition for the necessities of life. Out of this condition the struggle will grow naturally, as we shall see later.

Variation. No two living things are alike. No one ever saw even two blades of grass or two leaves on the same tree that were exactly alike. They may appear very similar; but when accurate weights and fine measurements are made, slight differences become evident. Even identical twins may resemble each other so strongly that the greatest of care is required on the part of the parents to distinguish them; yet if one weighs them, measures them, and scrutinizes their features carefully, disparities will be found.

These variations are of two kinds. One is due to differences in the environment, such as light, air, warmth, rainfall, food, or shelter. Such variations are called modifications. They are neither permanent nor inheritable. Contrasted with these there are other variations that go deeper and are due to natural, inborn differences which are found to exist in the egg cells and sperms. These are continuing and result in a fundamental difference between individuals. It is these innate variations that are significant in evolution.

The struggle for existence. The prodigality of nature literally fills the earth with potential offspring. The forest floor is covered with tree seeds, and the roadsides are thickly strewn with the reproductive structures of weeds, grass, and other plants. Even cultivated fields must be tilled with the greatest persistence to keep the crops from being smothered out by invading weeds. Insects lay eggs sometimes by the hundreds, and mammals produce from one to as high as a dozen or more offspring. The result of this constant overproduction is a continuous competition for room, light, food, shelter, and bodily protection. This competition, which becomes extremely intense, often assuming the character of a life-and-death struggle, may arise from three sources : (1) It may arise between individuals of the same species. This competition between animals and plants of the same kind is often the most crucial and deadly. In such cases all the organisms are competing not for slightly different elements of the environment but for exactly the same elements. What one gets, the other cannot ; therefore the chances of life for the second are reduced by just that amount. A square meter once laid off along the roadside in the spring was found to contain something like eight hundred young ragweed plants. Certainly not one fourth of these plants could ever grow to maturity under such terrific competition. (2) The struggle may arise between individuals of different species. Wolves prey upon the rancher's sheep, and the owner goes gunning for the wolves. Deer multiply in the mountain wilds, and each mountain lion infesting that region is estimated to capture on the average one deer a week. Insects lay a hundred eggs on a leaf, and a busy chickadee devours them all. An alfalfa field is purple with bloom, and the mowing machine lays it low. Throughout the world, both in water and on land, this competition prevails. (3) The struggle may also be between the individual and the environment. A Denver paper relates that a Colorado rancher saved five hundred bluebirds. The birds were on their northward migration and were overtaken by the worst March snowstorm known to that region in fifty years. This friend of the birds found them chilled and numbed by the cold. He gathered them into his barn and fed them until they revived and the storm abated. Hundreds of live stock are placed in jeopardy every time a blizzard sweeps the Western plains. Animals and plants, as well as people, are endangered by hailstorms and tornadoes. Every severe drought imposes a life-and-death test upon hundreds of organisms. So the story runs until it includes practically every living thing, be it tree, bird, or man.

The survival of the fittest. Although we have found nature prodigal, under normal conditions the number of any one species in a given region remains about the same. There may be a temporary fluctuation one way or the other, but in a number of years, except where destructive man interferes, the average varies but little. This arises from the fact that thousands of individuals of all kinds either die off or are overtaken by their enemies and are devoured for food.

But under normal competitive conditions this survival is not a fortuitous one. It is selective. Not just so many plants, or birds, or men succumb indiscriminately out of a thousand, let us say. On the contrary, variation has produced differences among individuals, and it is the least fitted that die and the best fitted that survive. The best fitted are not always the biggest or the strongest, or the most robust either. For instance, in a raging blizzard the buffalo with the longest hair and the toughest skin, irrespective of its size, is much more likely to outlast the giant of the herd who does not possess these qualities. Moreover, a pugilist may be much larger and stronger than a lawyer, but in a mortal crisis demanding wits as well as brawn the former is much more apt to find himself at a great disadvantage.

This sorting competition, to be sure, is much more acute among plants and lower animals than it is among men; but that it prevails among humankind too there can be no doubt. During the winter of 1931–1932 hundreds of thousands of Chinese peasants died of hunger and cold in the famine districts of northern China. In 1918 millions of Americans and Europeans were carried off by the epidemic of influenza that swept the army camps and homes of the general populace. Besides, the law of struggle and survival may work on a much higher mental and social level than it does among the creatures of the woods and fields. For example, when a desirable position opens, an applicant is likely to find that he has many keen competitors. The one who can present the best credentials will probably get the appointment. Of the runners in the mile race the athlete who has trained properly stands a much better chance of cutting the tape first. In the personnel of a bank the youth who, other things being equal, is a careful, persistent, and alert student of economic forces and finances and who keeps his body efficient and his head clear by proper living, is much more likely to find himself promoted to the president's chair twenty-five years hence than his fellow worker who wastes his time and dissipates his mental and physical powers through imprudent living.

The paragraph above is no attempt either to moralize or to preach. It is but the expression in concrete form of a universal law — the law of struggle, competition, and survival. Its application is softened somewhat by man's altruism and spiritual sense. But that the law works, nevertheless, to sort out and to elevate certain men, cannot be questioned.

Heredity. The fittest who survive in the struggle of life not only determine the character of the present generation, as Darwin believed, but also influence the nature of the generations to come. They are the ones who live to mature years and produce a full quota of offspring. But the progeny will also be fit because they will have inherited the same characteristics of fitness that marked their parents. Thus the fittest of each generation tend to beget their own kind, which comprise the next generation. In this way the successive generations of the fittest are perpetuated, and the unfit lose out in the struggle and disappear.

Evolution is conditioned by a changing environment. It is evident that a progressive change of any kind in living organisms could not continue indefinitely in a static environment. After a species became successfully adapted under such conditions, there would be no need for further change. Indeed, any wide variations that did occur would work to make the individual unfit rather than fitter to survive. Under these circumstances the new type would tend constantly to be eliminated. The total effect, then, of such elimination would be to hold the average of the species back to the dead level of its existent condition; in other words, it would remain stationary.

But all students of geology declare without the least hesitation that the earth has changed profoundly, especially at the surface, where life resides. This progressive change through which the earth has passed would not only make a parallel change in life possible, declare the evolutionists, but would actually make it imperative. Even if species at the outset had been created with a definite form and structure, they would have had to evolve into forms successfully adapted to the new conditions or they would have perished and become extinct long ago. To illustrate: Huge dinosaurs used to be the dominant form of life when the earth was warm and humid. As climatic changes occurred which resulted in a cooler, drier earth the vegetation became reduced in amount. Consequently these big hulks of animals, as some believe, could not get enough food to sustain their bodies and gave way to animals with more sagacity and lower food requirements.

Darwin and Wallace arrive at the same ideas regarding evolution. The name of Alfred Russel Wallace will always be associated with that of Darwin in connection with the theory of evolution. Wallace, some years younger than Darwin, went on an expedition to collect specimens. He too had read Lyell's *Principles of Geology* and Malthus's *On Population*. While at Sarawak, on the island of Borneo, he tells us, "I was quite alone, with one Malay boy as cook, and during the evenings and wet days I had nothing to do but to look over my books and ponder over the problem which was rarely absent from my thoughts." While pondering this matter the idea of evolution came to him as if by inspiration. As the complete idea took form in his mind he wrote it out and sent it to Darwin in June, 1858. Darwin was astonished to find that a man in that remote part of the earth had worked out a theory almost exactly like his own. He knew, too, that Wallace's work was original. Darwin had written out a short sketch of his theory in 1842 and a longer one in 1844, but had published neither of them. Lyell, who had become Darwin's close friend, had urged him to publish his papers, but the latter declined to do it because he felt his manuscript was still very incomplete. Of Wallace's paper, Darwin says in a letter to Lyell: "I never saw a more striking coincidence; if Wallace had my MS. sketch written out in 1842, he could not have made a better short abstract! Even his terms now stand as heads of my chapters." Darwin was then writing the *Origin of Species*.

Darwin was greatly troubled. He felt that although he clearly had priority claims he could not now publish without seeming great injustice to Wallace. His first generous impulse was to suppress his own theory, which had been the result of so many years of study, and to publish Wallace's paper. In his distress he wrote to Lyell explaining the situation. He said, "I would far rather burn my whole book, than that he or any other man should think that I have behaved in a paltry spirit." Lyell and Hooker, the latter another prominent scientific contemporary, refused to permit Darwin to do this, and at their suggestion the two papers were combined and published under the joint authorship of Darwin and Wallace on July 1, 1858.

Owing to this honesty and generosity a warm friendship subsequently sprang up between Wallace and Darwin. Moreover, Wallace always recognized Darwin as the better worker. In 1870 he wrote, "I have felt all my life and still feel the most sincere satisfaction that Mr. Darwin had been at work long before me, and that it was not left for me to attempt to write the *Origin of Species*." Later still he wrote, "I was then (and often since) the 'young man in a hurry,' he [Darwin] the painstaking student, seeking ever the full demonstration of the truth he had discovered, rather than to achieve immediate personal fame."

QUESTIONS FOR STUDY

1. Contrast the main features of the theory of special creation with those of the theory of evolution.

2. Make a brief outline of inorganic evolution and show in what way it agrees with the principle of organic evolution.

3. Show how evolution applies to geologic changes which occur upon the surface of the earth.

4. Give a brief historical account of the theory of evolution.

5. Explain Lamarck's conception of the principle of use and disuse in evolution.

6. What events in the life of Charles Darwin led him to study evolution?

7. Give a brief account of Darwin's theory of evolution as given in his *Origin of Species*.

8. What is meant by "environment"? Show how it could be effective in bringing about changes.

9. Who arrived at a theory of evolution similar to that of Darwin's, and what recognition was given to him by Darwin?

10. Explain why Lamarck's theory fell into disfavor.

11. What evidence is there that a changing environment may assist in making permanent changes in a species?

REFERENCES

BRADFORD, GAMALIEL. Darwin.
DARWIN, CHARLES. Origin of Species.
DARWIN, FRANCIS. Life and Letters of Charles Darwin, Vol. I.
FASTEN, N. Origin through Evolution, pp. 31–65.
HERBERT, S. The First Principles of Evolution, pp. 7–50.
NEWMAN, H. H. Evolution, Genetics, and Eugenics, pp. 10–27.
OSBORN, H. F. From the Greeks to Darwin, pp. 29–245.

Chapter VII · The Theory that Living Organisms have become successfully adapted through Change is supported by Many Evidences

Introduction. The first edition of the *Origin of Species* was issued on November 24, 1859. Although twelve hundred and fifty copies were printed, the edition was exhausted before night. This unexpected sale, which was a great surprise even to Darwin himself, indicates something of the degree of interest which had been aroused in this new theory. However, this interest must not be taken as an altogether friendly one. Many people bought the book quite as much from a hostile motive as they did from a sympathetic one. They wished to become acquainted with this dangerous doctrine in order that they might crush it root and branch. A prominent bishop in one of the leading churches, discussing the *Origin of Species* in a printed article, referred to Mr. Darwin as a "flighty" person, one who endeavors "to prop up his utterly rotten fabric of guess and speculation," and further denounced his "mode of dealing with nature" as "utterly dishonourable to Natural Science." Many scientists too were entirely unfriendly to the theory if, indeed, not in active opposition. Even Henslow, the great botanist whom Darwin said had been the kindest friend to him that ever man possessed and of whom Darwin's father said, "I fully believe a better man never walked this earth," found it difficult to accept the transmutation idea. He weakened somewhat near the close of his life and accepted evolution at minor points, but never could bring himself to embrace the theory as a whole. But, as time went on, Darwin's theory won followers. Such men as Hooker and Huxley in England, Haeckel in Germany, and Gray in America, as a result of their study and correspondence with Darwin, came to believe in the doctrine thoroughly. Even

212

clergymen came to consider with more interest and regard the great mass of carefully selected and intelligently arranged facts which Darwin had painstakingly gathered. The distinguished cleric Charles Kingsley wrote Darwin soon after the *Origin of Species* was published:

I have gradually learnt to see that it is just as noble a conception of Deity, to believe that he created primal forms capable of self-development into all forms needful . . . as to believe that He required a fresh act of intervention to supply the *lacunas* [gaps] which He Himself had made. I question whether the former be not the loftier thought.[1]

The years following brought increasing converts from all intellectual classes. Darwin's explanation of how evolution occurs has been seriously questioned at some points, but the central idea itself has found increasing favor and acceptance. Perhaps it would not be putting it too strong to say that by far the greater majority of persons today who have taken the pains to study the question seriously and in an unbiased manner have become adherents of the evolutionary principle.

What then, one may reasonably ask, are the compelling evidences that have led to this change in sentiment? What are the concrete proofs that living organisms have become adapted through an orderly process of gradual change? Let us examine them.

Morphology is an evidence of evolution. Morphology is the science of the form and structure of living things. If vertebrate animals, for example, be examined, it is surprising what a similarity of structure runs through the whole group from fish to man. *Homology* is the word used by scientists to express this similarity in structure. Morphology also offers evidence of another sort. Animals are found to have imperfectly formed organs at many points in their bodies. These structures are called vestigial organs, and they may function incompletely or not at all. If each of these two morphological evidences is discussed separately, its bearing on the truth or falsity of the evolutionary doctrine will become more evident.

[1] Francis Darwin, *Life and Letters of Charles Darwin*, Vol. II, p. 82.

Homology. Homologous organs, as used in this discussion, are organs similar in structure and origin and are similarly

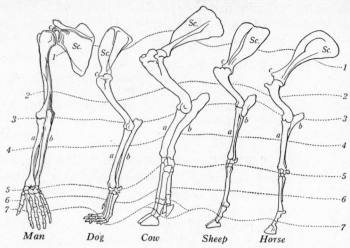

FIG. 83. Anterior Limbs of Vertebrates Compared[1]

placed in the body. Let us take, for instance, the foreleg of a dog. The most fundamental part of this organ is its bony structure. We find that the upper portion of the leg skeleton is composed of a single bone called the humerus, which is attached to the shoulder blade, or scapula. Below the humerus come the radius and the ulna, followed by that portion comparable to the wrist of man and with its group of similar carpal bones. Still further down are the metacarpals, and then the phalanges, which make up the bony divisions of the toes.

If comparison of this skeletal portion of the dog is made with the corresponding homologous organ of the hog, sheep, horse, and man, a striking similarity throughout the series is found. The main bones are comparable section for section. The only dissimilarity worth mentioning comes in the parts that correspond to the wrist, hand, and fingers in man. As we shall discover later, what is called the knee in the front leg of the horse is in reality its wrist. The whole hand has been

[1] From G. J. Romanes's *Darwin and After Darwin.* By permission, The Open Court Publishing Company.

reduced to one long, heavy "toe" terminated by a massive nail called the hoof. This modification has been necessitated by the use of the fore-leg as a walking organ which must sustain the horse's weight upon the hard earth. The "hands" of the sheep and the hog have been modified in a similar manner, but here the change has not gone so far, since two toes remain. In these animals the remnants of two other toes are still found. Man has taken the erect position and uses his hand both as a prehensile appendage and as a complex manipulatory organ. Had the series

FIG. 84. Wing of Bird and Bat

Note the similarity in structure [1]

been enlarged to include the fin of a fish and the wing of a bird, the homology would have still held. Though the whale is not a fish, by any means, yet it has adapted itself to life in the water, and its front appendage has become transformed into a finlike paddle (Fig. 86). The bird's wing has been modified into an organ of flight; yet the hand section, with several bones, still remains as a wing tip. In most birds even the remains of a separate finger with its rudimentary nail may be found. This nail is especially prominent in capons.

If, instead of considering the anterior appendage of vertebrates, other homologous organs be compared, the same likeness prevails. The leg of man is strikingly similar to the hind legs of other mammals. The homology holds likewise for the spinal column, the cranium, or the thoracic skeleton. If one takes systems such as the circulatory system, the digestive

[1] From G. J. Romanes's *Darwin and After Darwin*. By permission, The Open Court Publishing Company.

2

system, or the excretory system, which are composed of softer tissues, the homology is also strikingly evident. To a lesser degree the same principle is found in plants.

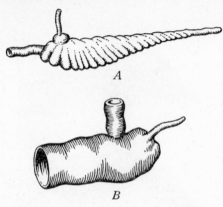

Fig. 85. Appendix of Rabbit and Man[1]

A, rabbit; B, man

This phase of morphology cannot be taken as final proof of evolution. Had organisms been the product of special creation homology might still have existed. It might have risen from the fact that a general plan of structure was held to throughout the various groups of plants and animals when creation was accomplished. On the other hand, if evolution *has* taken place, these fundamental similarities of structure would naturally be expected to exist. They would have persisted as a product of heredity. In the light of evidences to follow, honest, truth-seeking investigators have been compelled to take the latter view and to interpret homologous organs as an evidence of blood relationship running throughout the animal kingdom.

Vestigial organs. Vestigial organs are the rudimentary remains of organs that are believed to have been formerly completely formed and fully functional. In man the tonsils, the appendix, and the coccyx (or lower four or five joints of the spinal column, which is considered to be the remains of a tail) are examples.

A striking example of vestigial organs is found in the skeleton of the Greenland whale. Being a mammal whose remote ancestors are believed to have inhabited the land, the whale had little use for its hind legs when it took to the water. Consequently these limbs became abortive and are repre-

[1] From Francis Mason and others, *Creation by Evolution.* By permission of The Macmillan Company, publishers.

sented today by bony remnants which do not even protrude
outside the skin. The seal has its hind legs reduced and

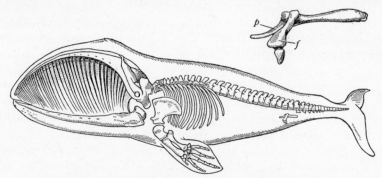

FIG. 86. The Greenland Whale possesses Vestiges of Hind Legs

p, pubis; f, femur [1]

incorporated, along with the caudal appendage, into a power-
fully efficient organ of propulsion. The python has remnants
of hind legs protruding as a couple of blunt, heavy spines

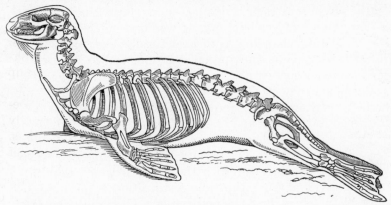

FIG. 87. The Seal's Hind Legs, incorporated with its Tail, form a Powerful
Swimming Organ [2]

among the ventral scales. In the inner corner of the eye of
practically all higher animals are found the remains of a mem-

[1] From G. J. Romanes's *Darwin and After Darwin*. By permission, The
Open Court Publishing Company.   [2] Ibid.

brane. In birds this forms a complete and perfectly functioning nictitating membrane by the blinking of which the eyeball can be moistened without closing the lids. Man has vestiges of three main ear muscles which are functional in many other mammals and serve to pull the ear forward, upward, and backward. A few individuals retain some power over these muscles and can move the ears to some perceptible degree. Wisdom teeth are rudimentary organs and do not appear at all in some persons. Adenoids are believed to be the remnants of a pharyngeal tonsil that formerly extended down into the throat.

Space will not permit of an extensive enumeration of vestigial organs, but there are literally hundreds of them. There is scarcely an animal whose body does not present a generous array of these tokens of the past, as they are believed to be. The plant kingdom too is replete with them, from the lazily floating algal forms to the sturdy trees of the forest. Incidentally it may be of interest to note that these vestigial organs in man are prolific sources of infection. Being more or less abortive, many of them have a poor circulation, which materially lessens their resistance to invading germs.

The import of vestigial organs would seem to be that living things are changing both structurally and functionally. If they are not, as the special creationists believe, why do these rudimentary organs exist? The only alternative to the evolutionary view would be that they were created in this imperfect, semifunctional condition. Such a position has failed to appeal to the logic and reason of thoughtful investigators.

Embryology furnishes evidence of evolution. Embryology is the science which treats of the early development of an organism. It is concerned with the changes which take place in an individual from the time fertilization is accomplished until birth (or, if it be the embryo of a bird, until the egg is hatched). The validity of the evidence furnished from this field rests upon the fact that the embryogeny of higher organisms appears to recapitulate the race history of its ancestors. The biologists condense the idea by saying that ontogeny (the development of the individual) recapitulates phylogeny (the development of the race).

Animal life in its simplest forms is unicellular. The amœba and the paramecium are examples. Other protozoa are found that prefer not to live alone but to aggregate into groups called colonies. The *Gonium* is such an organism. Each cell here is a distinct individual. It secures its own food and reproduces its kind quite independent of what the other members of the colony do; yet it prefers to live as a member of the colony. Above the *Gonium* the *Volvox*[1] appears. This peculiar hollow, ball-like colony is composed of a single layer of several thousand cells arranged in the form of a sphere. Here each cell may be thought of as a separate animal, but within the colony there is some degree of specialization, differentiation, and coördination. For instance, the spherical colony rotates about through the water. This movement is produced by the concerted action of many hairlike cilia that project in pairs from each cell. Certain cells may take upon themselves the task of dividing many times to produce a group of cells which, when freed, form a new colony. These particular cells are in a sense performing the work of reproducing for the other cells with which they are associated. Still other cells of the colony may assume the function of producing sperms. These sperms fertilize eggs grown in still other cells specialized for this purpose. After fertilization the eggs develop new colonies. Here we have a distinct differentiation and specialization for the purpose of sexual reproduction. In the *Hydra* a still more complex animal is encountered — one composed of two layers of saclike cells. At the free end of this animal an orifice is found through which food is taken and waste eliminated. In the *Hydra* also some cells are specialized for protection, some for securing food, and still others for reproductive purposes. Above the *Hydra* the starfishes, worms, and other organisms of growing complexity continue the lines of development.

Now let us return to the embryological development of a higher animal. After fertilization the single-celled egg grows into a cluster of undifferentiated cells called the *morula*, which

---

[1] The *Volvox* is classified by some scientists as an animal; by others as a plant.

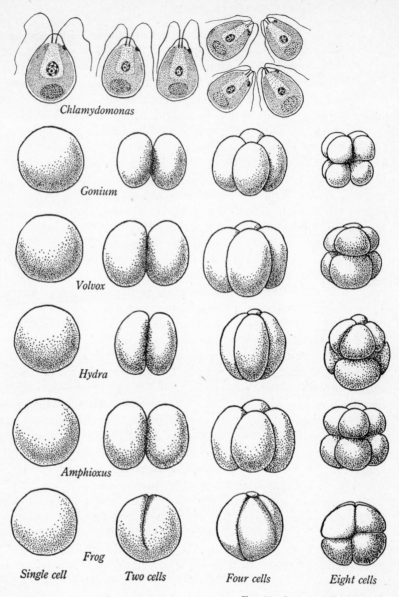

Chlamydomonas

Gonium

Volvox

Hydra

Amphioxus

Frog

| Single cell | Two cells | Four cells | Eight cells |

FIG. 88. Stages of Embryonic

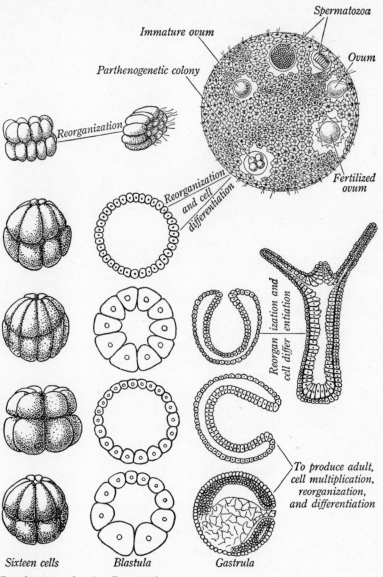

Immature ovum

Spermatozoa

Parthenogenetic colony

Ovum

Reorganization

Fertilized ovum

Reorganization and cell differentiation

Reorganization and cell differentiation

To produce adult, cell multiplication, reorganization, and differentiation

Sixteen cells   Blastula   Gastrula

Development showing Recapitulation

strongly resembles the *Gonium* colony. This continues to grow, a cavity forms at the center, and in this stage the

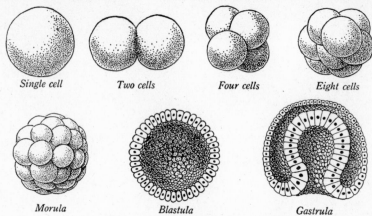

Single cell   Two cells   Four cells   Eight cells

Morula   Blastula   Gastrula

FIG. 89. Stages in the Development of an Embryo
Note their similarity to the forms in Fig. 88

blastula, as it is designated, bears a marked resemblance to the *Volvox*. Soon a form develops which is strikingly similar to the *Hydra*. Beyond this point the embryo develops rapidly into a much enlarged and highly complex structure.

The early stages of all vertebrate embryos are strikingly alike. It would take almost a specialist to distinguish between that of a fish and a man, for example. The difference becomes marked only as the fetus of the more advanced organism pulls away in its development from the lower type and approaches the state of the mature embryo. This general recapitulation can be explained only on the basis of a hereditary hang-over of the biological developmental patterns found in mature individuals down along the lines of ancestral development.

Let us now leave the evidence of evolution presented by general embryology, which could be pushed much farther if time permitted, and fasten our attention on specific instances of recapitulation. There are two species of European salamander, the black and the yellow. The first inhabits the higher Alpine regions; the second thrives in the lower reaches of the plains and valleys. Both species produce their young alive,

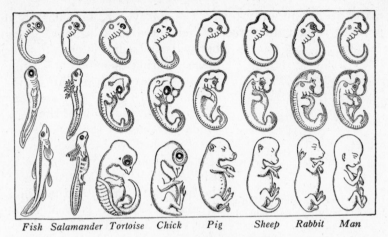

Fish Salamander Tortoise Chick Pig Sheep Rabbit Man

FIG. 90. A Series of Embryos from Different Animals arranged to show
Similarity in Development and Structure[1]

but the offspring of the yellow salamander is born with long
feathery gills and lives in the water for about the first six
months of its life. The young of the black salamander, on the
other hand, living in the higher stretches of the country,
where water is less abundant, are born without gills and are
ready to take up their abode on dry land immediately. An
examination of the oviduct of a pregnant black salamander,
however, discloses that at a certain period in their development these embryos too have feathery gills. Indeed, if an
embryo in this stage be removed from the body of the mother
and be placed in water, it will survive and develop much after
the fashion of the embryo of the yellow relative. The black
salamander probably either evolved from the yellow species,
or they both may have developed from a common ancestor.
The embryonic gills of the black embryos, then, are considered
to be a hereditary hang-over, so to speak, and indicate blood
relationship between the two species.

Similar examples could be pointed out in almost any group
of animals. Certain whale embryos begin the development of
teeth. The growth of these teeth is soon arrested and bony

[1] From B. C. Gruenberg's *Elementary Biology*. Ginn and Company, publishers.

plates are produced instead. The embryos of some birds also produce rudimentary teeth; but these soon disappear, and horny, beaklike lips later take their place. During their prenatal stage the embryos of all vertebrates (including man) show at least four gill arches on each side of the neck similar to those of the fish. These are furnished with blood vessels after the manner of gill arches in fish. At this stage the structure of the heart and, in short, the whole circulatory system is fishlike in character. Later the heart structure changes, and growth obliterates the gill pouches, with the exception of one which is modified into the Eustachian tube, connecting the middle ear with the throat cavity. The human embryo at one stage exhibits a tail longer than the legs and has webs between both fingers and toes. Finally, to mention but one more prenatal character, the body of a young human embryo is more or less covered with fine, short, silky hair up to about the sixth month of prenatal life. Most of this is subsequently shed, remnants of it being found floating in the placental fluid at birth. In rare cases it persists, and the whole body of the individual, including the face, is covered with hair like that of a furry animal.

FIG. 91. The Hair on this Man's Face is a Survival of a Prenatal Character [1]

Some enthusiastic biologists of the past have taken the extreme position that the whole racial history of animals may be traced through these prenatal characters. That view has been largely discredited. However, practically all reputable scientists in this field do regard these prenatal embryological structures as recapitulations of the evolutionary past. They are taken as evidence of phylogenetic relationship retained in this early stage of the individual by the influence of heredity.

[1] From R. S. Lull's *Organic Evolution*. By permission of The Macmillan Company, publishers.

The geographical distribution of plants and animals is believed to furnish evidence of evolution. Plants and animals are widely distributed over the earth. Yet, according to the "doctrine of specific centers," it is believed that any one particular species arose at some one definite place and then migrated out from there as a center. Overproduction and crowding would tend to promote these scattering movements. Sometimes a species migrated very widely and then, owing to some geological or climatic change, a barrier would occur of such a nature as to separate the two extremities of the occupied region. The barrier might be an arm of the sea, such as is known to have once extended from the Gulf of Mexico diagonally across the Rocky Mountain area to the arctic regions; it might be a mountain range elevated by geological shifts; it might be a great ice sheet, such as once covered the northern part of North America; or the barrier might be produced by some other condition. Once the extremes of the migration were separated, the barrier might continue for millions of years. Owing to somewhat different environmental conditions that would doubtless prevail in these two isolated extremes, the animal and plant life, respectively, of one region might gradually become different from those of the other region. If the two regions should develop quite different climates or other life conditions, then, assuming the truth of evolution, the animals or plants once alike and having the same ancestor might in the course of a few million years become decidedly unlike.

A good example of such change is believed to be found in the camel family. One branch of this group is the true camels, found to inhabit central Asia and Arabia; the other branch of the family is the much smaller, lighter colored, and graceful llamas and guanacos native to South America. No anatomist who has ever studied the structure of representatives of these two camel branches doubts for a minute the fact of their blood relationship. But how did they become so different?

At this point a study of the fossil remains of prehistoric camels comes to our rescue in a most striking manner. The fossils of these animals reveal beyond a doubt that the first camels appeared in North America. Fossils found in successive earth layers indicate also that the North American camel in its final form came about as a succession of changes from a much smaller ancestor.

From North America the camels migrated both northward and southward. Those going northward crossed a land bridge then existing between North America and Eurasia, and at length came to inhabit the regions they now occupy. Those going south passed into South America over the isthmus connecting the two continents. Later, for some reason not well understood, these ancestral camels, together with elephants, mastodons, mammoths, and many other prehistoric animals, completely disappeared from the North American continent. This left the two extremes of the camel family widely separated. Now these two branches of the camel family, as they migrated and during the succeeding millions of years that have intervened, have come to take on their present forms owing to the effect of different environmental influences. Scores of changes in other animals may be traced in a similar manner.

Such evidence, of course, is not beyond question. But the character of the fossils and their gradations of form as outlined can be found. The strong anatomical evidence as to relationship between the two living branches may also be seen by examination. To what conclusion do these evidences then point? The only rational conclusion to which painstaking students' of the matter have been able to come is that living camels have reached their present form through a slow, gradual evolutionary change as they adapted themselves to different conditions.

Islands also furnish important evidence of this distributional effect. The Galapagos Islands, of which we have already learned, may be taken as an example. It was a study of the animals and plants on this group of islands that perhaps did more than anything else to confirm Darwin in his belief as to the importance of evolution. The Galapagos are the weathered remains of volcanic peaks elevated out in the Pacific five hundred or six hundred miles from the west coast of equatorial South America. They are separated from the mainland by water varying from two to over three miles in depth; con-

Fig. 92. Descendants of North American Camels

Above, the Bactrian camel and the llama of South America, photographs by New York Zoölogical Society; below, the guanaco of South America, photograph by Colorado Natural History Museum

sequently there is little probability that they ever were connected with the mainland of the neighboring continent. In discussing the fauna and flora of this archipelago Darwin says:

Here almost every product of the land and of the water bears the unmistakable stamp of the American continent. There are twenty-six land birds; of these, twenty-one, or perhaps twenty-three, are ranked as distinct species, and would commonly be assumed to have been here created; yet the close affinity of most of these birds to American species is manifest in every character, in their habits, gestures, and tones of voice. So it is with the other animals, and with a large proportion of the plants, as shown by Dr. Hooker in his admirable Flora of this archipelago.[1]

[1] Charles Darwin, *Origin of Species*, Vol. II, p. 188.

The islands were found to have no vertebrates except a very few species of tortoises, lizards, and one or two kinds of snakes. In addition, there were several species of insects.

The only way Darwin could account for life on these islands was on the basis of one of two possible theories. Either the plant and animal species found had been created there or else they had migrated from the South American continent and had taken on certain new characters in response to the warm, moist, equable insular climate. But other islands have flora and fauna strikingly similar to those of the neighboring continent. The Cape Verde Islands, off the coast of Africa, have forms of life strikingly like those of the African mainland; the Bermudas, like those of North America; and St. Helena, like those of Africa again. It did not seem probable to Darwin nor has it to subsequent investigators that special creation should have taken place on all these isolated islands. Moreover, if it had taken place, it seemed even less probable that the living species called into being should have had such a striking similarity to the flora and fauna of the respectively neighboring continents. Biologists believe it much more probable that organisms migrated to these islands from the mainlands and then gradually changed as they successfully adapted themselves to the new conditions. The fact that the type of plants and animals found on the islands were in each case only those which could have migrated or have been blown there through the air, or which could have drifted in on logs or other floating débris, seemed to add strength to the supposition. In addition, it was noted that the older the islands were, as disclosed by the character of the geological formations, the greater the degree of difference between the insular organisms and those found on the adjacent mainland. So, taking all these facts into consideration, biologists can see no other plausible interpretation of them except to conclude that they indicate a definite change in living things; in other words, an evolution.

Blood tests offer evidence that evolution has taken place. There are several ways to make blood tests, but we shall outline the one followed by Dr. Nuttall of Cambridge University. It is called the precipitation method. As a usual thing chemists have been unable to determine the affinities of the blood of different animals on the basis of a chemical test. That there is a difference between the blood of different organisms, however, has been a matter of common knowledge for many years. In making human-blood transfusions the blood of a different animal cannot be used. If it were, serious consequences, if not death, would most likely follow. Even when human blood is used, a person is sought whose blood, on the basis of tests, shows the greatest degree of compatibility with that of the patient.

In making the blood test a rabbit or other similar experimental animal is injected with the clear, amber-colored serum drawn off from the coagulated blood of another animal, in this case let us say man. The rabbit, in response, at first runs a temperature and shows other symptoms of systemic disturbance. It recovers in a day or two, when a second dose of the serum, but larger than the first, is injected, with similar results in the rabbit's reaction. In this way successively larger doses of the serum are administered to the animal, which finally exhibits no reaction to the treatment at all. Now the rabbit is said to be *sensitized* to human blood and is hence immune to its otherwise deleterious effects. What really has happened is that the rabbit's blood has developed certain chemical substances called antibodies to protect the animal from the injurious effects of the human blood.

If, now, a small amount of blood serum from an *unsensitized* rabbit be added to a suitable portion of human-blood serum in a test tube, no chemical reaction whatever is observed; but if the sample of human-blood serum be treated with serum from the *sensitized* rabbit, a quite different and striking result follows. Instead of the mixture's remaining unchanged, as in the first case, a heavy white precipitate is thrown down, owing to the effect of the antibodies. The discovery of this blood reaction has proved quite important in both jurisprudence and in the administration of pure-food laws. If someone is suspected of a crime involving murder, and the suspect is found to have bloodstains on his clothing, the stains become strong incriminating circumstantial evidence; but if the suspect denies the charge and attributes the stains to the blood of a fowl or other animal which he claims to have slaughtered, this blood test may be invoked to prove the truth or the falsity of the claim. A weak solution is made by soaking the stains in salt water. The solution so obtained is purified by careful filtration and is then treated with serum from the blood of a guinea pig or rabbit kept sensitized to human blood. If no reaction follows, the bloodstains are invalidated as an evidence of guilt. If, on the other hand, the white precipitate is thrown down, the stains are then known to be formed by human blood, and the evidence becomes very incriminating. Even if the stains are old and putrid the test will still work. In European countries meat inspectors use the same test to determine the validity of the butcher's claims as to the contents of sausage and other similar meat products.

When these blood tests had been perfected, it was suggested that if evolution were true there might be a similarity of blood qualities between animals related by descent, as well as by morphological likeness. It was to test out this hypothesis that Dr. Nuttall did his work. His results were surprising. He found, for example, that when a rabbit was sensitized to the blood of one carnivore it gave a much more pronounced reaction with the blood of carnivora in general than it did with the blood of other mammals. Anti-llama serum gave a moderate reaction with the blood of camels. A very close blood relationship was found to exist between the deer family and antelopes, sheep, goats, and domestic cattle.

Strong anti-turtle serum gave a pronounced reaction with the blood of other species of turtles and with crocodiles but a very much weaker one with the blood of the more remote lizards and snakes. The embryology of the horseshoe crab had indicated its relation to certain land spiders rather than to marine crustacea. These blood tests confirmed the embryological evidence in a striking manner.

The anatomical evidence indicates that man is closely related to the apes, such as the orang-utan, the chimpanzee, and the gorilla, that he is less closely related to the Old World monkeys, and that he is remotely related to those of the New World. The blood tests revealed exactly the same kinship, the human anti-serum throwing down the heaviest precipitate with the man-like apes and the lightest with the South American monkeys.

From these examples it may be seen that blood tests have brought very strong confirmatory evidence to the theory of evolution. Biologists have found it very difficult to account for this blood relationship except on the basis of a common ancestral origin.

Domestication offers evidence of evolution. A study of the different types and breeds of domesticated animals presents strong evidence of evolution. The great range in size and form presented by the different breeds of horses is a striking feature at any horse show. The diminutive, long-haired Shetland pony looks like a pygmy beside the heavy-boned, strong-muscled, hairy-legged draft horse, and it in turn looks lumbering, indeed, when standing beside the thoroughbred race horse with its trim, rangy legs, alert eyes, and greyhound-like body. Yet it seems probable that all these types of horses have been developed from two or three species of wild horses. The horse has been domesticated so long that the history of its origin has been lost in antiquity, but the best records obtainable indicate only two or three ancestors at the outside. Cattle too show an equally large variety of species, as do swine and sheep. The exact ancestry of any of these animals is not positively known, but the number of the wild species is in each case relatively very small. Dogs, of which an almost monotonously long list of types has been developed by fanciers, are but the descendants of domesticated wolves. How many species of these ancestral wolves there were at the beginning it is impossible to say, but the list certainly could not have been half so long as the classification sheet presented at any modern dog show.

The domestic pigeon furnishes another classical example of derived types. They all seem to have come from a single wild rock pigeon. Darwin studied the origin of many domesticated animals, but with regard to the pigeon he tells us:

I have been led to study domestic pigeons with particular care, because the evidence that all the domestic races are descended from one known source is far clearer than with any other anciently domesticated animal. Secondly, because many treatises in several

2

FIG. 93. Many Types of Pigeons have been produced from a Single Ancestor, the Rock Pigeon[1]

languages, some of them old, have been written on the pigeon, so that we are enabled to trace the history of several breeds. . . . I have kept alive all the most distinct breeds, which I could procure in England or from the Continent; and have prepared skeletons of all. I have received skins from Persia, and a large number from India and other quarters of the world. Since my admission into two of the London pigeon clubs, I have received the kindest assistance from many of the most eminent amateurs.[2]

From this study, which is quoted in part to show how painstakingly careful Darwin was in his investigational methods, he tells us that the French writer C. L. Bonaparte lists two hundred and eighty-eight species of pigeons. Some "domestic races of the rock pigeon differ fully as much from each other

[1] Adapted from Romanes's *Darwin and After Darwin*. By permission, The Open Court Publishing Company.

[2] Charles Darwin, *Animals and Plants under Domestication*. By permission, D. Appleton and Company, publishers.

in external characters as do the most distinct natural gen-
era." [1] Yet the evidence all points to the wild rock pigeon as
their original ancestor.

Evidence that living things can change decidedly is furnished also by the Porto Santo rabbit. This case is really
more than an instance of change under domestication and
furnishes strong evidence that living animals may evolve into
other forms out in nature.

In 1418 or 1419 (the records are not quite clear) Gonzales
Zarco, a Portuguese navigator, liberated a female rabbit with
her litter of young on the island of Porto Santo, not far from
Madeira. The female was of the European domesticated
type and had produced the young during the voyage down to
the island. The insular climate was very favorable to the
rabbits, and there were no carnivorous enemies to prey upon
them. Consequently they multiplied rapidly and in less than
forty years were described as "innumerable." Four and a
half centuries later, when representative rabbits from the
island were taken back to Europe, they had changed so greatly
that the noted German zoölogist Haeckel described them as a
distinct species.

Although these rabbits had lived in a highly favorable
habitat, as is evidenced by their rapid increase in numbers,
nevertheless they had decreased markedly in size. They were
scarcely more than half the size of the European wild rabbits,
from which the European domesticated ancestral stock had
come. The fur on the back was redder, with fewer black-
tipped hairs, and instead of remaining pure white the throat
and belly had become gray. Both the tail and the ears had
lost their black tips. Darwin relates:

The two little Porto Santo rabbits, whilst alive in the Zoölogical
Gardens, had a remarkably different appearance from the common
kind. They were extraordinarily wild and active, so that many
person[s] exclaimed on seeing them that they were more like large
rats than rabbits. They were nocturnal to an unusual degree in
their habits, and their wildness was never in the least subdued;

[1] Charles Darwin, *Animals and Plants under Domestication*. By permission, D. Appleton and Company, publishers.

so that the superintendent, Mr. Bartlett, assured me that he had never had a wilder animal under his charge.

Moreover, it was a highly important fact that the superintendent "could never succeed in getting these two rabbits, which were both males, to associate or breed with the females of several breeds which were repeatedly placed with them."[1]

Changes in plants under cultivation have been as varied and pronounced as those described for animals. Naturally, then, if organisms can change under the influence of artificial selection, may they not similarly evolve into new forms under the guidance of natural forces? To most students of the problem it seems perfectly reasonable to suppose that they might.

Paleontology furnishes strong evidence of evolution. The fossilized remains of plants and animals — bones, teeth, and other preserved or petrified parts — throw considerable light upon the question of ancestral origin. It is well known, even to the casual observer, that the upper parts of the earth's crust have been deposited in layers. Railroad cuts, mine shafts, the walls of mountain cañons, and the preserved cores of oil wells all furnish abundant and conclusive evidence of this fact. The only places in which layers are not found are at those points where erosion has plowed off the sedimentary and metamorphic rocks and brought the surface down to the level of the original igneous material. These completely eroded areas, however, are of very limited extent, comparatively speaking, and by far the greater part of the earth's outer crust shows definite stratification, as was made clear in a former chapter.

In general it would seem perfectly safe to assume, too, that these strata must necessarily have been laid down in order, the bottom ones first and the upper ones last. Therefore the deeper down the layers are, the older they must be. Such an assumption could not be safely made, of course, for the strata of any one particular limited area. Magma intrusions, land elevations, and mountain-folding might easily confuse and destroy this successive orderly arrangement in any one locality. Indeed, in some places faulting and tipping might be so ex-

[1] Charles Darwin, *Animals and Plants under Domestication.* By permission, D. Appleton and Company, publishers.

treme as to elevate the strata on one side of the fault, tip them over, and cause them to fall back on one another. Other forces, such as erosion and deposit, might add their work to the scene and so confuse the structural formation as to make the record, when taken alone, utterly inconsistent and undependable. But when large areas are taken, despite limited discrepancies due to local disturbances, the orderly arrangement of the successive layers may be traced for long distances. Some strata may be traced for distances of five hundred miles or more. Under such conditions it is considered to be perfectly safe to assume successive arrangement in point of time from the lower strata on up.

Now the fossils of animals and plants are found deposited and preserved within the rocks of these successive layers. It would seem quite rational, then, to make the further assumption that those organisms whose fossils are found in any quantity must have lived and died in the period when the strata in which the remains are found were laid down. Such a conclusion appears to be the only one that could rationally be reached. With these two principles (order of strata and the appearance of fossils within them) in mind, then, it will be of interest to see just what the geological record reveals.

In order to talk and write intelligently about the earth's past history, the geologists have found it necessary to divide the geological strata into groups and subgroups and to name them. If one arranges these different periods in the form of a table which shows the types of plants and animals living at these particular times, together with the estimated age of the earth for each major period, it presents a most interesting and compelling story. It will be seen that there was a gradually growing complexity of both animals and plants from the earlier geological times up to the present. One type of animal or plant would appear in a primitive and usually diminutive form and increase in size and abundance until it became dominant for its particular age. Then, as geological changes made environmental conditions unfavorable and inhospitable for these organisms, they either perished or evolved into other forms better adapted to the new conditions.

TABLE II. SUBDIVISIONS OF EARTH HISTORY ACCORDING TO MEASUREMENTS OF TIME BASED ON RADIOACTIVITY

Era		Period	Millions of Years	Representative Animals	Representative Plants
Cenozoic	Quaternary Late	Holocene (the present)	1	Mammoths, mastodons, elephants, primitive man, present man	Present forms
	Quaternary Early	Pleistocene (the world-wide glaciation)			
	Tertiary Late	Pliocene	7	Three-toed horse, horse, lemurs, elephants, early camel, apes, Pithecanthropus	Deciduous trees increasing, lower flowering plants, many grasses
		Miocene	19		
	Tertiary Early	Oligocene	35	Ancestors of rhinoceros, primitive pigs, hippopotamus, horses size of rabbits	About the same as in the Cretaceous
		Eocene	55		
		Paleocene			
Mesozoic	Late	Upper Cretaceous	95	Reptile-like rhinoceros with three horns, crocodiles, land birds, few little mammals	Giant redwoods, flowering plants, as birch, maple, oaks
	Early	Lower Cretaceous	120		
	Late	Jurassic	155	Small to huge reptiles, flying reptiles, Archaeopteryx	Cycads, pines, arbor vitae
	Early	Triassic	190		
Paleozoic	Late	Permian	215 250	Giant insects, salamander-like amphibians, small reptiles 2 feet long, 5 inches tall	Seed-bearing ferns, cycads, and cone-bearing trees, perhaps ancestors of the redwoods
		Pennsylvanian ⎱ Carboniferous Mississippian ⎰	300		
	Middle	Devonian	350	Insects, spiders, scorpions, gill fish, lung fish	Lower land plants, giant lycopods, horsetails, simple ferns
	Early Late	Silurian	390		
	Early	Ordovician	480	Protozoans, sponges, corals, starfish, worms, Trilobites	Algal remains preserved
	Late	Cambrian	550		
Proterozoic	Early		1000	Perhaps protozoans and crustaceans — record not above question	Bacteria and lime-secreting algae whose simple, limy skeletons made fossil preservation possible
Archaeozoic			2000	No unquestioned evidence of life. The presence of graphite, limestone, and iron ore suggest the possible presence of bacteria and algal forms	
Cosmic					

To show more concretely and vividly how these changes gradually took place, let us examine the elephant series. It begins with a peculiar little animal scarcely three feet high that lived in northern Egypt. It possessed the beginnings of short tusks and had twenty-four teeth and a long head. The only sign of a trunk, as indicated by the shape of the nasal skeleton, was an elongated, supple upper lip that enabled the animal to pull in and bite off the forage with great facility. It was named *Mœritherium* (Greek *Mœris*, an ancient lake near which its remains were found, + *thērion*, "a wild beast"). It probably splashed about

Fig. 94. The *Mœritherium*, the Earliest Elephant Known

From a model

in the swamps of that region and fed on the tender, succulent vegetation. Its remains are found in the Eocene strata.

Up in the Oligocene the fossils of a more advanced type of elephant have been found in both Egypt and India, showing that this beast was rather widely distributed. It was somewhat

Fig. 95. The Head of *Paleomastodon*, an Early Type of Mastodon

From a model

larger than the *Mœritherium* and had distinct but short tusks, a shorter neck, and, strange to say, a considerably elongated lower jaw. The position of the nostrils indicates a short proboscis. It had several teeth, but not so many as the smaller animal. It was called *Paleomastodon* ("ancient breast tooth").

The Miocene yields fossils of a widely distributed animal almost as large as the modern elephant, called the *Trilophodon*

(three crests — from the three ridges across the teeth). It had enormously elongated lower jaws with two protruding tusks below as well as two above. It is probable that the lower jaw had become so long as to be somewhat of a handicap in browsing and obtaining food. The *Trilophodon* was a great emigrant. Its remains are found in Europe and Africa, and it was the first proboscidean to reach

FIG. 96. Trilophodon, One of the Stages in the Evolution of the Elephant[1]

North America. It is believed to have crossed over the land bridge uniting Asia and North America during the early Miocene. Its molars had become so massive that but two could be accommodated in a jaw at one time.

The mastodon (breast tooth) appeared in the Miocene and became abundant in the Pleistocene. It was the best known of the North American elephant series. This beast, which was as large as the Indian elephant, had greatly retracted the lower jaw and, for the most part, had lost the lower tusks, although some specimens still had remnants of them. The upper tusks had reached enormous proportions, sometimes exceeding nine feet in length. The proboscis had attained true elephant-trunk proportions, and the animal lived on the twigs of hemlock, spruce, and perhaps pine, together with other herbaceous vegetation. It had powerful legs and body and was widely distributed throughout the United States. It is said that almost every peat bog in New York, Indiana, Ohio, Illinois, Michigan, and Iowa (and there are thousands of them) contains the remains of one or more mastodons.

[1] From R. S. Lull's *Organic Evolution*. By permission of The Macmillan Company, publishers.

American Museum of Natural History

FIG. 97. *Mastodon Americanus*

Restoration by Charles R. Knight

The true elephants arose in the late Miocene and became dominant in the Pleistocene. Intermediate forms between these and a primitive mastodon type may be found. There were several species of elephants scattered over North America and Eurasia. Of these one arctic species had long black hair with a thick coat of brown wool beneath. The arctic tundras have yielded up frozen and well-preserved specimens of these animals. Of the North American species, all of which became extinct, some had huge tusks that curved inward and, in some cases, actually crossed at the tips so as to make them inefficient either as digging organs or as defense weapons. There are two species of the Old World elephants still living: *Elephas indicus* of India and *Loxodonta africanus* of Africa. The teeth are massive and, strange to say, continue to develop and take the place of those worn out, until the animal is somewhere near sixty years old. Of the two species the African elephant is the larger and differs from the Indian elephant markedly in both contour of head and size of ear.

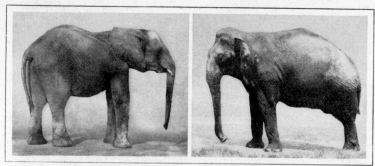

National Zoölogical Park

FIG. 98. The Only Types of Elephant now Living

Left, African elephant; right, Indian elephant

Other animals show a similar series of types in their family development. Among the most notable of these are the camels and the horses, both of which developed on the North American continent and both of which, for some inadequately explained reason, became extinct in the land of their origin. The wild horses still found in some sections of the southwestern United States are descendants of domestic horses that are believed to have escaped from the early Spanish explorers. The original horses were little animals scarcely larger than rabbits, with four toes in front and three behind. Modern horses still have remnants of some of these toes in the form of "splint bones," found on both the front legs and the hind legs.

Paleontology furnishes one of the very strongest evidences of evolution. Only one other theory has been proposed to attempt to explain the serial development of plants and animals from the lower to the higher, more complex forms, as revealed in the earth's strata. This is the theory of *catastrophism*. According to this view the marked changes in the earth's crust represented transformations produced by great geological disturbances, such as earthquakes, volcanic outbursts, floods, sudden elevations, and the like. During each of these great cataclysms it was believed by many that every living thing had been wiped off the face of the earth, leaving only their fossil remains to tell the story of their former existence. It

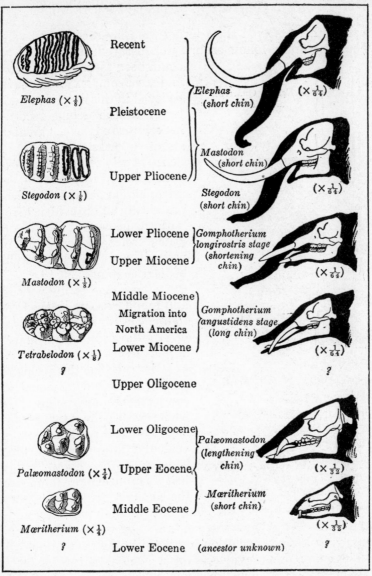

Elephas (× 1/8)

Recent

Pleistocene

Elephas
(short chin)

(× 1/64)

Stegodon (× 1/8)

Upper Pliocene

Mastodon
(short chin)

Stegodon
(short chin)

(× 1/64)

Mastodon (× 1/8)

Lower Pliocene

Upper Miocene

Gomphotherium
longirostris stage
(shortening
chin)

(× 1/64)

Tetrabelodon (× 1/8)
?

Middle Miocene

Migration into

North America

Lower Miocene

Upper Oligocene

Gomphotherium
angustidens stage
(long chin)

(× 1/64)

?

Palæomastodon (× 1/4)

Lower Oligocene

Upper Eocene

Middle Eocene

Palæomastodon
(lengthening
chin)

(× 1/32)

Mæritherium (× 1/4)
?

Mæritherium
(short chin)

Lower Eocene

(ancestor unknown)

(× 1/32)

?

FIG. 99. A Series of Diagrams showing the Evolution of the Head and Molar
Teeth of the Elephant[1]

[1] From H. W. Shimer's *An Introduction to Earth History*. Ginn and Company, publishers.

was further believed that during the quiescent period which would naturally follow each major disturbance, creation had again been resumed on a higher and more complex level. Cuvier believed in this doctrine and interpreted the deluge recorded in Scripture as the last great destructive catastrophe. No evidence at all conclusive has ever been adduced to prove the catastrophic hypothesis; indeed, practically all the evidence from both the geological and the biological sciences tends to disprove it. On this basis it would be next to impossible to explain the great range of life which is still found existing on the earth today, but which on the basis of evolution can be adequately accounted for.

It is true that the geological record is fragmentary and incomplete at many points. The whole earth is the extant record, and the fossil remains that reveal the story are scattered widely. No one knows, either, just where to explore to unearth the choicest pages. Yet each year bits of the history are being uncovered, a page here and a page there. When these fragments are assembled they all point to but one logical conclusion — origin through evolution.

There are other evidences of a general nature, such as classification and genetics, that add their weight to the evidences already cited in favor of evolution. However, their

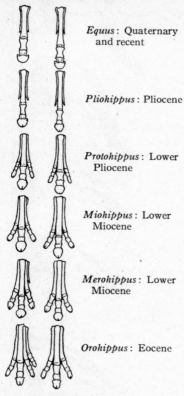

Equus: Quaternary and recent

Pliohippus: Pliocene

Protohippus: Lower Pliocene

Miohippus: Lower Miocene

Merohippus: Lower Miocene

Orohippus: Eocene

FIG. 100. The Evolution of the Feet of the Horse[1]

[1] From H. H. Newman's *Readings in Evolution, Genetics, and Eugenics.* By permission of The University of Chicago Press, publishers.

testimony is no more convincing than those already presented, and for this reason it will be omitted.

Specific adaptations indicate evolutionary changes. The end product of evolution is successful adaptation. This is the only reason why living things should change at all. The rocks bear silent testimony, however, that, with respect to the ability to make these changes, plants and animals of the past have fallen into two categories. Some have been plastic enough to respond in an effective manner to the selective force of new conditions and have succeeded; others have been too unyielding to respond in a suitable way and have become extinct. The elephant group already discussed furnishes excellent examples of both types. Altogether about three hundred and fifty species of prehistoric elephants have been identified. Most of these, unable to make suitable response to changing conditions, have perished and disappeared. A few ancestors, on the other hand, were plastic enough to evolve into the two species which survive today. But the environment is still changing. Civilized man has appeared on the scene. So far, the African elephant has been unable to evolve suitable weapons of defense against the darts and bullets of the greedy ivory-hunters. Consequently the living representatives of *Loxodonta* are rapidly diminishing. Lucas tells us that in a few years not a single *old* elephant of this genus with well-developed tusks will be left alive, and that unless they are protected by law they are doomed to a speedy extinction. The Asiatic elephant stands a better chance of survival, for a while at least, because his tusks do not yield such high-grade ivory and because he can be domesticated and utilized to carry logs in the teak yards of India.

But, to return to our main idea, some plants and animals have changed in a most striking way to meet new needs. Some structural and functional adaptations are so peculiarly worked out as to appear almost ingenious in their conception and perfection. They stand as one of the most gripping evidences that living things can and do change to meet new needs. There are thousands of illustrations, but we shall note only a few of the most striking.

Special plant adaptations. All plants are adapted to live successfully. They have roots to penetrate the soil in all directions, finely divided branches to form connections with large volumes of air, leaves admirably built to make food, and ingeniously constructed flowers to effect pollination in order to insure the production of seeds and fruits. Beyond this point, moreover, many plants have developed wonderfully effective disseminating organs to carry their offspring about over the earth to find new homes. Almost any plant provides for these functions. But the niceties of adaptation, the versatility of nature, and the limits to which she will go to insure the success of her children are seen more vividly in certain specific adaptations which we wish to note.

Some peculiar adaptations for water storage are found among the cactuses. Many of these plants grow where the total annual rainfall is not over four or five inches. Under these conditions they must be equipped to absorb quickly the water provided by the scanty shower and store it for use during the long dearth that often ensues between these periods of supply. Almost all cactuses have some water-storage provision. The barrel cactus of the Southwest, as its name suggests, is peculiarly equipped for this purpose. Indians and other desert travelers pressed for water often slash off the tops of these friendly plants and extract the water from the spongy tissue within for their relief.

The ability of cactuses to store water is shown by the following incident. A student of botany pressed and, as he thought, thoroughly dried a specimen of the little prairie cactus for a herbarium mount. When mounted, the sheet was filed away with other specimens. Imagine the student's astonishment when looking through his collection the next spring to find that this supposedly desiccated cactus had pushed out a good-sized bud, and, although stuck to the mounting sheet, had produced a fairly well-formed flower.

Dr. Gager tells us of a certain representative of the gourd family found in Mexico. A large root-stem tuber of this plant was placed in a glass case in the museum at the New York Botanical Garden in February, 1902. It was then so lifeless

Carson Studio

FIG. 101. The Barrel Cactus of the Arizona Desert

This cactus stores large quantities of water which can be used to quench the
thirst of the desert traveler

looking as to resemble a rock more than a plant. Yet each
summer, under these waterless conditions, this specimen persisted in pushing out a long, slender green shoot. The water
stored in the tissues, when the root-stem was brought from
the desert, permitted the shoot-bearing to continue until 1908.

Many other desert plants store water. Plants from these
regions are also notorious for their habit of producing long,
formidable thorns of one kind or another. This habit is
probably another adaptation to protect them from the ravages of thirsty animals that would otherwise destroy them.

There are some plants that supplement the food which they
make with protein products from captured insects. Some of
the leaves of the pitcher plant, *Nepenthes villosa*, have been
modified into remarkable pitcher-like receptacles. Each
pitcher is partly filled with a row of long, slender, smooth, stiff,
bristle-like hairs which curve downward toward the liquid.
Over the opening a leaflike covering is supported by a little
upright pedestal. "Curious" insects alight on the rim of the
pitcher and attempt to explore the interior by crawling down

the curved hairs. Once on these hairs, however, the insect is doomed. Their smooth surface affords no footing and, despite anything the victim can do, it slips further and further until it is finally precipitated headlong into the water contents below. The leaf secretes into the liquid a digesting enzyme which dissolves the softer parts of the insect's body. The digested products are then absorbed by the plant to supplement its diet.

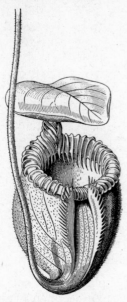

FIG. 102. The Pitcher of *Nepenthes villosa*, an Adaptation for Catching Insects[1]

Venus's-flytrap (*Dionæa*) is another interesting insectivorous plant. Its half-leaf blades are hinged on either side of the midrib like the jaws of a trap. The upper side of the blades is covered with a sticky, sirupy secretion, and the margin of the leaves is fitted with bristle-like teeth that curve upward. When a fly or other insect, attracted by the viscous bait, alights on the leaf, the halves slowly close like a trap. The insect, not noticing the impending peril, remains too long sipping up its newly found treasure. Once aware of its danger, the victim tries to escape. But it is too late. The incurving hairs at the leaf margins are now interlocked together so as to cut off all egress. The leaf halves tighten down on the victim, and a digestive enzyme is secreted to dissolve its body. After digestion and absorption the leaf opens for the next victim. If a bit of beef is substituted for the insect, the leaf follows its usual procedure; but if a tiny pebble is dropped on the leaf, it closes as usual but presently slowly opens again as if it had detected the trick.

The mangrove tree, found in the salty marshes of the West Indies and Florida, as well as in the tropics of the Old World,

[1] From A. Weismann's *The Theory of Evolution.* By permission of Edward Arnold, publisher.

presents some most striking special adaptations. Its roots, most often submerged in water, require special provision for an oxygen supply. To meet this need, the inundated roots send up heavy stubby branches that extend above the water's surface. These "knees," as they are called, are provided with lenticular structures and air-conducting tissue to carry oxygen down to the deeper submerged portions of the roots. In other words, the mangrove erects its own ventilating shafts. Because of this habit mangrove swamps are full of these "knees," which stick up out of the water like partly submerged fence posts.

FIG. 103. Venus's-Flytrap, an Adaptation of a Leaf used to catch Insects[1]

The mangrove also exhibits one of the most ingenious adaptations to insure the growth and establishment of its offspring to be found in the whole plant kingdom. The seeds appear in clusters, and they are not disseminated. Instead they normally germinate while still fastened to the parent tree. They push down long, pointed primary roots from four to ten inches in length and from one half to three fourths of an inch thick. These roots are heavy and somewhat enlarged near the pointed end to give the effect of a plumb bob. Above this peglike root, cotyledons expand. A tree in this stage, with all the dangling young roots, presents a strange spectacle. At the

[1] From Bergen and Caldwell's *Practical Botany*. Ginn and Company, publishers.

2

proper time the young plant is cut loose from the parent tree by an abscission layer which weakens the suspending tissues,

FIG. 104. Germinating Seeds of the Man-grove, an Adaptation to insure Planting[1]

and the enlarged embryo drops like a shot, with the point of the root directed toward the earth. When the soft, marshy soil is struck, the root is effectively driven into the mud, and the young tree is set for a successful and speedy growth. This tree was perhaps the original tree-planter. So effective is this embryo in maintaining the vertical direction while falling that two or three feet of intervening water may be penetrated, and still the root will stick into the muddy bottom in the upright position as designed by nature, and growth will thus be assured.

Special animal adaptations. Animals, like plants, have multitudes of the most striking adaptations. Take the whole form and structure of the bird, for instance. The general contour of its body indicates that it is of reptilian ancestry. Its tail, elongated neck, and feathers, which are but highly modified scales, point toward this one conclusion as to origin. Moreover, most significant of all, the reptile-like scales are still retained over its lower legs and feet. More recently this theory has been practically substantiated by the discovery of the paleontological remains of a prehistoric half-bird, half-reptile ancestor called *Archæopteryx*. This animal, as revealed by the

[1] From C. S. Gager's *General Botany*. By permission of P. Blakiston's Son and Company, publishers.

fossils, had a protruding face much like the bill of a bird, except that it was bony and covered with skin, and its jaws bore reptile-like teeth. The wings were feathered, rather short, and had three toes at the extremity instead of the remnant of one, as in modern birds. The body bore plumage, and the vertebral column instead of being shortened to its present proportions extended out into a long reptilian tail. Strange to say, too, each joint of the tail bore a pair of long coarse feathers arranged one on either side. This ancestor evidently used the nailed claws at the end of the front appendages to assist it in climbing trees, and then the wings were used as second-rate organs of flight to enable it to sail off rather clumsily from its elevated perch. This strange animal was somewhat smaller than a crow.

New York Botanical Garden

Fig. 105. Mangrove Trees Planted by the Adaptation of Nature

A, young trees; B, matured trees, showing roots

Yet from unattractive, intermediate creatures somewhat like the one just described, our modern birds, with all their speed, agility, grace, color, and song, are believed to have originated. This series of changes that have taken place in the structure and appearance of the bird is such as to adapt it to the life which it lives in the air and to make the male attractive for mating purposes.

FIG. 106. Restoration of *Archæopteryx*, an Extinct Reptile-like Bird[1]

Special adaptations are no more vividly shown among animals than they are in provisions which have been made for concealment and protection. In one of the well-known aquariums a large tank is arranged with two communicating

FIG. 107. Ptarmigan, showing Plumage Adaptation

Left, summer plumage ; right, winter plumage

compartments. One is floored with fine sand, and the other with well-variegated shingle. Within the tank are representatives of a type of fish called flounders. Visitors are invited

[1] From Knowlton's *Birds of the World*. By permission, Henry Holt and Company, publishers.

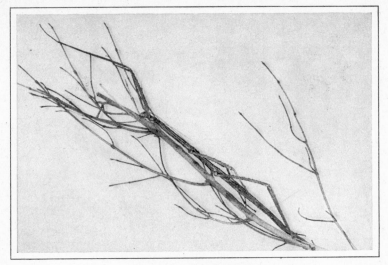

Fig. 108. A Walking Stick

Note the similarity to the twig on which it rests

to drive the fish from one compartment to the other and to watch the results. Over the shingled floor the fish are "coarsely blotched with dark brown, sandy-yellow, and pale cream" colors; but when driven into the other compartment, within five minutes they lose their blotched pattern and assume a uniformly distributed, fine-grained sandlike appearance.

Mountain ptarmigan are protected in a novel way by seasonal changes in the color of their plumage. During the summer they are more or less brown and mottled to conform to the general background of rocks, earth, sticks, and leaves. In the fall they molt and take on almost pure-white feathers to correspond to the snow-covered ground upon which they will live during the winter months. Weasels, small carnivorous animals related to the mink, exhibit the same protective change in the color of their fur.

Walking sticks escape devouring birds by an elongated body and slender limbs of the same color as the twigs and stems among which they live. The deception is further rendered effective by somewhat irregularly spaced dark bands around

FIG. 109. Kallima, the Dead-Leaf Butterfly

Note how much the underside of the wings resembles a leaf

the creature's body which simulate the nodal markings of the twigs themselves, making the insect extremely inconspicuous.

The prize mimicry protective provision, perhaps, is found in the Kallima, a butterfly of India. The wings are gracefully curved in outline and are highly and beautifully colored on the upper surface. On the underside they are a dull brownish gray and have well-marked main lines and branches that resemble the midrib and lateral veins of the leaf of the plant upon which the butterfly perches. When it alights, the creature brings its wings together above its back, as butterflies in general do. In this position the wings take on almost the exact appearance of the surrounding leaves. Only the sharpest eyes can detect the camouflage.

The larva of the curious puss moth achieves protection by a strange admixture of bluff and color. When disturbed, this really harmless caterpillar retracts its head into the first body-ring, with its inflated bright-red margin and its two intensely black eyespots. The whole appearance then becomes one of a terrifying flat face with huge black eyes and a threatening mouth.

Skunks go about their nocturnal tasks leisurely and grace-fully. They are seldom required to bring their protective scent bag into operation. If they are hard pressed, they spray out from glands under the tail a stream of nauseating liquid that is effective against almost any natural enemy. If the at-tacking party has had no con-tact with a skunk before, it is quite probable that one expe-rience will properly educate it. After that all the skunk has to do — and also what he gen-

Fig. 110. The Larva of the Puss Moth in its "Terrifying Attitude"[1]

erally does do — is to raise his banner-like tail as a warning to intruders. This sign, like the warning fire of a fighting craft, is usually effective. The heavier ordnance seldom has to be brought into play.

Many ants, as a sort of "live and let live" policy, have adopted a peculiar system of community-feeding. If some members of the group are more successful in their foraging expeditions than others of their own colony, they seem per-fectly willing to share their gastronomic booty with their less fortunate brothers. The hungry ant will approach one that is supplied with food and gently stroke its antennæ. This action is a distress sign and a solicitation for food. The appeal seems to be invariably effective in securing a generous response from the more prosperous ant. For now the two, beggar and donor, raise the fore parts of their bodies so as to bring their mouths close together. A droplet of liquid food is regurgitated by the one and is hastily swallowed by the other. This liquid comes from the crop of the ant with the well-supplied stomach; for, as Forel says, the ant's crop is a sort of "social stomach," and as long as one is supplied with food the others may share its store by simply making their needs known. This habit is an example of communism to an unstinted degree. Even for

---

[1] From E. B. Poulton's *Colours of Animals*. By permission, D. Appleton and Company, publishers.

the ant with the food supply none is digested and appropriated for its own use until peristaltic contractions push tiny droplets out of the crop into the more active divisions of the alimentary canal beyond.

As in plants, these special adaptations among animals might be cited by the thousands. There is scarcely a tree, a vacant city lot, a roadside, or a yard that does not present them by the score. They all stand out as achievements that have come about through change and permanent adjustment; for, if we are to repose any confidence in the record of the rocks, none of these organisms, either plant or animal, were on the earth when life first began.

Nor must one get the mistaken idea that these marked and rather spectacular adaptations are the only ones. Every organism presents a whole array of adaptations, not quite so striking perhaps, but just as effective and necessary in preserving the individual. In man, for example, the hand as a manipulating organ; the cranium as a bony protection for the brain; the sensitive adjustments of the eye; the correlation of the rate of breathing with the oxygen needs of the blood; the flow of gastric juice when food is in the stomach; growth in response to hormone action; and the elimination of carbon dioxide by the lungs, — these and hundreds of other structures and processes are adaptations. They too are indispensable in preserving the individual and insuring his success.

The evidence of evolution is cumulative. It is obvious that the evidences given in this chapter to support evolution are cumulative. If metallic zinc be treated with sulfuric acid, it can be proved conclusively that hydrogen is one of the products of the reaction; if an electric current be passed through the primary circuit of an induction coil, it can be shown beyond question that a current is induced on the secondary; and that exercise stimulates heart action in the human body can be irrefutably established. But the fact of evolution cannot be proved in any such a final and positive manner. The claim of evolution as a valid theory, then, must rest upon the cumulative effect of all the evidences supporting it rather than upon the decisiveness of any one of them.

For instance, morphology cannot be satisfactorily explained on any other basis than that living things have changed and that they are related by descent; the similarities found in the embryological development cannot be adequately accounted for on any grounds except that of hereditary relationship; the differences found between related groups of organisms, separated by a geographical barrier, cannot be explained except on the assumption of adaptive response to slightly different environmental factors; the ascending series of plant and animal remains preserved by the earth's crust points unmistakably to a gradual evolution of organic forms; and so on for all the other evidences. No one of them proves the case with finality, but they all point to the same conclusion. In a court of justice when a sufficiently large body of circumstantial evidence is accumulated against the accused, the charge is legally assumed to be proved, and the prisoner is adjudged to be guilty. So it is with evolution. Human life is so brief, and evolutionary changes are so slow and far-reaching, that no single evidence can be marshaled to prove the case beyond controversy. But all the sources elaborated in this chapter have presented such a mass of cumulative evidence that practically every one of the thousands of men and women who have studied the question seriously and in an unprejudiced manner have been compelled to accept the theory because of the sheer weight of the evidence piled up to support it.

QUESTIONS FOR STUDY

1. Make a list of the evidences of evolution and show how each contributed toward establishing the theory.

2. How does each of Darwin's factors of evolution contribute toward the movement of the process?

3. How is it possible to trace the age of fossil remains when found in sedimentary rock?

4. Trace, as nearly as possible, the steps which might have occurred resulting in the adaptation of desert plants to their environment.

5. Make a list of some of the special adaptations that may be observed in birds.

6. List some of the special food-getting adaptations of plants.

7. What effect will special adaptations have upon the probable survival of a species?

8. Describe the manner in which blood tests are made to show relationship, and interpret the results thus obtained.

9. Why may closely related plants and animals be found in such widely separated sections of the world, and how did they probably get there?

10. Which of the evidences of evolution is probably the most convincing? Why?

REFERENCES

DARWIN, CHARLES. Animals and Plants under Domestication, Vol. I, pp. 14–387.
DARWIN, CHARLES. Origin of Species, pp. 219–288.
FASTEN, NATHAN. Origin through Evolution, pp. 99–231.
HERBERT, S. The First Principles of Evolution, pp. 51–172.
LULL, R. S. Organic Evolution, pp. 29–640.
MASON, FRANCES, and others. Creation by Evolution, pp. 1–269, 355–370.
NEWMAN, H. H. Evolution, Genetics, and Eugenics, pp. 88–173.
SCOTT, W. B. The Theory of Evolution, pp. 27–172.
THOMSON, J. A. The Outline of Science, Vol. I, pp. 60–151.
WELLS, H. G., HUXLEY, J. S., and WELLS, G. P. The Science of Life, Vol. III, pp. 839–960; Vol. IV, pp. 1147–1199.

Chapter VIII · The Process of Evolution cannot yet be Satisfactorily Explained

Introduction. Among the people of the United States there has been a great deal of misunderstanding and uncertainty during the past few years as to the status of Darwin's theory of evolution. One writer declares his unconditioned acceptance of the doctrine; another states that Darwinism is dead; another proclaims evolution to be the only logical attempt to explain creation, and at the same instant still another assigns it to the limbo of passé and forgotten things. What, then, is the truth of the matter? Is the evolutionary concept losing favor?

The scientist's point of view differs from that held by many laymen. It is a fact that the Darwinian concept of evolution is being severely criticized these days by both scientists and laymen; but when these groups voice their objections, they have two entirely different ideas in mind.

By criticism the scientist does not mean to express a loss of faith and belief in the evolutionary process in any sense; indeed, the acceptance of the doctrine by scientific workers has been so general and compelling that today it would be difficult to name a single person of recognized standing in the whole scientific world who is not an adherent of the idea of creation by progressive change. A few years ago at a meeting of the American Association for the Advancement of Science, attended by thousands of the most eminent scientists from all parts of the United States and Canada, a resolution was drawn up and *unanimously* adopted to the effect that "the theory of evolution is one of the most valuable and fruitful of all scientific generalizations and is therefore accepted as valid."

The scientist does not mean to express a loss of faith in the validity of the concept when he criticizes the Darwinian theory,

257

but rather to state his disbelief in certain phases of Darwin's explanation as to *how the process goes on*. On the other hand, when the layman reëchoes the statement that "Darwinism is dead," he means an entirely different thing. What he has in mind (especially if he is untrained scientifically) is the idea that the whole concept of evolution has been disproved and discarded. Nothing could be farther from the truth.

An excellent example of the difference between the two points of view is furnished by the late Dr. William Bateson's address at Toronto, Canada, in 1921. The occasion was a meeting of eminent scientists, to which Dr. Bateson, then one of the leading scientists of England, had been given a special invitation as a speaker. He chose for his subject a sort of critique of the Darwinian theory. Dr. Bateson himself was a thoroughgoing evolutionist and had not the slightest intention of attacking the general validity of the concept of evolution. On the contrary, what he did have in mind was to present a frank, dispassionate, and unbiased discussion of Darwin's explanation of the evolutionary process and to point out what he considered to be its weaknesses and defects in the light of post-Darwinian discoveries. His address was to fellow scientists who understood thoroughly both his purpose and the import of what he said. Contrast with this the layman's complete misunderstanding of the meaning of Bateson's address as revealed by the newspaper report the following morning. It was issued as a signed letter bearing the caption "The Collapse of Darwinism." The following abstract is taken from Dr. Newman:

To an audience rarely paralleled in Canada for scientific eminence and influence, the famous Professor Bateson, with amazing frankness, removed one by one the props that have been considered the very pillars of Darwinism. A scientist of international repute, one of the leading if not *the* leading evolutionist, of the day, he exposed the weakness of many of the leading planks in *The Origin of Species*, and ruthlessly tore down one by one the once fondly believed links in the great chain of Darwinian evolution.[1]

[1] H. H. Newman, *The Gist of Evolution.* By permission of The Macmillan Company, publishers.

This address of Dr. Bateson's was hailed with delight by the anti-evolutionists, and, as Dr. Newman further says, "To the layman this statement seems to mean that a leader among the evolutionists has repudiated his evolutionary faith and has gone back to the traditional doctrine of special creation." But as the same writer correctly declares:

There is no more likelihood that any evolutionist will or can revert to the doctrine of special creation than that any geologist will revert to the biblical idea of a flat earth or that any astronomer will revert to the biblical idea of a stationary earth in the center of the universe with the sun, moon and stars revolving about it.[1]

But, to return to Darwin. What are the postulates which he made that have become unacceptable today? The reader will recall that the central ideas of his theory were a variation, or difference, among living things; a survival of the fittest of these variants due to a rigorous struggle; and, finally, the passing on to the offspring of these fittest characters through heredity. It is Darwin's theories as to the cause and nature of variation and heredity that have been most vulnerable to the attacks of his critics. Indeed, their shafts have proved so deadly as to enable one to say, without putting the case too strongly, that at these points Darwin's ideas have been completely overthrown. It will be of interest to see how this has come about.

Darwin believed in continuous variations. Naturally when Darwin made variation one of the five factors underlying the evolutionary process, it devolved upon him to explain how these differences occur. He found this most difficult. About the only explanation he could offer was that of the effect of the environment. For instance, if two kernels from the same ear of a pure strain of corn be planted, one in very sterile soil and the other in very fertile soil, the mature plants will present a decidedly different appearance. The well-nourished stalk may be two or three feet taller than the other, its blades may be longer and wider, and its ear may be several times as large as

[1] H. H. Newman, *The Gist of Evolution*. By permission of The Macmillan Company, publishers.

the nubbin produced by the less-favored plant. Now, as Darwin believed, natural selection worked upon these fluctuating environmental and other variations to preserve the more robust and successful individual.

Subsequent investigations, however, have tended to discredit Darwin's point of view as to how permanent, heritable variations arise. One instance will suffice for illustration. Johannsen, a Danish botanist, chose the common garden bean as good material to use in testing out the hypothesis that small environmental variations may be transmitted and become cumulative so as to change permanently the character of the offspring. The results obtained from beans ought to be reliable because these plants are closely self-fertilized, so there could be little danger of pollination from other sources. Johannsen chose the pure-line variety known as the Princess and began by selecting the smallest and largest beans that grew on a single plant. The average weight of the small beans was sixty centigrams; that of the large beans was seventy centigrams. He grew these two lines of beans separately for six successive years, selecting for seed each season the lightest beans that grew on the plants selected for low weight, and the heaviest beans that were produced by the plants selected for high weight. On the sixth year he computed the average weight of all the beans produced by the plants that had been grown successively year after year from the lighter seed and the average weight of all the beans produced by the plants that had been similarly grown from the heavier seeds. To his surprise the average weight of the beans from the lighter line was sixty-nine centigrams, while the average of those from the heavier line was sixty-eight centigrams. From this experiment Johannsen decided that such fluctuating variations found within a pure line are not inherited and therefore cannot change the character of the offspring permanently. Other investigators have corroborated Johannsen's findings by working with other plants.

These results, as can be clearly seen, cut the props from under Darwin's theory of variations completely, and it fell flat. To be fair to Darwin, however, we must remember that he himself was never fully convinced of the truth of his position regarding the matter. He pondered and studied long and hard on this baffling problem. In 1881, only a year before his death, he wrote to Professor Semper, a German scientific friend, regarding variation: "It is a most perplexing subject. I wish I were not so old, and had more strength, for I see lines of research to follow." So we must remember, then, that this theory of the heritability of fluctuating variations which has since been exploded, was not a settled conviction with Darwin; but, driven into a corner by the problem, it was the only explanation he could offer. From this situation his theory of variation has come to be considered an integral part of his doctrine of evolution.

George H. Shull

FIG. 111. Hugo De Vries, the Proponent of the Mutation Theory

Darwin's theory of variation has been superseded by that of variation through mutation. During the latter part of the nineteenth century, then, biologists everywhere were becoming skeptical of Darwin's explanation of variation. Darwin himself had recognized what he called "sports," or "saltations" (jumps), among plants and animals. These were individuals that for some reason, unexplained at that time, were more or less strikingly different from their parents. Dr. Bateson, with whom the reader is already acquainted, compiled such a list of them that he finally wrote a book on the subject which he called *Materials for the Study of Variation*.

About that time Hugo de Vries, director of the noted Botanical Garden at Amsterdam, Holland, became convinced that Darwin's point of view was wrong. He conceived the idea that variations which count in evolution arrive not by an accumulation of small environmental differences, but, like Athena, spring suddenly by a single bound from their parental type. Quantitatively the difference between this new form and the parent might be marked or it might be very small.

But when the change once occurred, the new type was permanent and was transmitted by heredity. It might vary in response to slightly different environmental influences too, but, without a reverse jump, it could never go back to the parental kind again. These variant forms came to be known as *mutants* (Latin *mutare*, "to change"). With this idea as a hypothesis De Vries began a systematic search for examples of plants in nature that were throwing out these new species. His industry led him to transplant several hundred specimens from the wilds into his garden in order that he might study them the more closely. Finally, in a field in the vicinity of Amsterdam he found an evening primrose (later known as Lamarck's evening primrose) that afforded him the very specimens for which he had been looking. De Vries said in describing it:

Lamarck's evening-primrose is a stately plant, with a stout stem, attaining often a height of 1.6 meters and more. When not crowded the main stem is surrounded by a large circle of smaller branches, growing upwards from its base so as often to form a dense brush. . . . The flowers are large and of a bright yellow color, attracting immediate attention, even from a distance. They open towards evening, as the name indicates, and are pollinated by humble-bees and moths.[1]

This primrose was first discovered in 1886. It was in a highly plastic condition with regard to producing mutants, and De Vries segregated a number of these derived species that bred true to form. For instance, one new species had a smooth leaf as contrasted with the more or less crinkly leaf of the parent stock; one had red veins and streaks in the seed pod; one was an especially heavy plant with dense, dark-green leaves; another was a little dwarf scarcely half as large as the parent form but producing the original-sized flower. Altogether De Vries discovered a dozen or more of these new mutant forms.

Subsequent investigation has cast some doubt as to whether all these so-called evening-primrose mutants of De Vries are

[1] Hugo De Vries, *Species and Varieties — Their Origin by Mutation*, pp. 523–524.

really mutants or not. The basis upon which these objections are made is too technical for us to consider here. But whether they were mutants or not does not matter much; for, by abundant examples of true-breeding variables from other plants and animals about which there can be no question as to the germinal purity of the parents, the theory of mutations has been thoroughly established. Dr. Morgan, formerly of Columbia University, and his coworkers have experimented in the field of heredity for years with a little fruit fly called *Drosophila*. Perhaps the reader has often seen its kind gathered about stale bananas or grapes, espe-

FIG. 112. A Mutation of the Evening Primrose discovered by Lamarck[1]

cially in the fall of the year. The original species of this fly used by Dr. Morgan has red eyes, a gray body, and long wings. Under his care *Drosophila* has thrown off a long list of mutants. Flies with white eyes, flies with yellow bodies, flies with mere vestiges of wings too short to fly, and flies with just a little bar of eye facets instead of a complete eye have been obtained. Most important of all, these forms have all bred true to type. From all sources, according to Walter, more than four hundred mutants from *Drosophila* have been reported.

Mutants are also abundant among other animals. In Massachusetts, in 1791, Seth Wright, a farmer, found among his flock a sheep with short legs and a sagging back. Recognizing at once the disadvantage for jumping fences which

[1] From W. E. Castle and others, *Heredity and Eugenics*. By permission of The University of Chicago Press, publishers.
2

nature had inflicted upon this short-legged sheep, its owner preserved it. In time it became the ancestor of the famous

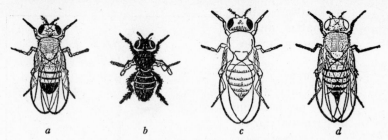

FIG. 113. Mutants of *Drosophila*, much Enlarged

a, wild type; *b*, black, vestigial wing; *c*, yellow body and red eye; *d*, gray body and white eye [1]

Ancon breed. In 1889 a breeder of Hereford cattle at Atchison, Kansas, discovered among his herd one animal without horns. Under his care this mutant became the progenitor of the well-known polled Herefords, which, shorn of their horns by nature, live together about the corrals and feed yards in bovine peace and amity. Mule-footed hogs with solid, uncloven hoofs have been obtained in the same way. White robins, spotted Negroes, pacing horses, pink-eyed guinea pigs, and waltzing mice are other examples of nature's versatility in this respect. Indeed, a complete list of known mutants would probably run into the thousands. It is now believed that most of our distinct, unhybridized forms of cattle, hogs, poultry, horses, pigeons, dogs, cats, garden vegetables, flowers, field crops, and horticultural fruits have been derived in this manner.

Just what causes these mutations to be thrown out by nature is not known. Certain factor changes and chromosomal aberrations are often associated with them. (The chromosomes, to be studied later, are tiny elongated bodies within the cells that contain the hereditary factors.) But just why the chromosomes should behave in this unusual manner is unknown. To some extent mutation is known to be influenced by external forces. For example, Dr. Muller of the University

[1] From T. H. Morgan's *Physical Basis of Heredity*. By permission of J. B. Lippincott Company, publishers.

of Texas has discovered that mutation in *Drosophila* can be speeded up one hundred and fifty times as fast as it normally occurs in this species by treating the parents with X rays while the germ cells are developing. But fruit flies in their original state, so far as we know, are not subject to X-ray irradiation. Still mutants are evolved. Since Dr. Muller's work it has been suggested that X rays emitted by the natural decomposition of uranium salts in the earth might be an active agent of mutation. This, however, must be taken only as a long-range guess. It remains for some energetic biologist with funds to go to a corner of the earth, such as is reported for northern Canada, where uranium disintegration is particularly heavy, to try the theory out.

Darwin's theory of pangenesis has also been overthrown. It is perfectly obvious that individual mutant plants and animals must be capable of transmitting the new, favorable characteristics to their offspring if evolution is to get anywhere. Otherwise each new variable character would die with its possessor, and the racial stock stand still. For this reason Darwin recognized heredity as one of his five evolutionary factors.

But how did hereditary transfer occur? In Darwin's day no one knew. So he advanced a hypothesis to explain it. He conceived all parts of the animal body — hands, feet, legs, arms, trunk, and head, or, as he said, "each cell or atom of tissue" — to be throwing off little microscopic "buds" which he called *gemmules*. These, he believed, were caught up by the blood stream and transported to the developing eggs or sperms in the body of the parent. Taking up their abode in these reproductive cells, the gemmule complements of both parents were brought together in the fertilized egg. These tiny bodies now became active, and as the embryo developed they stamped the characteristics of the parents upon the offspring. Darwin very appropriately called his hypothesis the theory of pangenesis (Greek *pan + genesis*, "growing from all parts"). He advanced it as a provisional theory only, but he came to have great faith that it would prove true. In a letter to his friend Asa Gray, dean of American botanists, he said, referring to a published article that Gray had written, "You give an excellent idea of Pangenesis — an infant cherished by few as yet, except his tender parent, but which will live a long life." [1]

But, unfortunately, Darwin's faith in his "infant," like the hopes of some other fond parents, was shortly to be dashed to the ground. His cousin, Francis Galton, who was also a devoted scientist, proposed to put the theory of pangenesis to the test. He reasoned after this fashion : if the blood of one animal is full of pangens gathered from all parts of the body, then, when that blood is transfused into the body of another animal, it ought to carry these pangens along with it. Consequently an offspring begotten in the body of the second animal immediately after transfusion should produce young showing the characteristics of the individual from which the transfused blood came. So Galton took, for example, a white female rabbit and transfused into its veins a considerable quantity of blood from the body of a large gray rabbit. Immediately thereafter the white rabbit was mated to an albino male, and the results of the experiment were awaited with eager anticipation. In due time a litter of young was produced, and every one of them was pure white, showing not the least sign of likeness to the gray rabbit.

Here, then, was another fallacy exploded by a single experimental result. Galton's method may have been open to criticism at some points, but its procedure was so striking and the results appealed so strongly to the imagination that the theory of pangenesis was never able to make much headway thereafter.

Subsequent to the statement of Darwin's theory, chromosomes, of which we shall hear much in later chapters, were discovered. Moreover, it has been established that these tiny chromatin bodies in each egg and sperm are the carriers of the hereditary factors instead of the pangens, as Darwin believed. For these reasons Darwin's theory of heredity — not the fact of heredity — has gone into the discard along with his idea as to how variations arise.

The efficient rôle of natural selection attributed by Darwin is being questioned these days. Darwin considered variations to

[1] Francis Darwin, *Life and Letters of Charles Darwin*, Vol. II, p. 266.

occur purely fortuitously. There was no rime nor reason to them. They might be better fitted to survive than their parents were, or they might be less fitted to succeed. Still others might be unlike the parents in some respects, but might be neither better nor more poorly equipped to live than they. Now upon this heterogeneous mass of organisms the forces of the environment made their selection. The least well-fitted would perish readily, those with intermediate chances of success would live longer and perhaps produce a few offspring, and the best-fitted would thrive and leave a full quota of progeny.

From Darwin's standpoint natural selection was a rather heartless sorting process. He warns us that animals and plants probably do not suffer so much in the sorting as men would, but even here there is suffering and pain. Cats compete with mice by eating them, hawks catch rodents, and the mountain lion makes his dinner off the fawn upon which he has pounced from the thicket. When we come to man there is also suffering and anguish, whether he is destroyed by a wild beast, whether he is killed off by a warring tribe, or whether he succumbs in a modern hospital to the onslaught of disease germs. Tennyson, in contemplating the whole question of evolution, cried out that Nature is "red in tooth and claw."

Many modern students of evolution (though not all, by any means) have come to question seriously whether this bloody selection as pictured by Darwin is as essential to evolution as he thought. That there are tragedies in nature everyone will admit. That the sorting effect of these tragedies has some bearing on organic change, too, cannot be gainsaid. When pleasure-seeking man shot the last passenger pigeon outside captivity in North America, to that extent he changed the life here. There can be no doubt about that. But as to whether these fortuitous tragedies found everywhere are the main essential factors in evolution, doubt can well be raised. They do exert a sorting influence, but they do not determine whether certain individuals that arise shall be fitted or not. Whether, then, evolution could not and would not go on without all this carnage, as necessarily implied in Darwin's point of view, is an open question. Some believe that it could; many believe that it could not; and still others are quite undecided.

Another objection that is being brought against the Darwinian theory of natural selection is that of probability: whether it is probable that mere chance and fortuitous variations could ever have brought together in one organism all the correlated parts and functions necessary for its success in a complicated environment. This objection really dates back to the time of Darwin himself, but it still continues to be urged against the theory with conviction and force. Let us take, for example, the blood system of man. It is very important and quite essential that the blood stream be kept at approximately the same pressure and corpuscular content. If it is not, the blood stream slows down, the food supplied to the tissues is insufficient, waste accumulates, acid conditions arise, and the oxygen carried to the cells is entirely inadequate. To prevent this, nature has devised a fine system of remedial correlations to be brought into play in case of an emergency.

The pressure may be kept up by quick clotting to prevent loss of blood after injury. Gray and Lint, working at Harvard, have found that the coagulation time of an experimental animal, before blood was withdrawn, was about seven minutes. Thirteen per cent of the blood was then withdrawn, and the clotting time as a protective measure was found to have been reduced to about two and a half minutes. When 10 per cent more of the blood was taken, the coagulation time was cut to about one minute.

The danger from hemorrhage is also safeguarded by the action of the spleen. The spleen is an organ lying back of the stomach and serves as a reservoir of blood in which the red corpuscles are much concentrated. When hemorrhage occurs, as from a wound, the spleen contracts, forcing its corpuscular content into the blood stream to help to compensate for the lost blood and to keep up the oxygen-carrying capacity of the circulatory system.

After injury the blood vessels are also found to contract automatically to raise the pressure of the blood and keep it coursing through the body at its proper rate. But this adjustment is at best only a quick, temporary expedient and cannot meet the needs of the organism very long. As a more effective adjustment the blood stream draws in water from the lymph in the tissue spaces. This passes through the thin capillary walls and tends at once to restore the normal pressure within the vessels by increasing the volume of blood. Incidentally, it is doubtless the copious withdrawal of water from the tissues that causes men wounded during a battle to suffer so acutely from thirst.

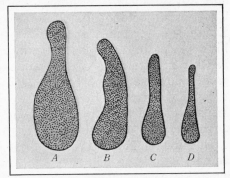

FIG. 114. Cat Spleen, showing Contraction after Hemorrhage

After Barcroft

Now, say the opponents of Darwinism, what is the probability that all this delicate and complicated chain of correlations, controlled largely by the autonomic nervous system, could have been organized and perfected by the forces of blind chance working alone? To believe this, they continue, throws too great a strain on one's credulity. But there are many other organs and systems in the body whose differentiations and correlations are just as complicated and delicate as the blood stream, or even more so. Are these the result of mere chance, too? The eye with its intricate adjustments, the ear, the kidneys, the liver, the heart, and the central nervous system are all familiar examples. These objectors refuse to believe that purely fortuitous variations acted upon by the sorting forces of survival could ever result in such profound effects of correlation as are found in higher plants and animals.

But practically all these same individuals are thorough believers in organic evolution. They say frankly that the evidences which exist about us every day — the witness of the rocks, embryonic changes, vestigial organs, homologous members, the variety of types, breeds of plants and animals, blood affinities, and a score of other evidences — all lead unmistakably to the conclusion that evolution is true. But if it is true, what brings about the proper variations for survival and advancement? Candidly, these people do not know. They frankly admit it. They accept with all their heart Darwin's central idea of evolution, but they cannot embrace his explanations as to just how nature achieved these results. Neither are students of the problem of evolution to be regarded as queer or unreasonable because they believe even though conclusive evidence as to method may be lacking. We accept a thousand things as fact every day that we cannot explain as to their intricate processes.

But, like all good scientists, these individuals have theories and hypotheses which they tentatively advance to explain evolution. We shall examine one or two of the leading ones.

Orthogenesis. As Burlingame and his associates have said, evolution may be likened to the growth of a tree in which the tips of the branches represent the species of plants and animals now found living. Just as there are innate powers within the tree to push the branches upward and outward in definite directions, so it is believed by the adherents of orthogenesis (right growing) that there are within living things definite forces which of themselves, or in combination with external factors, stimulate change and development along specific lines. For instance, they would say that the concurrent processes of shortening the elephant's neck and the extension of his upper lip and nose into a prehensile, food-securing proboscis, and the lengthening of the giraffe's forelegs and the elongation of his neck to reach the ground, furnish good examples. Many paleontologists, especially, hold this view. It may be seen at once that if orthogenesis were true it would make possible the elimination of much of the blind, fortuitous variation and seemingly cruel selection as advocated by Darwin.

Emergent evolution. More recently a new theory — the essential idea is known by different names, but we may call it *emergent evolution* — has been proposed. Jan Smuts, the eminent South-African statesman and patron of science, has written a book on the subject which he has called *Holism and Evolution.* Under Darwin's mechanistic theory the potentiality to vary was inherent in every living thing from the beginning. The selective forces of nature, then, as the reader will recall, acted ruthlessly upon this great mass of variations, some of which might be advantageous, others disadvantageous, and still others neutral. In this way the disadvantageous variations were killed off and the best-fitted individuals were preserved. From simple beginnings, then, through the survival of the fittest, nature has produced the long lines of divergent and progressive organisms which paleontology and morphology have shown to have inhabited the earth in the past or to be extant today.

The emergent evolutionists take an entirely different point of view regarding the whole process. They believe that as nature progressed in the production of new forms new potentialities also were added; that is, that plants and animals at different stages of their evolutionary changes took on entirely new powers and possibilities which they did not possess at the outset. These new potentialities were supervenient and enabled the organisms to go on to greater heights of development than would have been at all possible to the old qualities which they at first possessed. In other words, new powers had *emerged*; hence the term "emergent evolution."

These emerging factors are thought not only to have supervened in the case of organic things but to have been acquired by nonliving material as well. For example, hydrogen and oxygen are definite gases with well-defined specific characters. Combine them in the proper proportions, and water results. This new compound is now an entirely different substance. It has acquired new properties and potentialities not at all inherent in the uncombined gases. In short, *emergence* has taken place. Combine sodium and chlorine, and table salt, with new emergent powers, follows. Mix sulfur, charcoal, and saltpeter properly, and gunpowder is produced. Certainly, claim the advocates of this theory, the explosive properties of this new mixture were not in any sense inherent in the separate ingredients when standing alone.

As to just how emergent factors are added among living things is, of course, a theoretical matter. It has been suggested that mutations furnish one avenue. The causal factors which lead parental lines to throw out these mutants, as has already

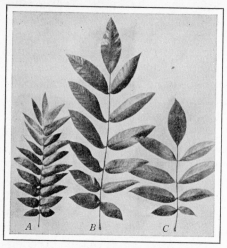

FIG. 115. Leaves of the Paradox Walnut and those of its Ancestors

A, common English walnut; *B*, the new hybrid; *C*, native California black walnut [1]

been made plain, are obscure. Yet they do occur. Along with them may emerge new potentialities. It has been suggested that the mental level of man, with no counterpart on the earth anywhere, may have emerged in this way. It has also been proposed that hybridization (the mating of two unlike parents) may furnish the mechanism through which emergence may take place. It is a well-known fact among breeders of plants and animals that hybrids (the offspring of unlike parents) often do possess characteristics entirely different from either of the parents when taken alone. Luther Burbank crossed an English walnut with a common California black walnut and for several generations selected the best of the progeny. In this way he developed a robust walnut which he called the Paradox. Its leaf character is very different from that of either of the parents, and it possesses remarkable, rapid-growing characteristics. Some of these trees were planted in hard earth along the street in front of Mr. Burbank's house. In fourteen years they had reached a height of almost eighty feet with a branch spread of seventy-five feet, although they had received neither cultivation nor irrigation.

As to whether there really are such things as emergent potentialities or not is a disputed question. Many biologists

[1] From W. S. Harwood's *New Creations in Plant Life.* By permission of The Macmillan Company, publishers.

today, perhaps the majority, are adherents of neither the orthogenic theory of evolution nor the emergent theory. It may not be going too far to say that they more nearly subscribe to the old Darwinian mechanistic view of natural selection than any other theory. Yet it must be remembered that Darwin's idea of how new forms—favorable variations—arise has been discredited. His selective agencies may in the end prove to be true. They certainly do eliminate those individuals "that are not good enough to live." But the problem as to how the fit arise has been answered only theoretically. None of the theories advanced up to the present may prove to be sound.

We have here, then, a strange situation. Practically every reputable biologist believes in the fact of evolution, yet not one of them is able to explain just how the necessary new forms arise. Dr. Nabours says, with reference to this problem, "We are in a morass"; and Dr. Osborn says: *"We are more at a loss than ever before to understand the causes of evolution. We await the man of genius to discover the causes . . . according to an entirely new concept."* [1]

Many people believe evolution and a religious faith to be antagonistic. Apparently many people believe the doctrine of evolution to be atheistic and irreligious. However, this does not necessarily follow. The reader should remember that even Darwin himself did not believe acceptance of the evolutionary idea to be incompatible with a religious faith. In a letter to an inquiring German student two years before his death, Darwin, who was then ill, had a member of his family write, "He considers that the theory of evolution is quite compatible with the belief in a God."

Much confusion seems to have risen over the habit of mind that puts religious faith and belief and the findings of science into two different air-tight — perhaps it would be more appropriate to say thought-tight — compartments and then tries to harmonize them. Certainly a sound philosophy and religious faith can never come out of such practice. Why should the

---

[1] H. F. Osborn, "The Nine Principles of Evolution Revealed by Paleontology," *The American Naturalist*, Vol. 66, p. 60.

full-blown rose, the birds in the trees, the beasts in the field, and the stately oaks standing in the forest not be considered to be as much a part of God's world as the subjects of which the Bible treats? The Psalmist reveals the Lord as saying:

> For every beast of the forest is mine,
> And the cattle upon a thousand hills.
>
> I know all the fowls of the mountains:
> And the wild beasts of the field are mine.

If this conception of the universe were kept in mind, it would obviate much strife and confusion. The scientist can make no distinction between the natural and the so-called supernatural. What man can study, experience, and learn about through his senses is the natural; the supernatural is that part of the universe which he has not yet been able to understand or for which his powers of comprehension are too limited. There is no difference between the two. The difference comes only in man himself.

Again the Bible must be taken as a book of religious instruction and not as a textbook of science. It was written by men who meditated upon the ethical and spiritual side of human nature — upon man's duties, his relation to the universe, and his destiny. If it involved science, the sacred writer set down what the men of his day believed and taught. He was not much concerned as to whether or not it expressed technical, scientific fact. He was concerned with teaching great ethical principles and spiritual truths. As a result, sometimes the science involved was true and sometimes it could not be substantiated experimentally. When Galileo was one time charged with teaching a dangerous and damnable heresy directly opposed to the authority of the Scriptures, he is reputed to have replied, "The Bible was given to tell men how to go to heaven, and not how the heavens go." That was a wise answer. As Conklin says, "This answer and all that it implies, if once accepted and believed, would go far to quiet the age-long controversy between science and theology."[1] The Bible must

[1] E. G. Conklin, *The Direction of Human Evolution.*

take first rank as a book of ethics and as a book that reveals man's evolving knowledge and sense of God; but it is not a reliable source of scientific truth.

Finally, it must be remembered that the theory of evolution does not attempt to say when, why, or by whom life was first produced upon the earth. The honest scientist when pressed for an answer will say candidly that he does not know. Evolution is not concerned primarily with the origin of life. It is concerned chiefly with the orderly course which life has followed since it appeared on the earth and with the laws and forces which govern its development. In other words, evolution is a *method of creation*. The Bible says that "God formed man of the dust of the ground." It does not attempt to say just how this creative act was done. That problem, on the other hand, is the very one with which evolution deals. It attempts to point out the long series of transitions from lower to higher forms, ending in man. It also tries to point out, wherever possible, the evidences of such transmutations and to assign factors whose interactions produced them.

So, then, since evolution neither denies the existence of God nor disclaims His directive influence over natural processes, it cannot be said to be necessarily atheistic. Some evolutionists are atheists, to be sure. On the other hand, many evolutionists are devout men. Those who have no confidence in the idea of a Divine control over nature may be disbelievers; but they are not disbelievers because of evolution, but because of other considerations.

QUESTIONS FOR STUDY

1. What postulates made by Darwin in his theory of evolution are not accepted today?

2. Make an outline of the several theories of variation that have been advanced.

3. What are mutations, and how are they believed to arise?

4. What is De Vries's theory of mutation, and what evidence do we have that it is true?

5. What was Darwin's idea of natural selection?

6. Explain the theory of orthogenesis.

7. Give the principal ideas involved in emergent evolution.

8. The sudden appearance of a white rabbit that bred true among a group of pure-line black rabbits would illustrate what?

9. In breeding *Drosophila* what results has Dr. Morgan obtained that support the principle of mutation?

10. Why is the effect which Darwin attributed to natural selection being questioned today?

11. Why does the theory of evolution not necessarily conflict with religious belief?

12. If the evolutionist cannot explain the original source of life, what does he attempt to say regarding it?

REFERENCES

CANNON, W. B. The Wisdom of the Body, pp. 27–60.

CONKLIN, E. G. The Direction of Human Evolution, pp. 155–247.

DARWIN, CHARLES. Origin of Species, pp. 219–261.

DARWIN, FRANCIS. Life and Letters of Charles Darwin, Vols. I and II.

DE VRIES, HUGO. Species and Varieties — Their Origin by Mutation, pp. 1–31, 516–575.

HERBERT, S. The First Principles of Evolution, pp. 172–224.

MASON, FRANCES, and others. Creation by Evolution, pp. 355–371.

NABOURS, R. K. "A Third Alternative: Emergent Evolution," *Scientific Monthly*, Vol. 31, p. 453.

NABOURS, R. K. "Emergent Evolution and Hybridism," *Scientific Monthly*, Vol. 71, p. 372.

NEWMAN, H. H. Readings in Evolution, Genetics, and Eugenics, pp. 237–314.

NEWMAN, H. H. The Gist of Evolution.

OSBORN, H. F. "The Nine Principles of Evolution Revealed by Paleontology," *The American Naturalist*, Vol. 66, p. 52.

SMUTS, JAN C. Holism and Evolution, pp. 85–223.

WELLS, H. G., HUXLEY, J. S., and WELLS, G. P. The Science of Life, Vol. II, pp. 600–643.

UNIT IV

The Adaptations of Plants and Animals determine their Habitat

TO THE casual observer plants and animals appear to be scattered promiscuously over the surface of the earth. There seems to be no reason, for the most part, why they grow where they do except that they have been able to gain a foothold. In this situation, however, general appearances fall far short of revealing the true picture.

Plants and animals are not scattered fortuitously over the earth. Bulrushes cannot grow on a high hill nor can buffalo grass grow in a swamp. Neither can an antelope thrive in the forest nor a bear find sustenance and shelter in the wide, open stretches.

Through ages of competition and struggle among themselves organisms have become adapted to live in particular habitats. There they can get food, shelter, water, and protection. There they thrive and succeed. It is this distribution of living organisms growing out of their competition with one another and their struggle with the forces of the environment that is to be studied in this unit.

CHAPTER IX · The Reaction between Organisms and the Factors of their Environment determines their Distribution

Introduction. Plants and animals literally fill the earth. The shallow seas teem with life; the fresh-water lakes abound with algæ and crustacea; the plains are covered with grasses and with herbivorous animals that get their sustenance there; the mountains are clothed with forests which provide shelter for deer and for the carnivora that prey upon them; the shallow soil of the tundras produces its lichens, quick-growing herbage, and arctic fox; even the deserts have their quota of insects, drought-resisting plants, and lizards. Wherever one goes, from the equator to the pole or from the lowest plain to the heights of the loftiest mountain, life abounds. This general question of where plants and animals live and why they live there is a most engaging one. Space will not permit us to present the matter in detail, but in this chapter we shall take a sort of bird's-eye view of the problem.

The distribution of man over the earth is determined by the environment. Human beings are not distributed over the earth in an accidental manner. Since man is found almost everywhere, at first glance one is likely to jump to the conclusion that his distribution is a matter of pure chance. Such, however, is not the case. There are definite reasons why he lives where he does. The Ethiopian, with his black skin, lives under the tropical sun and subsists largely on wild fruits and game; the Eskimo prefers the icy breath of the polar regions, where he travels by dog train and makes his dinner of blubber; the Caucasian is found in the temperate regions, where he is more comfortable and where the climate has been conducive to the establishment of an enterprising and versatile civilization.

2

Even within a single country, to be more specific, the people are not scattered about in a haphazard manner. They are distributed according to their needs and the opportunities to live successfully. The fisherman lives along the coast, where he may ply his trade; the rancher prefers the plains, where he may grow crops and raise herds on a grand scale; the lumberjack migrates to the forest; and the miner takes up his abode in mountainous regions where ores are found. The banker establishes his business not at a country crossroads, but in the city, where there are people to become his patrons. For the same reason the doctor and the lawyer hang out their shingles in populous districts.

Thus in human affairs we say that economic forces largely determine where people live and what they do. They must inhabit a region where they will be comfortable; they must make a living and support their families. Therefore they either go where they are acclimated and can follow their accustomed vocation, or they change their vocation to conform to the region in which they live. Aside from the strictly climatic factor, then, economic forces control to a very large extent the distribution of people. The very word "economics" is derived from the Greek word *oikos*, which means "home."

Plants and animals are definitely related to their habitat. Among plants and animals of the wild we find the very same conditions existing as among people. Each species is adapted, or is in the process of becoming adapted, to live successfully where it is found. The lowly earthworm inhabits the soil, where it can dig its burrow and eat decomposing leaves; the oyster lives attached to a rock in the sea, where the water currents bring it food; the antelope prefers the plains, which furnish it forage and over which it may bound away from danger unimpeded; the monkey likes the forest, which provides it with nuts and where it may nimbly escape its enemies; and the polar bear splashes about in the cold arctic waters, fattening on the fish and seal of these frigid regions.

In a very definite sense, then, plants and animals are disseminated over the earth in conformity to the same laws as those that control the distribution of peoples. Organisms are found where the climate permits them to thrive and where they can find food and shelter. Here, again, aside from the climatic factor, natural "economic" forces control them. For this reason the science which treats of the relations of plants and animals to the conditions in which they live is called "ecology" (Greek *oikos* + *logos*, "the science of home"). Then what we are really going to study in this chapter is ecology.

Plants and animals are believed to have migrated from definite points of origin. The evidence obtainable indicates that new species of plants and animals appear at some definite point and then spread out from that location as a center. The camel, it will be recalled, is thought to have originated in North America and then migrated south to South America and also northward over the land bridge to Eurasia. The elephant began with a little long-nosed ancestor in northern Africa, from whence it is believed to have spread to Asia and Europe. It subsequently crossed over the connecting land to North America, where, after an extensive development, it became extinct. The horse has had a similar history except that, like the camel, it seems to have arisen in North America. Plants, like animals, show the same tendency as to origin and migration. The giant Sequoias of California are but the cramped and gradually fading representatives of an illustrious ancestry that during prehistoric times spread far across the polar regions into Greenland and northern Europe. The fossil remains of these plants and animals, as well as many others, tell the story of the origin of living things and their subsequent migrations.

As soon as an organism has appeared at any specific point it begins its process of overproduction. This scatters its progeny outward in all directions. If there are other species in that region, the different species may crowd one another and give further impetus to this outward disseminating movement. Especially is this true of animals. Sometimes as species migrate from their centers of origin they meet midway between these points. When they encounter one another, there often follows such a competition as to lead to pronounced crowding. This may further accelerate spreading.

As a result of these migrations and crowded conditions, plants and animals tend to occupy the whole earth — shore line, mountain, plain, and valley. Moreover, they are pushed back into every nook and corner. The strong, well-fitted representatives, like prosperous people, occupy the best regions, where they flourish and thrive; the weaker organisms, less adapted to succeed under the pressure of sharp competition, like people ill equipped by nature, are forced into less-favored areas. In this way the whole face of the earth becomes occupied with living things. Wherever they can get a foothold they abound.

Within these habitats living organisms struggle for subsistence. Plants and animals can live only where conditions necessary to their particular needs may be found. Plants require soil nutrients, water, warmth, and sunlight. In general, animals must also have water, food, and a favorable temperature. In a sense, however, animals are less the victim of circumstances than are plants. Plants are stationary and must take whatever comes; animals can move about to get the necessities of life. They can also dig a burrow or seek shelter within a forest or behind a cliff. Because of these innate differences between animals and plants in relation to their habitat it will be necessary for us to discuss each one separately.

Plants must obtain light. Among plants in regions of dense vegetation, such as the tropics, the competition for light is most severe. Some plants require bright sunlight to flourish; others do not. For instance, the aspen, willow, bur oak, pigweed, and bean can scarcely stand shading at all. On the other hand, the linden, sugar maple, spruce, and wood nettle can tolerate a great deal of shade. The difference in this respect all depends upon the slow adaptations which the various species have made to this factor of their habitat.

This struggle for light leads to a layered arrangement of plants within a forest. This effect has been studied in one of the forests along the Missouri River, in southeastern Nebraska. The crowns of the red-oak and linden trees form the top story. In this position they are spread out to the exposure of direct sunshine as bright as that falling on the leaves of any cultivated crop. The tree tops are also subject to the action of wind, and altogether their habitat is not much different from that of the surrounding prairie.

Under the dense shade of the forest trees, however, conditions are entirely changed. The upper layer, shaded by the dense crowns of the trees, consists of such shrubs as ironwood, redbud, and a few woody climbers like the greenbrier, woodbine, and poison ivy. Down on the forest floor, screened first by the tree tops and then by the story of shrubs, is the third layer. This story has four fifths or more of the direct sunlight cut off. The plants consist mostly of shade-loving herbs.

This illustration furnishes a concrete example of how plants may respond to the crowding effects of competition. By adapting themselves to different degrees of subdued light both the succulent plants and the shrubs could retire into the otherwise unoccupied space under the forest crown and find a home. Indeed, they have changed the very nature of their demand for light to such an extent that they could no longer thrive if they should be pushed out into the direct sunlight.

In warmer regions with heavier rainfall this layering effect within forests is even more pronounced. The trees are taller and the crowns denser. Great trailing lianas, or climbing plants, creep far up the lofty tree trunks in order to bring their leaves into the light above. Sometimes these audacious climbers grow so luxuriantly as to overrun, weigh down, and kill off the helpless trees that support them. Below the crowns of the trees and lianas may be plants of lesser height that can endure shading. In the crotches of the trees, where bits of dead bark, decaying leaf fragments, and wind-blown débris accumulate as damp humus, the seeds of epiphytes (plants perched upon other plants) find lodgment and grow. Some of these epiphytes have blooms of exquisite beauty. Among these are the finest and most delicately tinted orchids. Down on the dark, perpetually damp forest floor the light is too dim for even the most shade-tolerant herbs. The ground is bare of every green thing, and only the colorless toadstools, mushrooms, and other fungi which do not need light can thrive.

Some plants solve the problem of light by developing early in the spring. As Huxley has said, the early bird catches the worm, but the early plant catches the light. Within the forest there are some plants that solve the light problem — the most pressing one of their existence — in just this manner. They shoot up from among the dead leaves with the first advent of spring. They spread out their foliage and bloom before the trees show scarcely any signs of awakening. Because these vernal flowers are among the first harbingers of spring they are therefore beloved by both man and child. Who of us has not thrilled at the first peeping violet, the nodding Dutchman's-breeches, the hepatica, or the spring lily? In our sentimental moods we may see in these modest flowers nature's signal of a new season. But the scientist, though he too may thrill, sees in these early plants a meaning of much deeper significance: he sees the unmistakable marks of the struggle for existence. These plants bloom early before the trees and shrubs leaf out because they must do this to succeed at all. Examination will show that practically all of them have fleshy roots or underground stems. From the nourishment stored in these storage organs the year before, leaves and flowers develop quickly the following spring. In this way flowers and seeds are produced and a store of food is laid up underground for the next year's growth, all before the trees leaf out and cast their dense shade. Later, cut off from the light by the forest canopy, the work of these vernal flowers for that season is then practically completed. They wither, dry up, and lie more or less dormant until the congenial conditions of the next spring induce a new period of activity.

Plant life often appears in zones. If one were to be elevated to a good vantage point and, from there, carefully observe the vegetation surrounding a shallow inland lake, it would not appear haphazard; on the contrary, it would present more or less definite zones of plant life. Farthest out in the water from the shore line would be found a zone of floating algæ extending completely around the lake. Behind that would occur a zone of rooted but mostly submerged plants, such as water lilies and pondweeds, whose leaves and flowers only

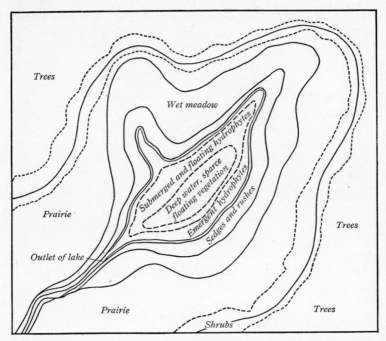

Fig. 116. Diagram showing the Zonation of Plants in a Lake Region of the Middle States

would emerge above the water. This, like the preceding zone and all successive zones, would probably extend ring-fashion around the whole lake. Next in order would come a zone of bulrushes and cat-tails standing in very shallow water or even extending back out of the water onto the low, flat, muddy ground. This zone would in turn be succeeded by one of slough grass, other coarse grasses, and sedges, which gradually give way to drier soil-inhabiting grasses, such as the bluestem. The grasses would also be interspersed with various herbs, such as the gentians and sour dock. In the rear the grassland would likely be followed by a zone of shrubs, which finally would give way to the uplands permanently occupied by forest trees.

If one could imagine a vast pinnacled mountain of solid rock standing for centuries unmolested in a plant-inhabited

Fig. 117. Zonation of Vegetation along the Banks of a Colorado River

Fig. 118. Zones of Grassland, Oak Shrub, and Yellow Pine in Colorado [1]

[1] From Weaver and Clements's *Plant Ecology*. By permission, McGraw-Hill Book Company, Inc., publishers.

region, he would find zonation about it also. After a time spores of lichens would blow in from the outside, and the rocks would become more or less incrusted with these pioneer plants. Farther down, around the base of the mountain, where the rocks had decomposed somewhat and talus had formed, a zone of dry-habitat mosses would very likely be found. These would be succeeded by a belt of drought-resistant grasses and herbs, followed by a zone of shrubs. Lastly, and merging with the

Fig. 119. A Lichen-Incrusted Granite Bowlder

permanently surrounding forest, would come the zone of trees. If the mountain were irregular and portions of talus slopes occurred interspersed throughout the cañons and along the base of rocks, it is perfectly evident that fragments of these zones might be found scattered all over the mountain's sides wherever conditions suitable for talus accumulation and soil-building occurred. It is also obvious that if either a lake or a stony mountain stood in a region where grassland is the permanent form of vegetation rather than forest, then both the shrubby belt and the tree area would probably drop out. The last zone to form in that event would be grassland.

The zones of vegetation surrounding a lake or a stony mountain change. If an observer from the vantage point suggested in the last section should remain to watch the vegetation about a lake or a stony mountain for, let us say, fifty or a hundred years, he would behold striking changes taking place in these zones of plant life. He would find that they are not stationary in any sense, but, on the contrary, are gradually shifting their positions like actors in a play. In fact, the spectator would find, perhaps much to his surprise, that a great drama of

FIG. 120. A Belt of Vegetation growing in Dry, Shallow, Rocky Soil (Blue-Grass Knotweed and Woolly Plantain)[1]

plant life, just as real and vivid as was ever enacted before the footlights, was taking place before his eyes. The tempo and movement of the play would be slow, to be sure; but great, conflicting forces of nature would be involved; competition between the plant communities, as actors, would be as keen and decisive as those in human life; and the outcome of the struggle between these natural components of the habitat would be as real and inevitable as is the outcome of a gripping tragedy upon the stage. Let us analyze this plant drama briefly to see the course it follows.

Reverting to the lake, the zone of algæ grows in great profusion. Its filaments die, sink, and tend to build up the bottom gradually and to make the water shallower. Soon the seeds of the water lilies and pondweeds are carried to this point and find lodgment in the submerged mud. They grow and, competing with the algæ, overtop it, cut off the light, and shade it out. In a few years the decaying débris of the water lilies and pondweeds fills up the lake some more at this zone, making it less favorable for the occupants and more

[1] From Weaver and Clements's *Plant Ecology*. By permission, McGraw-Hill Book Company, Inc., publishers.

conducive to the bulrushes and cat-tails. The seeds of the latter preëmpt this area sooner or later, and the plants which they produce finally overtop and crowd out the water lilies and pondweeds. In the Northern states and Canada, and also in similar latitudes of Eurasia, the cat-tail and rush stages, as well as the early phases of the next zone, are often heavily represented by peat mosses. Indeed, as the lake approaches extinction the central portion of it becomes filled with a peaty deposit. These deposits are often extensive in area and form great, water-soaked, quaking bogs. In time, however, the peat near the surface will decompose sufficiently to permit either grasses or shrubs to take possession. So the drama moves along slowly but inevitably, each plant zone, or community, changing the habitat and preparing it for the occupant's own downfall and displacement. This course continues, each successive zone evicting the one before it, and it in turn being supplanted by the zone behind. Finally the lake may become completely filled up in this manner and its site be occupied by a closed covering of vegetation. The course is so regular and undeviating that ecologists have called the whole process *plant succession*. The last actor to take the stage and triumph over all his precursors is the form of vegetation permanent to that region. If in a moist climate of suitable temperature, it will be a forest of some description; if in a drier climate, it will be grassland. This permanent form of plant life is called the *climax community* because it is the end, or climax, toward which all the successive vegetational changes are proceeding.

If at any time, after this great natural drama is completed, the soil is again denuded of vegetation by any agent, a sort of abbreviated succession takes place to restore the climax community. For instance, if an avalanche plows the vegetation off a mountain side, if a forest fire destroys the timber, or if a farmer breaks up the prairie sod, and these areas are left unmolested, plants from the outside crowd in. In the forest, as in most other cases, the pioneers are usually plants with wind-blown seed that can be easily carried in from the sides. Classical examples are the aspen and the fireweed. The aspen,

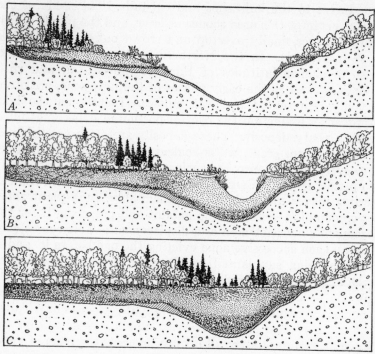

Fig. 121. Diagram showing how Lakes may be filled by Vegetation and how Forest Trees may later occupy the Site (after Dachnowski)[1]

of little commercial value itself, nevertheless is of great value to man in promoting plant succession; the fireweed receives its name from its ability to appear profusely in damp sites, even the first season after a fire.

These pioneer plants shade the ground, lower the temperature, add humus, and increase the moisture. In the coniferous regions this changed soil condition makes it possible for the next stage in the recovery to take place. In the Rocky Mountain region it will probably be the more shade-tolerant lodgepole pines. Once these pines are established and in possession of the field, they shade out and kill off the intolerant aspens and herbs. The pines continue to make the soil cooler

[1] From *Bulletin 16*, Geological Survey of Ohio.

Fig. 122. An Aspen-Covered Area

and moister, and for this reason are in turn succeeded by the Douglas fir, if the altitude is favorable. In the course of many years the Douglas firs, which can stand more shading, become the dominant vegetation, and nature's abbreviated drama, caused by the forest fire, is then complete.

In prairie regions a similar short succession will take place, but the plants involved will, of course, be suited to prairie conditions. The series usually begins with wind-blown weeds and proceeds from these annuals through various kinds of temporary grasses and herbs up to the climax grasses, whatever they may be. In the prairies of the north-central states it is likely to be bluestem and needle grasses; farther west, on the arid plains, it would be grama and buffalo grasses.

H. G. Wells and his associates give us an example of plant succession enacted on a grand scale. This is taking place on the volcanic island of Krakatoa. By three terrific explosions in August, 1883, this island in the East Indies literally blew

Fig. 123.　Lodge-Pole Pine succeeding Aspen in a Burned-Over Area

itself to pieces.　More than four cubic miles of volcanic dust and débris were blown into the air.　The quantity was so great that the dust drifted entirely around the earth.　The European sunsets, even at that great distance, were tinted a deeper red for many months by the volcanic dust floating in the atmosphere.　On the island itself, as well as on two small neighboring islands, every vestige of life, both plant and animal, was destroyed.　The nearest point from which life could come was a couple of small islands about fifteen miles distant. Java and Sumatra were each situated about ten miles farther away.

Less than three years after this cataclysm a Dutch botanist visited the island for purposes of observation.　He found that in this moist climate the ashy soil from shore line to peak was covered with a gelatinous layer of bacteria, blue-green algæ, and diatoms intermingled.　Dotted about in this layer several species of moss and eleven kinds of ferns were found growing.

Fig. 124. A Douglas-Fir Forest near Glen Haven, Colorado

In some places the ferns were profuse. The spores of these plants had all been blown in from the neighboring islands. In addition a few kinds of flowering plants had arrived. These were species with either wind-blown or water-borne seeds, and the individual plants were widely scattered. There were no shrubs or trees.

Ten years later, when the island was visited for the second time, flowering plants were in the ascendant. Fifty species of them had been carried in. In some places the individual clumps of ferns and flowering plants had spread so as to close their ranks and completely occupy the soil. Stretches farther inland were entirely covered with heavy growths of tall grasses. But the forests had not yet arrived.

In 1906, twenty-three years after the explosion, Professor Ernst and a party visited the island, which now bore a closed, luxuriant vegetation. Along the shore a strand-forest had grown up. Among these were several kinds of coconut-palms and figs, many of which bore ripe fruit. Animal life was well represented by species that could either fly or drift in on floating logs. Among these were mosquitoes, ants, wasps, and birds. Fruit bats were also found, as well as some lizards, one specimen of which had reached a length of three feet.

Despite its luxuriance, however, neither the plant nor the animal life had reached its permanent stages. New species were still arriving each year, and Ernst estimated that it would take another fifty years for the tropical vegetation to establish itself in its full, dominant state.

Wherever the permanent vegetation is disturbed by any agent, even in small areas, hints of this abbreviated drama may be detected. A gopher hill thrown up in the prairies will show traces of it for a few years until the tiny mound again becomes sodded with the climax grass. The farmer, who must battle so incessantly in the early part of the season to prevent the weeds from taking his crop, is but defending himself against the initial actors who would make his field the stage-setting for a great plant drama. Once these weeds got started, the play would proceed with dead certainty to the final climax unless man's industry and determination led him to interfere. Farmers often say that abandoned farm land will "go back to prairie," which is true; but in most cases they do not understand the great successional changes involved.

Owing to vegetational succession animal life also may exhibit a succession. It has already been made clear that animal life is dependent upon plant life for its subsistence. It is true that the carnivora prey upon other animals, but their victims, in turn, live upon plants. So, in the end, it all comes back to the plant.

Different animals prefer to live among different kinds of plant communities. This comes about because they vary as to taste and as to the kind of shelter and protection which their needs demand. The antelope and the buffalo like the plains, with their broad expanse and abundance of nutritious grass. The coyote also likes this habitat because a young buffalo calf or a baby antelope makes a fine dinner. If these choice foods happen to be scarce or the mothers too effective in the defense of their young, he can piece out his scanty fare with ground squirrels and other rodents. Quail and rabbits prefer brush and scrubby growth. Here the quail build their nests along the margin, huddle behind a clump during a snow-storm, and pick up seeds and bugs from among the leaves when the snow is off. The rabbit finds juicy herbs, and not only shelter from the elements but a barrier against enemies that would pursue him. Deer like the mountains, with their alternating meadow and forest growth, where they may feed and find protection. The bear also likes the mountain fast-ness. Here he can find a shelter and a mixed diet of flesh, berries, herbs, and honey. The panther and the mountain lion are attracted to the forest because from the lower limbs of a tree or from a clump of shrubs they may easily spring upon the back of the luckless deer as he passes along. So, as has already been stated, the kind of animal life found in any region depends primarily upon the kinds of plants that grow there.

If this be true, then one would naturally expect a shifting of animal life to accompany a successional change in vegeta-tion. Animal ecologists tell us that this is a fact. Whenever vegetational changes occur which make the habitat less con-genial to the animals that live there than it was before, they promptly begin to move. This more agreeable home is usually found among the plants of the next earlier stage of the suc-cession. In this way an accompanying animal succession al-ways follows a series of vegetational changes. About the only fundamental difference between plants and animals in this case is in the fate of the individual. Plants, being rooted to the soil, are immobile and are literally killed off by their successors; animals, on the other hand, are usually not killed off by the changing forces of the environment, but simply migrate into the more agreeable surroundings. To make the picture more complete and vivid, let us briefly follow such a sequence. For purposes of clarity we may go back to the vegetational succession which we traced about the margin of the inland lake as the plant life slowly closed in and filled it up. Let us assume that the lake is fairly extensive and in a wooded region.

2

Animal succession about a large disappearing inland fresh-water lake. The center of the lake may be occupied by fish of different kinds, depending upon its depth and size. If the lake is not too large and is situated in the Northern states, pickerel, muskellunge, and pike may abound. Nearer the shore, in the zone of algæ, millions of crustacea besides many other animals find both food and shelter. Pickerel, bass, and smaller fish upon which the larger feed, may be abundant also. In the pond-lily zone live many small fish, together with crustacea of various kinds. The larger fish, such as the black bass, often pass out into this belt to feed and to secure protection. Near the middle of a hot summer's day a few years ago in central Minnesota, two fishermen quietly and slowly pushed their rowboat in from deeper water through the pond-lily belt in order to observe the life found there. Insects of many kinds were swarming about the gorgeous floating flowers. As the boat edged into the leaf-covered surface, black bass shot out from under the shade of the lily pads into deeper water. While the craft was passing shoreward through this particular zone of vegetation, which was about a hundred feet wide, there was an almost continuous series of disturbances in the water as the larger fish, startled by the approach of the boat, plunged into the deeper water below.

In the bulrush and peaty-moss zone of that lake, which lay inland from the water-lily zone, was in the main another group of animals. Here beautifully colored and variegated damsel flies and dragon flies hovered over the reeds or darted through the air. Occasionally a female dragon fly would be seen to alight on the stem of a reed near the water's edge. Presently she would lower the posterior end of her long, slender body below the water line and, puncturing the stem at this point with her ovipositor, would deposit her eggs. The stems of the reeds six or eight inches above the water line bore many hideous-looking faded brownish nymph skins split down the back. From these perhaps the very same gaudily colored dragon flies that were then swarming through the air had emerged just a few days before. In the water tadpoles and pollywogs were busily hunting food and apparently enjoying the swim. Little open patches of water were full of wriggling mosquito larvæ. Occasionally a water snake could be seen darting through the reeds when startled or, when undisturbed, slowly crawling about through the rushes seeking its prey. At dusk the rippling water sometimes betrayed the presence of the silent but busy muskrat, and once in a while his houses could be found piled high above the water line.

Beyond the water's edge, in the mixture of peat-moss sedges and coarse water grass, were frogs leaping about or warily waiting to lap up any unsuspecting insects that might be crawling along. Here garter snakes could also be found. Farther out, in the dry land grasses, ground rodents of many kinds were abundant, while hawks hovered above ready to pounce upon any of these little creatures that might be caught too far from their burrows. Here at night the coyotes also roamed looking for food.

Farther back, in the brush and timber, rabbits were found and squirrels skipped about through the trees. Birds flitted among the branches. Timber wolves were often found, and an occasional deer could be started.

So, to epitomize, from the lake's center to the deep woods various zones of animal life intervened. In general the type found at any point depended largely upon the kind of vegetation which then occupied that particular belt. The animals, being mobile, had less sharply defined lines of occupancy than did the plants, but that they did tend to live within a specific type of vegetational zone was perfectly obvious. So it is all over the earth. Owing to migration and mobility the separating lines may overlap and be indistinctly defined. But, in a general way, the animal life of any large region may be mapped and charted. This is a profound fact of animal distribution.

Plants and animals are also distributed according to latitude. We have just portrayed life changes that may take place in any given area. Migrant plants and animals move into a denuded region and pass through a series of progressive changes. In the end they reach the stable, climax state which is suited to that particular set of conditions. In this section we shall attempt to point out more clearly what is probably already known in a general way; namely, that temperature changes, often accompanied by a varying moisture supply, profoundly affect the distribution of life from the equatorial belt poleward.

The representative equatorial areas are regions of heavy rainfall and hot, moist atmosphere. Under such conditions vegetation is luxuriant. Heavy forests of broad-leaved trees abound, with dense lianas stretching out through their crowns. In the open spaces are great jungles of bamboo and cane. Along the margins and in the less densely shaded portions of the forest's interior rank growths of ferns, bananas, and other herbaceous plants abound. These tropical forests furnish an ideal home for bats, parrots, humming birds, sloths, armadillos, monkeys, and reptiles.

In the subtropical belts both north and south of the equator, although the rainfall is reduced, the climate is still warm. Here the trees often break their ranks, and the area in between is occupied by grassland and scrub. The result is a beautiful parklike effect called a savannah. In these regions, favorable both for food and for protection, an abundant animal life is found. Zebras and antelopes roam in great herds. Giraffes, elephants, rhinoceroses, peccaries, and wild hogs are represented. Here too, in Africa, the anthropoid baboon finds his home. Along with these herb-loving animals come the flesh-eaters that prey upon them. These are typified by wolves, lions, pumas, tiger cats, hyenas, and wild dogs.

With increase of latitude the world's two most pronounced desert belts are encountered. These are represented in the Northern Hemisphere by the great Gobi Desert in Asia, the Sahara in Africa, and the arid regions of southwestern United States and northern Mexico; in the Southern Hemisphere, at a corresponding latitude, the desert areas of Chile, Bolivia, and Brazil in South America, the Kalahari Desert of South Africa, together with the Australian desert, are found. In some cases these are regions of extreme desiccation. The forest and grassland of the subtropical belt gradually give way to leathery-leaved scrub; to trees reduced to shrublike dimensions and often covered with thorns; to cacti, sage-brush, and drought-resisting bunch grasses. Because of lack of moisture to support vegetation the individual plants and tufts of grass are scattered. Much of the bare sandy surface in between them is exposed to the baking effect of the hot desert sun. In some sections, such as the Sahara, water is too scarce to produce any plant life whatever. Here forage is so scarce that few if any large animals can find sufficient sustenance to survive. Since there are no large herbivora to supply quantities of flesh, there are no large carnivora either. Such animal life as does exist there is reduced in size and capable of enduring extremely arid conditions. Many insects are found, together with centipedes, desert birds, rodents, and drought-loving reptiles.

In the next higher latitudes of both hemispheres lies a moister zone with a temperate climate. There are semidesert areas occasionally found, to be sure, but such as do exist are limited in extent and are due largely to local physiographic conditions. In this zone the inland parts of the continents are covered with grassland, which for the most part forms a closed stand. Owing to differences in moisture these areas may range from the short-grass plains to the tall grasses of prairie and pampas. It is these regions that support the extensive stock-growing and agricultural interests of the world. Along the coasts and lower river basins extensive forests occur such as are found along the Pacific and Atlantic coasts and in the northeastern states of the United States. Of these forests the drier and the sandier soils are occupied by conifers, represented especially by the pines, while the more fertile regions of heavier rainfall are covered with deciduous trees. In these temperate belts the buffalo, antelopes, deer, rodents, a wide variety of birds, together with wolves, coyotes, pampas cats, bobcats, minks, foxes, and skunks find their homes. In the upper lakes of this zone flocks of waterfowl, such as ducks, geese, brants, and cranes, nest and rear their young.

In latitudes higher than these temperate belts, land areas in the Southern Hemisphere practically drop out. Some islands are found above this latitude, and the southern point of South America extends farther to the south; but for the most part land gives way to water. In the Northern Hemisphere a very different situation exists. Here great forests of coniferous trees, with interspersed broad-leaved varieties, cover the ground. Represented in these forests are spruces, firs, tamarack, and cedars. In North America a broad zone of Canadian forest, interrupted here and there by grassland and lake, extends entirely across the continent from the Atlantic to the Pacific. In northern Europe and in Asia, wherever the moisture supply will permit, a corresponding belt of coniferous forest is developed. Animals commonly inhabiting these regions are the moose, elk, grizzly bear, wolf, bobcat, and beaver. Many waterfowl of various kinds swarm into the lake and swampy areas of this belt, where, during the summer months, they breed and multiply by the thousands.

North of this great forest zone, except in favored localities, the short summers and long winters gradually reduce the size of the native trees and other vegetation. As one approaches the arctic circle the tundra regions are encountered. In North America this belt extends from the mountainous parts of northern Alaska along the Arctic Ocean to the Atlantic. It is from a few hundred to a thousand miles in width, and in some places is even wider. In winter it is a region of snow and ice; in summer it is one of lakes and streams. The ground is frozen to great depths and, during the short summer, thaws only on the surface. The plant life here is represented by dwarf willows and alders, reindeer "moss," and a profusion of short-season grasses, sedges, herbs, and small shrubs. The animal life comprises the polar bear, the walrus, the seal, the arctic hare, and the arctic fox, which lives chiefly upon the hare. The shaggy musk ox is found, and in some places great herds of caribou graze upon the scanty vegetation. Migratory birds, such as eider ducks, snow geese, swans, curlews, golden plover, and arctic terns, migrate during the summer to these regions to nest and rear their young. Similar conditions and life regions are also represented in the northernmost parts of Europe and Asia.

Life in mountain ranges is also zoned. In general, then, we have seen that both plants and animals are distributed in great life belts that extend from the equator to the poles. They overlap at many points, and there are no sharp lines of demarcation between them; but when one compares the total life of any particular belt with that of the adjacent belts, the difference is easily seen. Somewhat the same zoning of life is found on a small scale as one proceeds from the foothills of a mountain range to the crests of the taller peaks. Elevation, in general, means a cooler climate. So then, roughly speaking, the same zones of mean temperature are found to exist between the base and the summit of a mountain range as exist between the temperate belt and the arctic regions three thousand miles farther north.

Let us take a cross section of the Rocky Mountains in Colorado as an example. Along the foothills of this cross belt of the range the plant life is little different from that of the surrounding plains. In addition to the stretches mostly covered with grass, clumps of scrub may be found, together with a few scattered gnarled yellow pines. At the higher elevations of the range, up near the timber line, the forest is largely one of Engelmann's spruce and balsam fir, together with interspersed areas of Douglas fir, depending upon the exposure and the supply of moisture. Swampy areas within this belt are occupied by alpine meadows that rival the very tundras themselves in both the species found and the profusion of their bloom. Along the timber-line margin heavy growths of dwarf willows similar to those of the arctic regions are also found. Between the lower and upper zones of plant life, park-like stands of yellow pine abound, and still higher up come stretches of the more or less dominant Douglas fir. Variation in exposure, slope, and soil moisture are all factors that tend to obscure sharp lines of demarcation and mix up the species; but, in general, these zones may be very clearly observed.

The animal life is much less sharply delimited. The distance from the foothills to the peaks is usually not more than thirty to forty miles, and the larger animals can easily range over this distance; in fact, altitudinal migration is quite common. In the summer months the mountain sheep, deer, and elk live among the grassy areas and alpine meadows of the higher altitudes; during the winter months they descend to lower elevations to escape the severe cold found higher up and to secure both food and shelter. Beasts of prey, such as the mountain lion, move back and forth with the deer, upon which they feed. The black-speckled native trout like the colder waters of the higher ranges, while the rainbows accommodate themselves to both the upper and the lower stretches of the mountain streams.

The migrations of animals affect their distribution. It is a well-known fact that some animals have a thoroughly established migratory instinct. In some cases migration occurs but once or twice in a lifetime, in other cases it is an undeviating annual affair, and in still other cases it is an irregular and unpredictable movement depending upon fortuitous circumstances.

The migration of the eel and the salmon. The eels that inhabit the fresh-water streams of North America are hatched out in the waters of the Pacific and the Atlantic. When they reach a certain size they then migrate from the marine waters up the inland rivers. Here they may live as fresh-water fish and grow for several years. As the mating season approaches they become restless in their fresh-water habitat and begin their long journey out into the ocean to spawn. Once the eggs are laid and fertilized, the old eels become sluggish and are believed to die, leaving the young to take their place.

The salmon, on the other hand, pass most of their adult life in marine water. Here they range widely. They may even swim from one side of the Pacific to the other, as has been determined by investigations based on tagging records and reports. When they become sexually mature they grow restive in their salt-water habitat and head for the streams of fresh inland waters. They swarm into these rivers in almost unbelievable numbers. They crowd one another, and if an obstruction of any kind impedes their progress they may almost dam up the stream. On they go upstream, leaping low waterfalls and floundering through shallow rapids. At length they reach the spawning grounds, near the head waters. Here they lay their eggs. Once this procreative duty has been discharged, the old fish deteriorate rapidly. They grow thin, their fins become ragged, and they may even become blind. In a short time the forces of senility overcome them and they die off, leaving the eggs that they have just laid to bring forth a new generation, which will return to the ocean.

The migrations of birds. One of the greatest mysteries of bird life is their annual migrations during different seasons into widely separated parts of the globe. Some of the migratory birds that nest in the northern United States and southern Canada pass their winters in the Gulf states and Mexico. Others travel much farther to spend their winters. The blackpoll warbler nests and rears its young in Alaska. As autumn approaches it takes an almost beeline course southward across Cuba and Jamaica to the north-central part of South America, where it spends its winters. The next spring it returns to the Alaskan regions to carry on its nesting operations again.

The golden plover has similar distant migratory habits. It spends the summer months along the northern shores of Alaska and Canada. With the approach of fall many of them take up their long flight directly across the Pacific Ocean to the Hawaiian Islands. They must make this distant journey of two thousand miles without either food or water, for there are no intermediate stopping places. How they find the way to their winter quarters on these relatively small islands no one knows. A divergence of a degree or so from the straight line at the start would carry the birds far to the side of their destination. Yet they seem never to miss it.

The palm for globe-trotting, however, must go to the arctic tern. It nests during the summer along the most northern limit of land bordering the Arctic Ocean in North America. It arrives in May, and the season is so late there that snow must often be dug out of the nest; indeed, the first nest found by man in that region, only seven and one-half degrees from the pole, contained a downy chick circled by a bank of newly fallen snow that had been removed from the nest by the parent. When the young are capable of sustained flight, these birds leave their summer home. A few months later they are found eleven thousand miles away, skirting the edge of the Antarctic continent south of South America.

How birds acquired the migratory instinct and what forces set them in motion are both profound mysteries. It is believed that this instinct has been produced by natural selective forces, and that it therefore has survival value. But why the birds in the Northern Hemisphere should go south in the autumn instead of east or west, or why they should travel northward in the spring, is not clear. Several theories have been advanced, but none of them are conclusive. One theory holds that the condition of the food supply is the principal factor that initiates these movements; another maintains that temperature is the main consideration; still another, of more recent origin, is inclined to the point of view that the length of day is an important influence if not the controlling one. There are other theories, but these are the main ones. To the student of the whole question it seems probable that no single factor is the cause of the migratory movements. It is more likely that when the problem is finally solved, if it ever is, it will be found that the response is due to a complex of environmental forces. Temperature, food supply, daily light-duration, the nesting urge, as well as many other factors, may be involved.

Regardless of the external force or forces concerned with migration, the immediate cause of the response on the part of the organism is likely to be found to be a hormone secretion. This point of view is far from established yet, but there are evidences pointing to that conclusion. This may possibly be true of the fishes as well as of the birds.

Fortuitous animal migrations. Almost everywhere in nature there seem to be two opposing forces working. One is overproduction, which tends to increase the number of individuals of any one species; the other is destruction, which tends to limit the number of any one kind and to hold them in check. Hawks, for instance, are enemies of rodents and tend to hold their numbers down. Let anything happen to the hawks such as an epidemic of disease, or a temporary destructive crusade against them by man, and their number would be reduced.

With the danger from their enemies diminished, the rodents would increase. Now the few hawks left surviving would find food abundant and life relatively easy because of the increased number of rodents. Under these conditions the hawks themselves would multiply rapidly and this, in turn, would tend to reduce the number of rodents again. This condition of equilibrium which seems to prevail in a very complex manner among all living things is known by students of nature as the *balance in nature*.

Let man interfere with this balance in nature, and serious consequences may follow. The English sparrow was introduced into this country from Europe. Since there are no natural enemies here to hold it effectively in check, it has completely overrun the country. The prickly pear was taken into Australia by the hand of man. So favorable did it find the arid conditions there that it responded luxuriantly; in fact, it was in a fair way to preëmpt the whole land when a natural enemy in the form of the cochineal bug was found to counteract its spread. When man is contemplating taking a strange species of any kind into a new land, he ought always to count the costs. Without its old enemies that naturally hold it in check, it may become a veritable pest.

It sometimes happens that a natural check is removed from a species, even though it remains in its own native environment. This may arise by the destruction of its enemies through an epidemic, a storm, a drought, or the onslaughts of a second enemy that preys upon the first enemy. Or it may come about through an increased food supply or temporary favorable climatic conditions. Now, with this natural check off, the species under consideration increases enormously. It may even multiply to the point where it consumes the whole food supply, overruns the country, and, by sheer hunger, is forced to migrate. This condition then leads to what may be termed fortuitous migrations. These have no directional objective, or course. In fact, the migrating animals will proceed straight into the face of disaster if they happen to be headed that way. Their movements are an attempt to escape a peril of some kind, and they are as likely to go in one direction as another. In *The Science of Life* the authors tell us that before game in Africa was much interfered with by man, settlers in South Africa periodically used to see hundreds of thousands of springbok, a kind of antelope, trekking southward. Owing to overproduction several hundred thousand might be in sight at the same moment, and the movement might continue for days at a time. One horde was estimated to be fifteen miles wide and a hundred and forty miles long. Under the play of their innocent and unrestrained instincts, the authors tell us, all these animals were trekking to certain misery and death.

The same authors are authority for the following statement regarding the lemmings, little ratlike creatures which inhabit the moors of Scandinavia and farther north. Periodically they multiply so enormously that they invade the lowlands in huge swarms, moving mostly by night. They appear so suddenly that Olaus Magnus, telling of them in the sixteenth century, was convinced that they had fallen from the clouds. They are said to climb walls barring their progress and to swim rivers, losing many of their number every day. Finally the survivors reach the sea, which they doubtless take for a stream, and plunging in swim until they drown. Collett relates that one ship steamed for a quarter of an hour through miles of swimming lemmings. This migration in reality results in no benefit to the animals themselves, for in the end not a single one of all that leave their homes returns.

Early settlers vividly portray the ravages of grasshopper migrations in the pioneer days of the Western states. In Nevada the mouse plague that occurred in 1907 was estimated to have destroyed three fourths of the whole alfalfa acreage of the state.

Many plants and animals are associated with each other purely on the basis of individual needs. At this point in our discussion of the ecological relations of plants and animals, we come to some peculiar types of association. Here the association is more independent of the physical factors of the habitat and rests more on considerations of food supply and, to a lesser extent, of shelter, reproduction, and protection. The closeness of the relationship varies. In some cases the relation is a loose one, resulting in mutual benefit; in others it is more one-sided; in still other cases not only does all the advantage accrue to one member of the combination, but the second member is often parasitized to the point of death itself. No exhaustive discussion will be attempted. Space will permit the mention of but three or four types of these associations.

Colonial organisms. Some organisms live together in a colonial state. This is a condition of mutual aid and permits of a degree of specialization and differentiation which is sometimes remarkable. Honeybees, wasps, hornets, and ants are examples.

At the mating season among ants, for instance, when weather conditions are favorable, winged males and females swarm out of colonies adjacent to each other. They issue by the thousands, leaving the excited and distracted workers rushing aimlessly about the entrance to their burrows. Once above the earth and in the open air, the sexes come together in small compact swarms, where they mate. The polygamous females may consort with several different males, which promotes cross-fertilization. When the nuptial flight is over, the males drift off at random to die of starvation. Some of the queens return to their original homes to continue the life of the colony. Others become separated and, finding a suitable spot or opening in the ground, creep in. Here, by autosurgery, they break off their wings at a preformed plane of weakness. For days they take no food, but live, mature their first eggs, and feed the first batch of young from their degenerating wing muscles. As soon as the first young are hatched, the neuter workers take over all the care of the colony. They excavate burrows, prepare nests, clean out waste, and provide food for both the female and the young. From that time on the queen retires from household labor and spends her energies in laying eggs to increase the colony. These differentiations of body form and specializations of task seem to be as beneficial to these lowly creatures as is the differentiation and specialization of labor among human beings.

Symbiosis. Symbiosis, as we have already learned in connection with the nodules that grow on legume roots, is an association of two different species of organisms for mutual

FIG. 125. A Symbiotic Relation between the Hermit Crab and a Hydroid Colony[1]

aid. We may take certain species of ants again as an illustration. These ants capture plant lice, protect them, and use them as "cows." They transport the aphids about from one plant to another to find food and shelter. They may even take them underground and dig extensive tunnels along the roots of the plants which are to provide the juices for the lice. Under these favorable conditions the aphids thrive and multiply rapidly. The ants, in turn, exact their tribute from the lice in the form of a sweet exudation. They approach the aphids, stroke them gently with their antennæ, and the honeylike secretion issues in tiny drops from the louse's body. The sweet exudate is now eaten with great relish by the ants.

Other examples of symbiosis are furnished by the hermit crab and certain marine hydroids. The crab takes up its abode in the empty shell of a snail, from which it feeds by extruding its pincers and the fore part of its body. The hydroids attach themselves to the shell and form a furry colony that almost completely covers and obscures it. The crab leaves particles of rejected food which the hydroids gather up. Moreover, as the crab projects its front legs and propels the shells about from place to place, the hydroids too are transported to more bounteous feeding grounds. It is probable also that the stinging cells on the tentacles of the incrusting animals afford the crab considerable immunity from attack. Among plants the lichens, in which algal cells are incased by fungal filaments, furnish a typical example of symbiosis.

[1] From Jordan and Kellogg's *Evolution and Animal Life*. By permission, D. Appleton and Company, publishers.

FIG. 126. Left, Young Cowbird in Warbler's Nest; right, Cowbird Egg in
Warbler's Nest[1]

Parasitism. Parasitism is a one-sided association in which
one organism lives at the expense of the other. The dependent
member is called the parasite; the sustaining one is termed
the host. There are thousands of this type of association
found in both the plant and the animal kingdom. Sometimes
the effect of the parasite on the host is less drastic and may be
tolerated almost indefinitely; in other cases the attack of the
parasite is so devastating and deadly as to result in the speedy
death of its victim.

The cowbird furnishes an example of mild parasitism. This
bird, a little smaller than the blackbird, does not build a nest
of its own. It makes its unfortunate victim act as incubator,
nurse, and provider for its young. The parent watches its
opportunity and deposits its eggs in the nests of other birds.
The host birds are generally smaller than the cowbird itself,
and eggs of the latter usually hatch a short time before the
others. This premature hatching gives the young cowbird

[1] From Herbert Friedmann's *The Cowbirds.* By permission of Charles C.
Thomas, publisher.

a decided advantage in growth. Being lustier it snatches and gobbles up most of the food brought in by the parents.

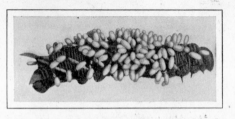

Then, because of superior strength and size, it may in due time actually crowd its helpless competitors out of the nest entirely, and they subsequently perish. Now, through its wicked instincts, becoming the sole recipient of the bountiful provision of the hard-working foster parents, it has an easy time and clear sailing in the matter of growth and development. The parent birds may suspect that something has gone radically wrong in nature. Nevertheless, true to their blind instincts, they continue to feed the intruder just as if it were their own offspring.

FIG. 127. A Sphinx Caterpillar covered with Cocoons of a Parasitic Insect[1]

The ichneumon flies present a more severe type of parasitism. This group of insects, of which there are a great number of species, derives its name from the so-called Egyptian ichneumon, or Pharaoh's rat, which devours the eggs and young of the crocodile. Most of these flies have piercing ovipositors (egg placers), by means of which they puncture the skin of other larvæ and deposit their eggs. Sometimes the eggs are injected into the cocoon either within the body of the larva or between the body wall and the cocoon shell. These parasitic larvæ hatch quickly and at once begin to devour the body of their host. The invading larvæ grow rapidly and soon themselves come to the surface of their riddled victim and spin little cocoons. Oftentimes the back of a dead caterpillar will fairly bristle with these tiny parasitic cocoons standing on end. These cocoons presently hatch out into parasitic flies again.

One of the largest and most imposing parasites of this general group is the Thalessa fly. It is parasitic on the larvæ

[1] From Sanderson and Jackson's *Elementary Entomology*. Ginn and Company, publishers.

of the common wood borer. The eggs of the host are deposited in the woody tissues by the parent with her boring ovipositor.

Here the larvæ hatch and go boring indiscriminately through the woody trunk of the tree, feeding on the wood as they go. Their tunnels are about the size of a lead pencil. About this time the Thalessa flies are also ready for business. The fly alights on a tree over one of these tunnels. How it determines the proper place is a mystery. It now curves its astonishingly

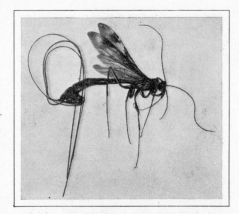

Fig. 128. The Thalessa Fly

long ovipositor up over its back, brings it down against the tree trunk, and gradually forces it in. This is a laborious task; and often a female Thalessa, with her ovipositor sunk deep into a tree, may be found in an almost exhausted condition. Finally, however, her ovipositor penetrates the wood borer's tunnel, and in this burrow she lays her eggs. In a few days the Thalessa eggs hatch, and their larvæ start on the hunt through the tunnel for the larval wood borer. Coming upon their host, they attack it and feed upon its soft, pulpy parts. In due time the Thalessa larvæ themselves pupate and later emerge as full-grown flies.

The plant kingdom too is well represented among the parasites. Grain rusts and smuts, downy mildews of grapes, chestnut blights, apple canker, melon wilt, and pear blight are all examples. Most of the plant parasites belong to the fungous and bacterial groups. Some of these parasitic bacteria are the cause of many of man's most dangerous diseases. However, the plant parasites are not confined wholly to these lowly forms. Some of them are flowering plants. Indeed, the largest flower known is the bloom of the Malayan parasite Rafflesia.

2

It may be over a yard across, and presents a combination of yellow, blue, red, and white colors. It sometimes weighs as much as twenty-five pounds. Rafflesia is parasitic on the roots of other plants.

Ewing Galloway

FIG. 129. The Rafflesia Flower

One of the striking features of many obligate parasites, both plant and animal, is their degenerate bodily structure. In their extreme form parasites have become wholly dependent upon their hosts for food. Accordingly, relieved of all necessity either to make food or to pursue it, they have lost all or most of the organs required for such activities. Among plants, parasites have lost most of their leaves and true roots; their stems too are often abortive and weakened. Nothing remains but parasitic roots with which to suck the very lifeblood of their victims and, generally, a profusion of flowers by which to reproduce their own scourging kind. In the animal kingdom parasitism has had a similar blighting effect. The parasite living within the body of its host has no need either for wings or legs to seek its food. Hence the parasite has lost them. Its comely body has disappeared; and, in extreme cases, nothing remains but some kind of deadly hooks or suckers by which it clings to its victim, supplemented by organs of reproduction to procreate its kind. The tapeworm has lost not only all organs of locomotion but its whole digestive system as well. Its only semblance of head, if, indeed, it can be called a head at all, is a ring of hooks by which it attaches itself to the sides of the intestine. It absorbs its nutriment directly through the body walls from the digested contents of the host's alimentary canal. The body of the worm behind the so-called head consists of a series of joints which may easily be broken off and separated. Each segment has both male and female elements and is capable, within itself, of producing a whole mass of parasitic eggs to infest other hosts.

So parasitism is a relative matter. It may run the whole gamut from slight dependence down to the most obligate parasite that cannot be separated from its host for an hour without disaster. The climbing rose is contracting bad habits, for it leans on other plants and depends upon them for support. The mistletoe is farther along the path of degeneracy. Its sickly colored leaves make carbohydrates, but it depends upon its host for water and minerals. The grain rusts have gone all the way and are high-handed parasites of the most pronounced kind. Animals present a similar parallel. The owl that lives in the prairie-dog burrow shows the first signs of dependency; the lion that requires his fifty zebras a year has gone halfway along the path of retrogression, but he still retains his mighty paws and well-developed legs; while Sacculina, the parasitic barnacle, has gone to the end of the parasitic road. It lives attached to the abdomen of the crab. Its body has the appearance of a swollen sac, and it absorbs its nourishment from its helpless victim through rootlike structures that ramify through the tissues of its host. But it has paid the price of its predatory habits. The eyes, the mouth parts, the thoracic appendages, the segmented structure, and many other organs found in the typical barnacle have all disappeared. It has reached the extreme stage of degeneracy.

In closing, parasitism teaches some vivid facts regarding evolution. It seems to be generally believed that evolution is always in the direction of an advance; that, when a living organism has evolved, it has always changed into some form which is more complex and highly specialized and which could be recognized at once as a more advanced type. But such is not the case. Change may go backward as well as forward. It may be retrogression as well as advancement. In short, an organic change may be de-volution as well as e-volution.

QUESTIONS FOR STUDY

1. What are some of the factors that are important in determining the habitat of man?

2. Assign some reasons why animals may migrate fortuitously.

3. Explain why different plants in a forest bloom at different times.

4. What factors may account for the migration of plants, and how do these operate?

5. Describe the effect of increase in altitude upon plant life and compare it with the increase in latitude.

6. What factors are believed to contribute to the migratory habits of animals?

7. What factors contribute to plant succession?

8. Describe an example of symbiosis that is of value to man.

9. What adaptations aid plants in migration?

10. Explain what effect parasitism may have upon the body of the parasite itself.

REFERENCES

BURLINGAME, L. L., and others. General Biology (second edition), pp. 481–513.

CLEMENTS, F. E. Plant Succession.

COWLES, H. C. "The Ever-Changing Landscape," *Scientific Monthly*, Vol. 34, pp. 457–459.

GADOW, HANS. The Wanderings of Animals.

PEARSE, A. S. Animal Ecology.

SHULL, A. F., and others. Principles of Animal Biology, pp. 275–309.

WEAVER, J. E., and CLEMENTS, F. E. Plant Ecology.

WELLS, H. G., HUXLEY, J. S., and WELLS, G. P. The Science of Life, Vol. III, pp. 961–1011; Vol. IV, pp. 1147–1199.

UNIT V

Mendel made Fundamental Discoveries which have made Possible Improved Plants and Animals

MENDEL'S discovery of the basic laws of heredity was a great contribution to human knowledge. Centuries before, man had recognized the fact of heredity, but he knew practically nothing of the fundamental principles which it followed in its expression. Breeders previously pursuing an empirical trial-and-error selection had made considerable progress in improving plants and animals. But now advancement was to be much more certain and rapid. For, by taking the earlier types and breeds and by following Mendelian principles, great things were to be accomplished. Flowers with more beautiful parts and delicate perfumes were to be obtained; more luscious fruits and heavier-yielding crop plants were to be evolved. Animals likewise were to be developed according to man's desires, whether he wished a midget Pekingese dog or a powerful draft horse; whether he wanted a slim-bodied racing hound or a dairy cow that would be a heavy producer of butter fat.

With these domesticated plants and animals as material to work on, and with the laws of genetics at his command, there appear to be practically no limits to the results man may get by breeding. Within natural limits he will be able to produce just about any type that he may desire.

CHAPTER X · The Characteristics of Plants and Animals are transmitted to their Offspring according to Definite Biological Laws

Introduction. Darwin, as we have found, made heredity one of the five factors underlying his theory of evolution. When the successful among plants and animals once arrive, their superior qualities must be passed on to the offspring if the fittest types are to continue and finally win out over the unfit ones. But how is this transmission of parental characters accomplished? Darwin candidly admitted that he did not know. But, true to his scientific point of view, he advanced a theory. That theory, we will remember, was pangenesis. Pangenesis, however, received a body blow when he and Galton tried out the transfusion of blood in female rabbits with negative results in the offspring.

For thirty years or more after this test men were still very much in the dark as to the causes lying back of heredity. To be sure, Mendel had determined much about the quantitative results of heredity as expressed numerically in the offspring, but he too had to speculate when it came to assigning real causes.

Although a great deal yet remains to be discovered about heredity, in the past thirty-five years much has been learned about the subject that is of inestimable value to society. This knowledge is influencing all breeding operations in both the plant and the animal kingdom. It is even hoped that an intelligent observance of this knowledge will reflect itself in a better human race.

Heredity has been found to be a matter of cell behavior; therefore, before we can understand it we must know how cells in higher plants and animals behave when they divide to form other cells. This knowledge is fundamental as a starting point.

317

Mitosis is a process by which the two newly formed cells receive identical genes. The term "mitosis" may be defined as indirect cell division. The adjective *indirect* is employed to distinguish this type of cell division from that found among the simpler one-celled plants and animals farther down the scale of life. Apparently, among some of these lowly organisms, one cell simply pinches in two directly to form two individuals, much as a piece of dough may be pinched in two to make two biscuits. Among the higher organisms, however, cell division becomes a very much more complicated affair. There are several intermediate stages, and it is for this reason that the process is described as indirect.

Mitosis normally occurs in embryonic tissue where growth is taking place. To be specific, we may trace it in the fertilized egg of *Ascaris*, a parasitic worm very commonly found in the intestines of hogs. This parasite is not uncommon in the alimentary tracts of children who get the eggs by putting dirty objects from the region of pigpens into their mouths.

When the fertilized egg of *Ascaris* is ready to divide, the chromatin of the nucleus resolves itself into a long, tangled, ribbon-like structure called a skein or spireme. This occurs within the nuclear membrane. The skein then shrinks up, becoming shorter and thicker, and subsequently breaks transversely into four separate parts called chromosomes (Greek *chroma + soma*, "color body"). About this time the nuclear membrane dissolves away, leaving the chromosomes lying free within the cytoplasm. The chromosomes are usually bent into a more or less V-shape or U-shape, instead of being straight and rodlike. We should pause here to emphasize that the chromosome number is *specific*. By that we mean that all individuals of any one species, disregarding sex differences, have the same number of chromosomes. The number of chromosomes in different species, however, may vary widely. For example, as reported, one species of *Ascaris* has 4; the pea, 14; field corn, 20; cattle, 16; the tomato, 24; the dog, 22; man, 48; and the monkey, 54. The number in some species is believed to exceed a hundred.

While the changes referred to are going on, the centrosome, a body outside the nucleus, divides, and the two halves pass

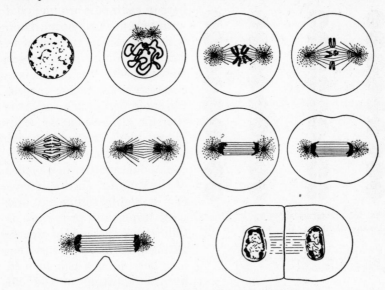

FIG. 130. Diagrams showing Stages in Mitosis

to opposite sides of the nucleus. Between these two poles formed by the centrosomes, fine protoplasmic fibers extend to complete a fusiform body called the spindle. About this time the chromosomes move up to the equatorial plane of the spindle, forming a rather flat group which is represented more or less diagrammatically in Fig. 130. Now the chromosomes each split longitudinally into two equal halves. This splitting is a most important process and needs further explanation.

It is believed that the hereditary characteristics possessed by the individual are determined by tiny factors in the chromosomes, called *genes*. These genes are ultra-microscopic. By employing a special technic Professor Oswald Blackwood of the University of Pittsburgh has recently reported them to be $\frac{1}{3,200,000,000}$ of an inch in diameter. Moreover, investigational results show, beyond much doubt, that these genes are arranged longitudinally along the chromosome, much like beads on a string. Now when the chromosomes all split, there are just twice as many of them as before, and the two new members of each pair are identical with the old chromosome from which they came in respect to the determining genes which they possess. It is in this precise way that nature insures the same hereditary constitution for all cells in the individual plant or animal body.

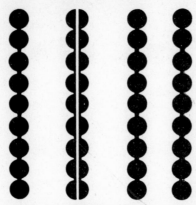

FIG. 131. Diagram showing the Linear Arrangement of the Genes in the Chromosomes

Note how each gene is divided equally when a chromosome splits during mitosis

After splitting, the chromosomes migrate to the two poles of the spindle. In this migration a highly significant and very important procedure is followed. Instead of segregating at random, the two members of each pair formed by the splitting of the old chromosome pass to opposite poles of the spindle. This insures that the aggregation of genes contained by the chromosomes at one pole will each have an exact counterpart in the genes contained by the chromosomes at the other pole.

At the poles the chromosomes become joined together again, end to end, and are drawn out into a long thread, much like the one formed by the parent nucleus in the early stages of cell division. The spindle disappears; and a new nuclear membrane forms around the chromatin thread at each pole, but does not include the centrosome, which is left outside in the cytoplasm. The chromatin now becomes arranged as a network in the nucleus. Meanwhile a deep cleavage furrow has been pinching in midway between the two forming nuclei which finally cuts the original cytoplasm into halves and completes the division of the old cell, forming two new ones.

In principle the mitotic stages traced in the dividing egg of *Ascaris* are essentially the same for the cells of all higher plants and animals. Certain differences in other species may be found. For instance, in many cases the chromosomes apparently do not unite end to end to form a skein, but remain separate and distinct from the beginning. In plants, instead of a cleavage furrow pinching the newly formed cells apart, a plate develops through the old cell to separate the two newly formed ones. This plate begins as knotlike developments on the spindle fibers, which enlarge and finally coalesce to form the dividing wall. But these differences, it will be readily seen, are not fundamental. They are unimportant variations that do not affect the general principles necessary to insure two new cells exactly like the old one in respect to the genes which they contain.

Heredity works largely through sexual reproduction. Reproduction is the production of distinct individual offspring by the parent organisms. As has already been made plain in a former chapter, there are two general types of reproduction: asexual and sexual. Asexual reproduction is accomplished either by a unicellular organism dividing to make two or else by a fragment of the plant or animal parent becoming detached and then developing into a new individual. In some cases this reproductive fragment may be a microscopic spore, or it may be a foot or more long, as is the case of a grapevine cutting.

With a few exceptions which we need not note for our purpose, it is evident that an offspring produced asexually will be quite like the parent. The reproductive fragment will have exactly the same chromosomes and the same gene complex. Therefore any characters developing in it which are different from the parent will be very largely, if not wholly, the result of variable environmental factors.

But when we turn to sexual reproduction the case is a very different one. Here a male cell from one parent unites with a female cell from the other parent to produce the young. If the parents are exactly alike (which, in the strict sense, may seldom occur), the offspring would be expected to be exactly like the parent line; but, if the parents differ by so much as a single gene, in that event we have far-reaching possibilities for

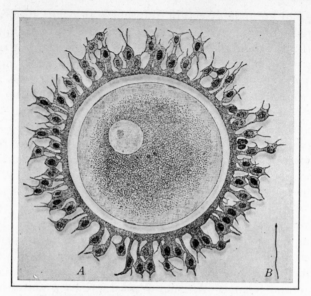

FIG. 132. *A*, the Egg, and *B*, the Sperm of Man[1]

differences in the offspring. It is in sexual reproduction, then, that the laws of heredity have a chance to express themselves.

Sexual reproduction is marked by fertilization and subsequent growth. In outlining sexual reproduction we might start at any one of several points in the cycle. It will perhaps meet our needs best if we begin with the reproductive cells themselves.

The female reproductive cell is called the egg. The eggs are relatively large nonmotile cells and are often furnished with considerable food material. They are more or less spherical in form. The male reproductive cell is called the sperm and is, in relation to the egg, very minute in size. The sperm carries little or no stored food, and the animal sperm is provided with a long, whiplike tail which propels it actively through the liquid in which it is released. The human egg, for example, is about $\frac{1}{125}$ of an inch in diameter

[1] From L. B. Arey's *Developmental Anatomy*. By permission, W. B. Saunders Company, publishers.

and may just be seen with the naked eye. In comparison with the human sperm, the egg is a veritable giant.

By appropriate mating habits the sperms are delivered where they may reach the eggs, so that fertilization, or the union of the two cells, may be accomplished. The two sexual cells taken collectively are called gametes (Greek *gamos*, "marriage"). When gametes fuse in fertilization a body called the zygote is formed. The zygote marks the first stage of embryonic development. Just how nature works at this point is unknown. But in some way the set of genes coming from the father and those coming from the mother act in such a way as to stamp the characteristics of the parents on the offspring. Sometimes the young, as we shall see later, may resemble only the father in some particular character; in other characters it may resemble only the mother; and in still other characters it may represent a sort of compromise due to the mutual action of the two sets of parental genes.

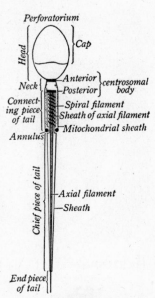

FIG. 133. Diagram of the Sperm of Man[1]

During the early period of embryonic development the cells of the embryo multiply rapidly by mitosis. Up to this point the cells apparently are all alike. They, as well as the original zygote and the gametes, are composed of special reproductive protoplasm called *germ plasm*. The mass of undifferentiated cells arising from the zygote (assuming that we are talking about animals) do not long remain alike. Rather early a differentiation of this germ plasm begins. Boveri states that in *Ascaris* it may be noted by the sixteen-cell stage. At this point one or more cells continue unchanged for

[1] From L. B. Arey's *Developmental Anatomy*. By permission, W. B. Saunders Company, publishers.

the specific purpose of forming gametes for the next generation. Because of the reproductive function to be assumed,

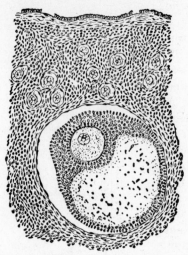

these cells do not change their
character but remain as germ
plasm whose gene complement
was contributed by both the
father and the mother. As the
embryo develops, this isolated
germ plasm does not remain
inactive, however. It proceeds
to multiply by mitosis to
form either unripe eggs or unripe sperms, depending upon
whether the embryo is to be a
female or a male.

In addition to this tiny mass
of germ plasm isolated early
for reproductive purposes, the
much larger mass of hitherto
undifferentiated germ plasm
is simultaneously set aside to

FIG. 134. Diagram of Unripe Eggs
in the Ovaries [1]

develop the body tissues of the young embryo. For this reason
these cells are no longer called germ plasm, but somatoplasm
(Greek *soma* + *plasma*, "body form"). The somatoplasm
cells continue to multiply rapidly by mitosis and then to
differentiate into bone, muscle, cartilage, nerve, skin, blood,
and other tissues of the young embryo. At birth or hatching time (if it be an egg-laying animal) the embryonic animal is all formed so far as tissues and organs are concerned.
Among these organs are ovaries in the female and testes in
the male.

In the meantime, while the body of the embryo was developing, the number of unripe eggs or sperms continued to
increase by mitosis. In the case of the female it is believed
by many biologists that all the unripe eggs which that individual ever will produce are formed before birth. They

[1] From L. B. Arey's *Developmental Anatomy*. By permission, W. B.
Saunders Company, publishers.

are formed in the ovaries, and each is inclosed in a little compartment called a primordial follicle. These unripe eggs

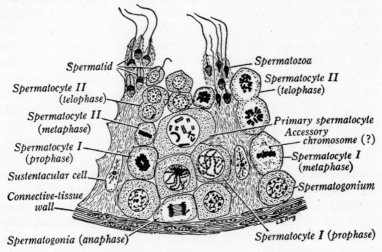

FIG. 135. Diagram of Unripe Sperms in Testes[1]

continue to reside in the ovaries until the fetus is born and grows to sexual maturity. In the male the sex cells are formed in the testes. Here, too, unripe sperms are formed before birth, but they may also continue to be produced after birth. In the female either one egg is released at a time or, at the most, comparatively few; but in the male a great number of sperms are delivered at once. For instance, in one of the higher animals one egg is normally released at a time, while in the male as many as 300,000,000 sperms are estimated to be released simultaneously. Since so many sperms are lost entirely, this great number is necessary to insure fertilization. It is doubtless for this reason that unripe sperms continue to develop after birth.

A stream of germ plasm extends from generation to generation. A hundred years ago our grandmothers leavened their bread

[1] From L. B. Arey's *Developmental Anatomy*. By permission, W. B. Saunders Company, publishers.

with home-grown yeast. In those days a continuous supply of fresh, compressed yeast was not delivered to the stores in every city and hamlet. In some instances yeast even in the dry cake form was hard to get. Consequently housewives kept a home culture of yeast for their needs. In general this culture consisted of a little flour and sugar, mixed with water until it was of a liquid consistency, into which yeast cells were introduced. Here the yeast cells would multiply and flourish. On baking days most of the culture would be taken out and used to leaven the bread. A small portion of the culture, however, would be retained as a "starter." To this starter a little more flour, sugar, and water would be added. As a result, the yeast would continue to multiply and grow until the next baking day; then the same practice of removing most of the culture for the bread and retaining a small quantity as a starter would again be followed. In this way there was a continuous stream of the same yeast culture passing from baking day to baking day for months, or even years. On the other hand, that part which was removed each time to leaven the bread served its purpose as a rising agent, perished, and disappeared as the bread was baked and consumed.

In a similar manner the germ plasm of animals passes on in a stream from generation to generation. After a tiny mass of germ-plasm cells are formed by a series of cell divisions following fertilization, a functional separation takes place. Most of the cells become somatoplasm, which in the end must perish when the individual body which they form dies. A few of the original germ-plasm cells, however, are retained as a starter to form gametes for the next generation. In this way there is a continuous stream of undifferentiated germ plasm passing from generation to generation. Germ-plasm elements from both parents are combined at each fertilization, to be sure, but this does not in any way impair the continuous flow of the reproductive life stream.

From this it may be seen that each generation of young men and young women is the custodian of the germ plasm for succeeding generations. For a time this precious stuff is given over by nature to their care and keeping. Fortunately

what these custodians do does not seem to change the hereditary character of the genes directly. What they do, however,

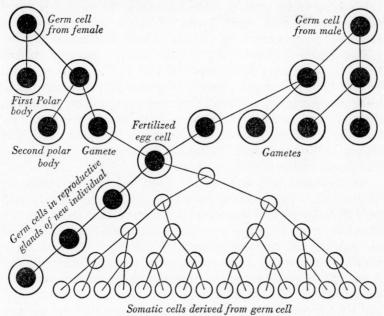

FIG. 136. The Continuity of Germ Plasm

may profoundly affect the physiological strength and vigor
of this germ plasm. If we are to rely upon the evidence furnished by observation and certain experiments performed on
animals, the wardens of this precious life stream hold a sacred
trust. By clean living and proper health habits the highest
degree of physiological potentiality may be insured to the
future offspring. On the other hand, unworthy living, the
excessive use of alcoholic liquors and drugs, or the contraction
of venereal diseases on the part of the parents may tragically
blast and blight the health of the offspring throughout their
whole lives.

Unripe eggs and sperms pass through an important ripening
process called maturation. In a former section it was pointed
out that during the embryological period the unripe eggs and

2

the unripe sperms are formed in the sex organs and remain there until the individual is sexually mature. The length of this period varies greatly among animals. In mice it is but a few weeks; in man, from twelve to fifteen years; and in the elephant, from twelve to sixteen years. But when the organism reaches sexual maturity the unripe eggs and sperms begin to function. The advent of this period is marked by certain bodily changes. Specific ductless glands begin to release their secretions. Their hormones arouse other organs to activity, and definite physical changes occur. In fact, biologically speaking, these developments, together with the ability to function sexually which accompanies them, is the full import of the adolescent period.

At this time eggs and sperms ripen and become ready for fertilization. The process by which this change occurs is a very definite one and is scientifically known as *maturation*. To illustrate it we may use a series of diagrams (Fig. 137). Let us consider the ripening process of the egg first. The early stages of maturation are essentially the same as those followed in mitosis. Chromosomes form from the chromatin, a spindle with its poles develops, the nuclear membrane disappears, and the chromosomes move down to the equatorial region of the spindle, just as they did in mitosis. From this point on, however, very important differences in chromosome behavior occur that must be followed closely. The one who wishes even an elementary understanding of how higher plants and animals, including man himself, inherit must master the following steps completely.

At the equator the chromosomes pair in a definite manner. The result is that, omitting the sex chromosomes, the two members of each pair come originally one from the paternal line and the other from the maternal line. These two chromosomes contain genes that control the development of identical bodily characters. Because they always associate in this manner in maturation they are called a pair of *homologous* chromosomes. This homologous arrangement may be traced definitely under the microscope for organisms such as *Drosophila*, in which there is a wide variation in chromosome length.

Fig. 137, *B*, shows the chromosomes paired at the equator. The reader must remember that these figures are pure diagrams and that the early stages of this cell division have been omitted. No attempt is made to preserve either the actual position of the chromosomes or their proportions within the cell. We are interested only in the principles involved. The theoretical case chosen for discussion represents a species that has four chromosomes only. If the species possessed ten chromosomes, there would obviously be five pairs of homologous chromosomes; if it possessed forty-eight, as in man, there would then be twenty-four pairs.

After the chromosomes have been closely associated as a pair for a brief period, they then begin to

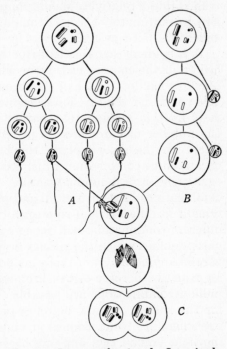

FIG. 137. Diagram showing the Steps in the Maturation of the Egg and Sperm

A, maturation of the sperm; *B*, maturation of the egg; *C*, fertilization

separate. No splitting whatever occurs in this case, as it does in mitosis. One of each pair leaves the other and moves out toward the periphery of the cell. Which one stays and which one moves out is apparently a matter of pure chance. At one time it may be the maternal chromosome and at another time it may be the paternal. Simultaneously the members of the other pairs of homologous chromosomes separate in a similar manner. In each case, however, the separation occurs absolutely without reference to what has occurred in any other pair. In one case the paternal chromosome may be left, while in the very next pair the maternal one may remain behind, as is shown in the diagram. The departing chromosomes are at length all ex-truded into a relatively small group called the *first polar body,* which is ultimately lost. The main portion of the cell may be called the half-ripe egg.

It is evident that this half-ripe egg is now a very different cell from the mother cell, from which it came. In the first place, it has lost just half its chromosomes. For this reason this particular cell division is known as the *reduction division.* It is by this method that nature prevents the doubling of the number of chromosomes in fertilization, as will be seen later. In the second place, the half-ripe egg has a certain chromosome combination which may vary widely as different eggs in the same individual mature. It is evident that different combinations of the maternal and paternal chromosomes may remain in different half-ripe eggs. Moreover, the number of different combinations will depend wholly upon the number of chromosomes which the species in question possesses. In the accompanying diagram there are four possible combinations. In man, with forty-eight chromosomes, more than sixteen million combinations are possible. The same would be true for the sperm. It now becomes perfectly clear that part of one's heredity is determined at maturation. If the egg ripen-ing be a human egg, and one parent had blue eyes and the other black eyes, for instance, in about 50 per cent of the eggs formed the chromosome with the blue-determining fac-tor would be retained, and in the other 50 per cent the one with the black-determining factor. This relation would hold for all other chromosomes carrying contrasted character-determining genes.

The half-ripe egg now passes through a regular mitotic division. In this case, again, instead of forming two cells, one member of each pair formed by the split chromosomes is pushed out into a *second polar body.* This second polar body, like the first, disappears leaving the single ripe egg as the end product of the maturation process.

The ripening of the primary sperm cell follows essentially the same stages of development (Fig. 137, *A*). The first division is a reduction division; the second is an equational, or mitotic, one. The two cells formed by the reduction division, however, are equal in size and are both saved. Each of these two divides again mitotically, resulting in four ripe sperms from the single original unripe cell. From this it may be seen that each sperm mother cell by maturation produces four ripe sperms, whereas the maturation of a single egg mother cell produces but one ripe egg. This difference in the maturation plan, it is evident, results in a correspondingly greater number of sperms, needed to insure fertilization.

In some animals and plants the course of maturation does not follow completely the exact ideal stages of development as outlined above. However, the essential principles involved are the same. So, for our purpose, the exceptions may be reserved for more technical studies.

In fertilization the egg contributes half the specific number of chromosomes, as does also the sperm. The result is a full restoration in the zygote of the chromosome number for that particular species (Fig. 137, C). Since the corresponding homologous chromosomes carried by the egg and the sperm, respectively, may have different character-determining genes, it is evident that one's heredity is partly determined too by what occurs in fertilization; that is, by the different combinations formed. For example, if a chromosome in the egg having a blue-eye determining gene were matched by one in the sperm carrying a black-eye determining gene, the eye color of the offspring would be different from that produced if the sperm also carried a blue-eye determining gene. So, then, one's heredity is determined by what occurs at two points. One of these is the reduction division in maturation; the other is fertilization.

Heredity has been recognized since the days of antiquity. In the days of the ancients it had been noticed that among living things like tends to beget like. But the fact alone had been recognized. No one knew how heredity proceeded nor the principles through which it was expressed. Indeed, a beginning on the true aspects of heredity was not made until the seventeenth century ushered in the scientific age.

Even then, for two hundred years afterwards, the efforts of observers in this field were fragmentary and unrelated. Here and there an important fact was discovered. But these isolated facts gave no more indication of the fundamental laws of heredity than a stray brick gives evidence of the stately cathedral into which it may be built when it is properly combined with other bricks. It was not until the middle of the nineteenth century that the genius of heredity arrived. He had the rare ability both to discover facts and to put them together so that they would mean something. In the history of science it is surprising how often highly important isolated facts in many fields have been discovered. But because these facts lacked the touch of a master hand to relate them and to perceive their true significance, nothing came of the more or less isolated discoveries. Many men had seen bacteria previously, but it took Pasteur to deduce the truth that diseases are caused by them; other physicists had observed that a magnetic needle shifts in opposite directions when under the influence of alternating magnetic poles, but it took a Faraday to see that this principle might be utilized to build an electric motor. In the field of heredity the man who was able to combine isolated facts in such a way as to lead to the deduction of fundamental genetic laws was Johann Mendel. Because of the far-reaching results of his work it will be of interest to note a few of the main facts connected with his life and work.

Mendel was born on July 22, 1822. He was a very industrious boy and took to school work readily. After finishing the lower schools of his time, he wished to pursue his studies further, but lacked funds. Acting on the advice of some of his friends, he decided to enter a holy order so that he might continue his education under the support of the church. He became a candidate for the priesthood in the order of St. Augustine, at Brünn, Austria. His initiation into this order was marked by the addition of the name Gregor, and he became known thereafter by the full name Gregor Johann Mendel. This, in common usage, was shortened to Gregor Mendel.

Mendel's work in the monastery was chiefly that of a teacher in the high school, known as the Brünn Technical

School. He began as a supply teacher, and while in the high school was never promoted beyond that rank. His acceptance as a regular teacher depended upon his passing certain examinations. He failed two of these examinations, which prevented his advancement. In fairness to Mendel it must be said, however, that his failures were in no sense due to lack of industry. Without tutoring he prepared for these examinations himself. Despite his failure in the first examination, he continued as a supply teacher.

FIG. 138. Gregor Johann Mendel

Mendel attempted his second examination in May, 1856, at Vienna. While the reports are incomplete, his biographer states it must be assumed that he failed in this attempt also. Whether he actually tried the examination and was unsuccessful, or whether, for some reason, he retired from the test is not clear. The report current at the time, however, was that he had a sharp difference of opinion with one of the examiners on some point of botany, and that as a result of the stubborn argument which followed he retired from the room. Nowotny, who was one of Mendel's associates at the Brünn school, believed that the dispute occurred, and that his disagreement with the examiner set Mendel off on his long study of heredity in peas to justify his position. At any rate Mendel's famous experiments began that very year. If this report is true, it is one of the few instances in history when a heated argument led to valuable results. Doubtless in Mendel's case the resolve to vindicate himself would not have yielded such valuable knowledge had not his determination been coupled with scientific ability of a high order and a burning desire to know the truth about the controverted point.

Mendel selected pure lines of peas. Although Mendel, in the end, worked with many different species of plants, his fundamental laws were worked out by using garden peas. The prelate granted him a plot of ground about one hundred and twenty feet long and a little over twenty feet wide in the monastery garden. Upon this tiny plot he carried on the investigations that have made his name famous.

To begin with, Mendel collected thirty-four different kinds of peas. Of these some were tall and some were short; some had flowers along the sides of the stem, and others produced their flowers at the end of the stem; some had wrinkled seeds and some had smooth seeds. So the characterization ran for the whole group. For two years Mendel tested these varieties in the monastery garden. He finally selected certain strains from the lot which always transmitted their particular character to the offspring. Thus, the tall peas always produced tall offspring, and so on for the other characters. Now, concluded he, since these characters are always passed on from generation to generation, they must be inherited. His problem, then, was to determine just how this transmission occurred.

Mendel's work was characterized by exact records based on careful analysis. Mendel finally selected fourteen different varieties of peas, each with a single character which he wished to study. In this respect his procedure differed radically from that of other men of his time. They attempted to study the whole individual plant or animal, with its entire complex of characters, as a unit. Mendel centered his attention on a single character only, in each variety. This plan simplified the problem enormously. Moreover, the pairs of characters Mendel selected were in each case contrasted the one with the other. For instance, round-seeded peas were contrasted with a wrinkled-seeded variety, green-colored pods with a yellow-colored pod, and so on for all the seven pairs.

Mendel proceeded by the method of hybridization. If he were studying pure yellow-seeded and pure green-seeded peas, for example, he would open up a bud on one variety just before it bloomed, deftly snip out the stamens, and then carefully transfer pollen to this bloom from a flower of the opposite variety. He now bagged the artificially pollinated flower with a bit of paper or calico to keep bees from carrying in pollens from other sources.

Mendel found that seed produced by these hand-pollinated flowers always grew plants — in this case, with yellow seeds.

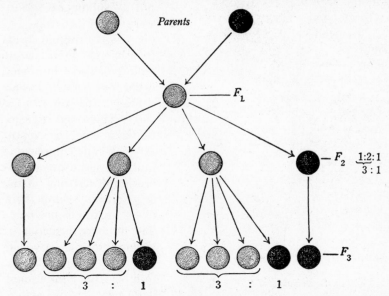

Parents

F_1

F_2 $\dfrac{1:2:1}{3:1}$

F_3

3 : 1   3 : 1

FIG. 139. Diagram illustrating the Results of a Cross between Yellow-Seeded Peas and Green-Seeded Peas[1]

Why this should be he did not know. Neither do we know today. Evidently this F_1 generation of *hybrids*, as they are called, had a seed-color-determining gene from both the yellow-seeded and the green-seeded parent. Yet it exhibited the character of the yellow-seeded parent only. Because this yellow-seeded character had covered up and dominated the green color in the F_1 hybrids, Mendel called it the dominant character. The contrasted green-seeded character, which had been hidden in the F_1's, he called the recessive character. For each of the contrasted pairs of characters studied, the dominant character alone appeared in the first hybrid generation.

Mendel now grew the F_2 generation of hybrids from the F_1's. When he did this a definite result followed which was revealed

[1] From Hugo Iltis's *The Life of Mendel*. By permission, W. W. Norton and Company, Inc., publishers.

only by the careful and accurate numerical records that he kept. In this case the second generation of hybrids were

FIG. 140. Segregation of F_2 Peas in Pod

not all yellow-seeded, as were the F_1's. Instead, on the average, three fourths of them were yellow-seeded, but the other fourth were green-seeded. In some way the green-seeded character had emerged again after being completely hidden in the first generation of hybrids. This segregation appeared even in the same pod. The numerical ratio $3:1$ might not occur, of course, if only a small number of individuals were grown; but when approximately a thousand F_2's were grown, Mendel found that the ratio of yellow-seeded to green-seeded peas came out almost exactly in this manner.

The three yellow-seeded peas, however, were not all alike. One out of the three was found always to produce yellow-seeded offspring like itself; in other words, it had taken on the pure form again. The one-fourth green-seeded peas were also found to be pure and, when planted, to produce nothing but plants of their own kind. The remaining two yellow-seeded peas, on the other hand, proved to be dominant hybrids exactly like the F_1's. When these produced the F_3's, the result was a segregation into a $3:1$ ratio again, as in the F_2's.

The fact that a pair of dominant and recessive genes may be combined in a hybrid and then segregate again in their original pure form was a new and highly important discovery. Genes may be associated in an individual, but they do not amalgamate. Instead, they retain their individual identity.

How persistent this characteristic of retaining their identity is has been revealed by Marshall and Muller. They kept the banana fly, *Drosophila*, in the hybrid form, in respect to three recessive characters, for something like seventy-five generations. Then, when permitted to segregate, the recessive traits came out apparently in exactly the same form in which they had gone into the cross in the first place.

The fact that genes when associated in a hybrid do not mix but come out again in their pure form is of great import to society. Every young person who contemplates marriage into a hereditarily defective family should remember that recessively defective individuals may appear again in the future generations of children. There are also individuals who, impelled by the best motives in the world, declare that the United States has no right "to build a fence around itself." They maintain that our immigration laws should be much more liberal than they are. The question may be debatable. But if the United States does lower its immigration bars to the point where hereditary defectives would be admitted, it should be thoroughly acquainted with the consequences. We can assimilate people as to manner of dress, personal tastes, and social customs, but we cannot change their genes.

Mendel found some cases of heredity much more complex. The form of hybrid which we have just been discussing was a *monohybrid*. It is so called because the original parents differed in respect to but one character. A monohybrid is the simplest hybrid that can be produced. Mendel worked with more complex hybrids. He crossed pure-line peas that were unlike in two characters. The F_1's exhibited both dominant characters, and were called *dihybrids*. For instance, he hybridized a pea having a smooth yellow seed with another pea having a wrinkled green seed. The F_1 hybrids were all smooth yellow-seeded peas. This showed that these two characters were dominant. When the F_2's were grown, however, they showed a $9:3:3:1$ ratio instead of a $3:1$ ratio as in the monohybrid. We may show by means of a checkerboard devised by Dr. Punnett of England how these ratios for the F_2's of both hybrids come about.

In the monohybrid the yellow-seeded character was dominant. The initial letter of each character may be used to

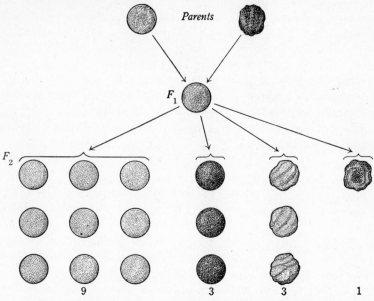

FIG. 141. Diagram illustrating the F_2 Results of a Dihybrid[1]

represent both the character and the gene which produced that particular character. Then:

$yy =$ the pure yellow-seeded parent, and y the single gene in its gamete.

$gg =$ the pure green-seeded parent, and g the single gene in its gamete.

The formula for the F_1 hybrid, then, would be yg. When this hybrid produced gametes, they would be either y or g because of the reduction division. Dr. Punnett, by forming a checkerboard as on page 339, with all the possible eggs placed in a row at the top and all the possible sperms arranged at the left side, represented all possible individuals that could be produced in the F_2 generation.

[1] From Hugo Iltis's *The Life of Mendel*. By permission, W. W. Norton and Company, Inc., publishers.

	y	g
y	yy 1	yg 2
g	yg 3	gg 4

In this checkerboard the pure individuals, like the original parents, are represented by squares 1 and 4. The hybrids are shown by squares 2 and 3.

Passing now to the dihybrids, the parents may be represented by the letters

$ssyy$ = the smooth yellow-seeded parent.
$wwgg$ = the wrinkled green-seeded parent.

The parental gametes would then be

sy and wg,

the F_1 would be

$ygsw$,

and the gametes produced by the F_1 would be

$ys, yw, gs, gw.$

Now, arranging the checkerboard:

	ys	yw	gs	gw
ys	yy ss 1	yy sw 2	gy ss 3	gy sw 4
yw	yy sw 5	yy ww 6	gy sw 7	gy ww 8
gs	gy ss 9	gy sw 10	gg ss 11	gg sw 12
gw	gy sw 13	gy ww 14	gg sw 15	gg ww 16

In this checkerboard squares 1, 2, 3, 4, 5, 7, 9, 10, and 13 carry both dominant genes and would therefore represent smooth yellow-seeded peas; squares 6, 8, and 14, wrinkled yellow-seeded peas; squares 11, 12, and 15, smooth green-seeded peas; and square 16, the only wrinkled green-seeded pea of the lot. Adding, and placing as a ratio series, the 9 : 3 : 3 : 1 result is obtained.

In the squares numbered 6 and 11 two new *pure-line* individuals are represented. These peas are different from the original parents, however, in that they have inherited a character from each of the parents. The result is a pure-line wrinkled yellow-seeded pea and a pure smooth green-seeded pea. These individuals, as is obvious, would always breed true. It is in this way, by hybridizing and selecting, that the desirable characters of two or more parents may be combined in a single offspring.

Mendel formulated his results into laws. For between seven and eight years (1856–1864) Mendel worked away with his peas. He had no associates in his endeavors, but was driven on by the sheer impetus of his own interest. Even if Father Nowotny was right in his surmises and Mendel was working to vindicate his contention in the botany examination, still his persistence shows a high degree of resolution. Such dogged determination is invariably the mark of genius. Finally, as a result of the long pages of careful records which he had meticulously kept, Mendel was able to abstract some general principles. For our purpose they may be stated as follows:

1. Unit characters (Mendel believed each character in the individual to be produced by the action of a single pair of hereditary factors).
2. Contrasted pairs (dominant and recessive).
3. Dominance (one factor overpowers the other when associated in the chromosomes of the same individual).
4. Segregation (the segregation of genes that occurs in the reduction division. Mendel advanced the idea of the segregation of the hereditary factors in some way, but he knew nothing of the reduction division).
5. Recombination (the recombination that occurs in fertilization).

Mendel's work when reported was not appreciated. In February, 1865, there was to be a meeting of the Brünn Society for the Study of Natural Science. Gregor Mendel's name appeared on the program. He planned to give an account of his work and report the general principles of heredity which he had discovered.

One can well imagine the enthusiasm with which he prepared his manuscript. He had worked over seven years on the single problem of heredity. Thousands of individual plants had been grown. His experiments had been planned and carried out with scrupulous care. Nothing, either of labor or of pains, which he considered necessary to make his results reliable had been spared. Some of the monohybrids he had carried through even the sixth and seventh generations, only to find that they still segregated in the 3 : 1 ratio, as at the outset. Mendel doubtless knew that he had discovered principles of tremendous importance.

The paper was presented clearly and simply but with force and conviction. But, strange to say, it fell on deaf ears. If the ears were not deaf, they certainly had no conception as to the great significance of the truths which they were hearing. Mendel, alas, like many geniuses who had preceded him, was a man ahead of his day. The minutes of the meeting simply record the fact of Mendel's appearance, followed by the laconic comment that there "were neither questions nor discussion."

Perhaps the chief reason why the results of Mendel's labors were not appreciated at that time may be laid at the door of Charles Darwin. Six years before Darwin had published his *Origin of Species*. This book had created a great stir in both scientific and theological circles, as the reader will remember. For several years after 1859 the subject of evolution took the center of the stage at almost every biological meeting; indeed, the minutes of the very meeting at which Mendel read his paper record that "Professor Alexander Makowsky, one of the leading members of the society, refers with the utmost enthusiasm to Darwin's theory of the origin of species."

Mendel retired to the monastery a disappointed man, but

with no misgivings as to the importance of the truths which he had presented. Father von Niessl, an associate of Mendel's at the monastery, tells us that the latter, in discussing the reception of his paper one day, uttered these prophetic words, "meine Zeit wird schon kommen" (my time will come after a while). After the meeting Mendel's report was published in a small local scientific paper. Unfortunately, however, the copies of this edition seem to have become as completely forgotten as the original manuscript itself.

Mendel's time did finally come; but, sad to relate, it did not come until after his death. At the time his historic paper was presented Mendel was working on the problem of inheritance in the hawkweed. This study was continued for several years. Hawkweed, however, has small flowers, and to work with such tiny flowers imposed a severe strain upon Mendel's eyes. His vision began to fail. In 1868 he was raised to the prelateship of the monastery. His enlarged ecclesiastical duties, together with his failing eyesight, caused him in 1871 to discontinue his experimental work in the field of heredity.

Another unfortunate circumstance arose which cast a shadow upon Mendel's last years. Soon after he was made abbot he became involved in a sharp and long-drawn-out controversy with the state over the right of the government to tax the monasterial property. As a result of this conflict Mendel became embittered and uncommunicative. Doubtless because of his heavy responsibilities as prelate and the strain of his quarrel with the temporal authorities, he fell a victim to Bright's disease with complications. He passed to his final rest on January 6, 1884. The notice of his death in the Brünn *Tagesbote* concludes with these words: "His death deprives the poor of a benefactor, and mankind at large of a man of the noblest character, one who was a warm friend, a promoter of the natural sciences, and an exemplary priest."[1]

The Mendelian laws are rediscovered. Eleven years before Mendel's death a German by the name of Anton Schneider was peering into a microscope examining certain cells. He

[1] Hugo Iltis, *Life of Mendel*, p. 278. Translation by Eden and Cedar Paul.

discovered, among other details, some peculiar bodies which were afterwards called chromosomes. In 1883, less than a year before Mendel's death, a Dutch scientist, van Beneden, saw chromosomes segregating in the reduction division. The strange behavior of these bodies at once challenged the attention of scientists. It suggested that chromosomes might have something to do with the process of heredity.

Interest in chromosomes and heredity was now given a great impetus. By 1900 De Vries at Amsterdam, the proponent of the mutation theory, Correns in Germany, and Tschermak in Austria had just about worked out the funda-

G. H. Shull

FIG. 142. Monument erected to the Memory of Mendel in the Monastery Garden by his Admirers

mental laws of heredity anew. Then it was recalled that an obscure monk, thirty-five years before, had presented a paper at Brünn postulating some revolutionary conclusions as to heredity. This recollection led to a search for Mendel's paper, which was rediscovered in 1900. Although these three men, working independently and without the assistance of Mendel's report, had arrived at the very same conclusions, they all renounced their claims of discovery. In a fine spirit of self-abnegation they all accorded Mendel the honor. Thereafter, then, these hereditary principles became known as Mendel's laws in honor of the man who first discovered them. Mendel had been right. His time did come. But the tragedy of it lay in the fact that the recognition of his work did not come until after his death.

2

The knowledge of heredity has grown since the days of Mendel. Since the rediscovery of Mendel's laws our knowledge of the processes of heredity has grown greatly. Investigators have had to modify Mendel's conclusions at some points and to extend them greatly at others. For example, we now know that dominance is in some instances imperfect. Correns crossed red and white four-o'clocks and, instead of getting red F_1's, got pink offspring. We also know that, instead of unit characters, a single character may be the result of several pairs of genes working together. Many other modifications and extensions of Mendel's laws have been made; but many of them are complicated and must be left for more advanced study. However, there is one thing that should be kept in mind: that subsequent discoveries and additions to hereditary science have not abrogated Mendel's results in the least. His principles still remain as the fundamental laws of heredity. They form the very basis of genetic science today and doubtless always will.

The applications of hereditary laws in plant and animal breeding have resulted in a multitude of new types and forms. Their monetary value is almost incalculable. The observation of these laws in human society also has been highly beneficial in some quarters and holds much promise for the future. But considerations as to the practical value of the laws of heredity when applied must be reserved for later chapters.

QUESTIONS FOR STUDY

1. What is the biological basis of heredity?

2. Contrast the process of fission with that of mitosis.

3. Draw diagrams and describe the various stages in mitosis.

4. What is meant by the statement that the number of chromosomes is specific? Give some examples of this.

5. What is fertilization, and why is it necessary?

6. Very early in the developing of the zygote cell differentiation takes place. What is its function?

7. Draw a diagram to illustrate the continuity of the germ plasm and explain it in detail.

8. Explain the process of maturation of the unripe egg in detail.

9. Contrast the maturation of the unripe egg with that of the unripe sperm.

10. How can chance chromosome segregation during maturation and chance recombination in fertilization affect one's heredity?

11. What are the hereditary factors called, and how are they arranged in the chromosomes?

12. Why may there be wide differences in the characteristics of children of the same family?

13. If, in cattle, one crossed a pure-line red (recessive), polled (dominant) female with a pure-line black (dominant), horned (recessive) male, what would be the characteristics of the two new pure-line individuals developed in the F_2 generation?

REFERENCES

ILTIS, HUGO. Life of Mendel. (Translation by Eden and Cedar Paul.)
JONES, D. F. Genetics, pp. 52–74.
LOCY, W. A. Biology and Its Makers, pp. 315–319.
SHULL, A. F. Heredity, pp. 1–93.
WALTER, H. E. Genetics, pp. 1–16, 93–120, 208–224.

Chapter XI · Through Selection and the Scientific Application of the Laws of Genetics Man has greatly improved his Domestic Plants and Animals

Introduction. The account as to how and where man first acquired most of his domesticated plants and animals is either entirely unknown or is exceedingly vague and fragmentary. It is probable, however, that domestication grew naturally out of primitive man's needs and necessities. Moreover, it is quite certain that he had made marked progress along these lines before he began to keep dependable records of his activities and achievements.

It is true that the wild life of almost every continent offers favorable opportunities for domestication. Many species of both our tame plants and animals seem to have come originally from either Asia, North Africa, or southeastern Europe. This condition, though, probably arose not so much because these regions offered better wild forms from which to begin, but rather because primitive man in these quarters seems to have been more enterprising and ingenious. At any rate, domestication on an extensive scale appears to have received its start in these parts of the globe.

The history of early domestication is fragmentary. As primitive people became more thickly populated the supply of game and other wild life grew proportionally scarce and difficult to obtain. Man met this emergency by taming some of the wild animals best adapted to meet his needs for food and clothing. Later he added other species to assist him in travel and to become his beasts of burden. At the outset he probably captured the young of these animals and nourished and cared for them until, having lost some of their fear, they would stay about his early habitation. As these semi-domesticated animals grew in numbers and kinds, their owners needed more forage to feed them. At first this exigency may have been met by a nomadic life, in which the tribes roamed about from place to place, stopping only where the grazing was good and where shelter could be obtained.

In the course of time primitive man discovered that grains, grasses, and fruits could be domesticated also. Gradually, out of his necessity, he learned that by planting the seeds of these wild plants in favorable places and by digging about them with a stick or other crude implement he could stimulate their growth and increase the yield. By this means he could not only increase the supply of grain and forage for his animals but could also enlarge the variety and quantity of his own food.

By some such experiences as these primitive man made his first beginnings at animal husbandry and agriculture. At the outset the newly domesticated plants and animals were changed but little from those remaining in the wild state. Such characteristics as they possessed were those which enabled them to survive and succeed under the conditions of untamed competition. But now man, by his care and husbandry, had removed them from their original habitat. He increasingly extended to them more care and protection. Under these circumstances the wild characteristics which had enabled these plants and animals to succeed out in plain, forest, and thicket were not needed. On the contrary, man desired in them other qualities to supplant these wild ones — characteristics which would enable the animals to produce for him larger bodies for food, a heavier milk supply, greater size and strength for burden-bearing, or a longer, better fleece for weaving. From the plants he desired forms that would yield larger grains, more forage, tougher fibers, more delicious fruits, and later, as his æsthetic nature developed, more beautiful foliage and flowers.

Man had made much progress in improving his plants and animals before the science of genetics was developed. Long before the laws of heredity were discovered man learned that he could improve his crops and herds by selective breeding. Just why this was possible he, of course, did not know. He only knew that like begets like and that good parental lines are much more likely to produce desirable offspring than poor ones. Pictorial tablets reveal that even the Egyptians had learned to improve their domestic animals by breeding and selection.

Most of our definitely established breeds of domestic animals had been derived long before Mendel's laws were rediscovered in 1900. Years of careful selection and, in many instances, the most persistent and deliberate inbreeding had accomplished this. These breeders were (in an empirical manner, of course) applying the laws of genetics. That is the reason why in many instances their success was so marked. Their progress had been comparatively slow, however, because they were proceeding in a sort of hit-and-miss fashion. To segregate out and to recombine into a single plant or animal the desired traits of excellence found scattered through several lines is a difficult matter, and especially so when the husbandman is proceeding by the trial-and-error method. Yet, in time, patience and interest can achieve much. But to realize definite goals in breeding by the shortest and most direct route is a task not for the untutored mind but for the trained scientist. He must know not only the type of inheritance with which he is dealing but also how to segregate and recombine desired factors through hybridization.

But whether the results were achieved by these early breeders whose intense interest led them to proceed empirically, or by more recent breeders who have followed definite genetic laws, it is surprising what has been accomplished in the improvement of plants and animals. These attainments have resulted in horses fitted for every purpose, from the heavy draft type to the spirited racer; in cattle ranging from the large beef animals to the premium butter-fat producers; in chickens raised for food and those raised specifically for egg-production; in farm crops yielding several times the amount of grain afforded by the older varieties; in fruits whose appetizing appearance and delicious flavors are famous the world round; and in flowers whose fragrance and beauty enable them to grace a flower show or to adorn the humble cottage. Nothing that man has done in all the realm of nature has resulted in greater achievements. Neither has anything else done more to relieve him from the strain and stress of living and to free him from the exigencies and uncertainties of rainfall and climate. To make these facts clear we shall select just a few instances from the many thousands that might be taken, and try to show what breeding in these typical instances has meant to man.

How plant-breeders developed the Marquis wheat. Wheat is regarded as being the greatest single food of man. Samples of a small-grained variety have been taken from the pyramids of Egypt, dating back more than three thousand years before Christ. Through the intervening years many varieties have been derived from what were doubtless mutations and hybrid races. Varieties well adapted to one region have been found to be wholly unsuited to regions with different climates.

The northwestern United States and the prairie provinces of Canada require a spring wheat. The winters are too severe to permit the growth of the fall-sown strains. In these regions, too, an early-maturing wheat is desirable. Occasionally great grain-rust epidemics sweep over this part of the country, often resulting in millions of dollars' worth of damage if, indeed, not the loss of the entire crop. The earlier the wheat matures, the more likely it is to escape the ravages of the rust parasite, which thrives especially in warm, moist weather.

The Marquis wheat is a hybrid spring grain developed by A. P. and C. E. Saunders, cerealists of the Central Experimental Farm at Ottawa, Canada. The parents of this wheat were the early-ripening, good-milling Red Calcutta, which had been imported from India, and the Red Fife, which, up to that time, had proved to be the highest-yielding and most popular wheat grown in the prairie provinces of Canada.

The different hybrids resulting from this cross were separated by the Saunderses at the Ottawa station in 1903. One of these selections appeared to have especially desirable characteristics, and was named Marquis sometime between 1905 and 1907.

Marquis is a beardless spring wheat with a hard red kernel and a stiff straw which enables it to stand up well until it is dead-ripe and ready for the combine harvester. In general appearance the plant strongly resembles its Fife parent, but it possesses the early-ripening, heavy-yielding, and excellent milling qualities of its Red Calcutta ancestor. Ripening earlier than the Fife, the Marquis often escapes the extreme rust attacks and blighting dry weather that frequently overtake the former. On an eight-year test in Canada its average yield was from 13.5 to 38 per cent heavier than that of the native Red Fife parent.

FIG. 143. Parents and Progeny, Marquis Wheat

A, Hard Red Calcutta, female parent; B, Red Fife, male parent; C, Marquis. (Courtesy of Central Experimental Farm, Ottawa, Canada)

The farmers of both Canada and the northern United States immediately adopted Marquis wheat as a spring crop. Today it is grown on something like fifteen million acres; and this hybrid plant, perfected by plant scientists, has been worth literally hundreds of millions of dollars to the farmers of the two countries.

Burbank developed a remarkable potato. Luther Burbank, the plant wizard, while carrying on his plant-breeding work became interested in Irish potatoes. Commercially this plant is propagated by means of sections cut from its tubers. When grown in the northern part of the United States it has practically lost its power to produce seeds. Occasionally, however, a seed ball may be found. When the seed balls do appear they are about the size of marbles and present somewhat the appearance of a miniature tomato.

One day Mr. Burbank was walking among the potato vines in his garden when his eye chanced to fall upon a seed ball growing on one of the plants. His interest was challenged at once. It must be remembered that in 1872, or more than a half-century ago, potato development had not reached the degree of perfection to which varieties of the present have been brought. The tubers were smaller then and much less attractive in appearance. For some time previous to this

FIG. 144. Potato Seed Balls

Note the similarity of these fruits to tomatoes

incident Mr. Burbank had been thinking about the potato and the possibilities for its improvement; so, before his eye caught sight of that lone potato seed ball, reflection had already made his mind fertile ground for a new idea. At once the thought occurred, "I wonder what would happen if I should save these seeds and plant them." His curiosity developed into a plan for action, and he decided to gather the fruit when it was ripe and save the seeds for the next year's experimental planting.

But disaster almost cut his plans short. He visited the plant several times thereafter to make sure that no harm should come to his coveted prize. Finally one day, judging the fruit to be ripe, he went to pick it. But, alas, there was no fruit there! He searched diligently all about the plant that bore it, but was unable to find it anywhere. For several successive days he returned to search for it. Finally his persistence was rewarded. He found it lying at some distance from the plant which bore it, whither it had probably been carried by some bird or rodent. The twenty-three seeds which it produced were carefully extracted and preserved until the following year.

FIG. 145. The Burbank Potato (Left) and its Ancestor, the Early Rose

At the proper time the next season the seeds were carefully planted and their growth and development watched with great anticipation. At length harvest time arrived, and the tubers of each plant were dug separately. To Mr. Burbank's surprise every plant had produced potatoes of a different kind. Most of them were small, knotty, or deep-eyed and of little value. Among them, however, were two vines that produced fine large white potatoes of great promise. They were much superior, in fact, to the parental Early Rose variety, from which the seed originally came. Later the tubers from one of the two plants proved to be superior, and the others were discarded.

From the tubers on this one plant as a start the famous Burbank variety of potato took its origin. The potato proved to be far superior in every way to any other potato grown at that time. As more tubers became available the United States government assisted in the distribution and introduction of this potato into new localities. At the height of its production, in 1906, over six million bushels of this famous variety were produced on the Pacific coast alone. At that time the annual value of the Burbank potato was estimated to have been $17,000,000. Subsequently other good varieties have been developed, but the Burbank still remains a choice potato among growers.

But to return to Mr. Burbank's seed ball: just how did it come to yield seeds that produced different varieties? We are not positive; but Mr. Burbank's theory was that the Early Rose potato itself was a hybrid and was continued in its germinally impure state by the method of asexual propagation followed in potato culture. He believed, moreover, that the flower which produced his original seed ball was fertilized by pollen from another bloom, transferred perhaps by an insect. By this hybridization process, then, just as in the case of the Marquis wheat, the good qualities of the various ancestors came to be recombined in that one precious seed out of the twenty-three produced by the original seed ball.

Domestic animals have been greatly modified by breeding and selecting. All domestic animals have been greatly changed and developed into diverse breeds and types by the hand of man. Of the hundreds of species that have been thoroughly tamed, there is perhaps not a single exception to this rule. Consequently it will be necessary for our purpose to discuss three or four groups only. Taking these as examples it may easily be seen how the principles of heredity have been and may continue to be employed by man to mold and fashion living things according to his desire.

Domestic dogs are believed to be the descendants of wolves. Dogs have been the companions of man from the days of savagery. Among the remains of ancient cave-dwellers the partly preserved bones of man and of dogs are found lying together. Among the excavated ruins of Nineveh, tablets have been unearthed which portray the heavy hunting mastiffs straining at the leash to go on the hunt. These tablets antedate Christ by several centuries. So dogs were domesticated early, and before the Christian Era they had already begun to be diversified into different types according to the wishes of man. The figures on Egyptian monuments of five thousand years ago present the forms of several widely different breeds.

Although their original pedigree is unknown, dogs are believed to have come from wolves. Every continent has its wolves and jackals, and wherever primitive man has been

FIG. 146. The North American Coyote

FIG. 147. The Timber Wolf

National Zoölogical Park

FIG. 148. The Dingo

found he had already tamed the wolf and adopted it as a pet, a sledge animal, or a companion of the chase. Darwin says:

It is highly probable that the domestic dogs of the world are descended from two well-defined species of wolf (viz. *C. lupus* and *C. latrans*) and from two or three other doubtful species (namely, the European, Indian, and North African wolves); from at least one or two South American canine species; from several races or species of jackals; and perhaps from one or more extinct species.[1]

This, of course, is almost the equivalent of saying that primitive man for the most part, wherever he lived, domesticated the wolves of that region and finally developed them into dogs.

It is believed by some that the aborigines' motive in taming the dog at the outset was in most cases to obtain a hunting partner. The savage, devoid of horses and with but very crude and ineffective weapons, had observed how successfully wolves hunted in packs. With his mind stimulated by hunger and necessity his imagination

FIG. 149. A Great Dane, Champion Attilla Hexengold

Owned by Charles Ludwig. Photograph by courtesy of Sally May, Secretary, The Great Dane Club of America

was not slow in picturing the advantage of having such an ally. True, the tamed wolf might kill the prey and get his dinner

[1] Charles Darwin, *The Variation of Animals and Plants under Domestication*. D. Appleton and Company, publishers.

first; but taking second turn at a carcass would doubtless not be so distasteful to a savage with a painfully empty stomach as it would be to modern man.

Mrs. R. C. Clarke

FIG. 150. The Chihuahua

This dog weighs only two pounds at five years of age

In some cases the wolf was obviously tamed to make a beast of burden. The Eskimo dogs of Greenland and Alaska are believed today to be nothing more than tamed wolves. Captain E. Parry, an explorer of northern North America, tells in one of his journals of a pack of thirteen wolves that came boldly close up to his vessel, the *Fury*. Yet his men dared not shoot lest the wolves might prove to be "huskies" belonging to the friendly neighboring Eskimos.

The modern wolf that most resembles the tame dog is the so-called *dingo* of Australia. It is red or tawny and may have been the wild ancestor of some of our modern dogs. It is easily tamed when taken as a puppy and becomes very playful and companionable. The dingo, as well as all other wolves and jackals, mates freely with dogs.

After wolves were tamed and became domesticated dogs (*Canis familiaris*) they were al-

Tauskey

FIG. 151. The English Bulldog

most like clay in the potter's hands. By new mutations, doubtless, by cross-breeding, and by selection man has been able to develop almost any type he wanted and to fit it for any purpose. Some breeds, such as the Great Dane, are almost gigantic in size, weighing in some cases three hundred pounds or more;

others, such as the tiny Chihuahua, weigh scarcely one and
one-half pounds and may be supported on the outstretched palm

John Sinnott

FIG. 152. The Greyhound

of the hand. The former
are excellent watchdogs.
Fuertes and Haynes tell
us that a few years ago
a burglar sought to enter
a house by a window.
Near the window he met
a Great Dane. Failing
to judge the dog's in-
tent, he proceeded. The
next morning his body
was found where he had
been strangled to death
by the dog. The minia-
ture Chihuahuas, on the other hand, have no value except
to be coddled as pets. Graceful, intelligent collies have a
strong pastoral instinct; the English bulldog is savage and
pugnacious. Greyhounds have little keenness of scent, but are

as swift as the wind; the
long-eared bloodhounds
have an almost incredi-
bly keen nose when on the
trail. The great, noble,
and kindly St. Bernard
is kept by the monas-
teries in Switzerland to
rescue travelers lost in
the cold and snow. One
of the most famous of
these modern dogs was
Barry. Among the forty

St. Elmo Bloodhound Kennel

FIG. 153. The Bloodhound

lives saved by him, one was a child that he found in the
snow. It was already overcome by the drowsiness that pre-
cedes death. The dog licked the child's face to warm it, then
crouched in the snow beside it so that it might fasten its
arms about his neck and be carried back in safety to the

monastery. The epitaph over Barry's grave reads: "Barry the heroic. Saved forty persons and was killed by the forty-first." His tragic end was the result of a mistake. He was attempting to rescue a lost traveler who mistook his friendly advances for an attack.

Altogether there are probably one hundred breeds and types of dogs. They vary from the tall, lithe Russian wolfhound to the long-bodied dachshund with its short, bench-crooked legs. Strange to say, too, they exhibit as much difference in disposition and mental traits as they do in bodily characteristics; and these less tangible nervous and mental traits are also matters of breeding and selection quite as much as are the physical ones.

Rural St. Bernard Kennels

Fig. 154. The St. Bernard

The horse has been greatly improved by breeding and selection. Unlike the dog, there is no living wild horse that seems to be the direct ancestor of the domesticated horse. The wild ass and the zebra belong to the same genus, but they are not the progenitors of the highly diversified and specialized horses of today. The fossil records reveal that in prehistoric times wild horses were scattered over North America. They were also found in South America, Europe, Asia, and in Africa. The history of the horse's ascent from a little five-toed primitive ancestor about the size of a house cat is clearly depicted by the fossil deposits in North America. Later the equine strains that appear to have been in the direct line of ascent became extinct on all these continents, save perhaps Asia alone. One wild representative, Prejvalsky's horse, is still found in the desert stretches of western Mongolia. This animal is a long-haired, rather thick-headed pony. Some authorities have taken the position that it is the probable ancestor of the modern horse; others hold it to be but a retrograde feral type that

New York Zoölogical Gardens

FIG. 155. Prejvalsky Ponies, the Wild Horses of Asia

escaped from the domesticated condition centuries ago. So at the outset it must be said frankly that the horse's true ancestry is buried in the past. Conflicting views are found, so that what shall be said on this question must be regarded as but carefully formulated opinion.

Domesticated horses are known to have been used in Babylonia about 2000 B.C. They were introduced into Lower Egypt during the seventeenth century B.C. by the Hyksos, or shepherd kings, who came from the north and east of Syria. These facts would seem to indicate that the horse's first domestication probably occurred somewhere in Asia. It may be that nomadic tribes learned the enormous advantage of using the horse as a convenient and swift method of individual transportation. Horses were later yoked to chariots and became valuable adjuncts in war. The use of the horse, however, as a common, workaday beast of burden and for tillage is of comparatively recent origin. In Britain oxen served as general plow animals up to almost the end of the eighteenth century. In America, although some horses were used from the outset, oxen were extensively employed as heavy draft animals up to a generation ago.

2

Mrs. Elizabeth Bryce

FIG. 156. The American Saddle Horse

The light-legged breeds of western Europe and America are believed to have come in a large measure from a single Asiatic ancestor. These breeds include the saddle and coach horses and the more specialized types reared for racing. The blending of blood from the Arabian, Barb, and Turk strains of horses has probably had much to do with the development of these lighter-bodied, speedy types. By careful selection horses have been derived from these strains which were able to cover a mile in considerably less than two minutes. Before the advent of the automobile it was this branch of the horse family that furnished the prized coach and carriage horses of the nobility.

Some writers have taken the position that the heavy draft horses were probably domesticated from a large-boned, broad-footed, heavy-set forest type of ancestor which was found either in Europe or in western Asia. Wild horses were

W. R. Brown

FIG. 157. An Arabian Stallion

abundant in Europe in Paleolithic times and were used extensively for food by the men of the Old Stone Age. Stacked before one cave used as a place of human habitation were found the bones of "several tens of thousands of horses," all

L. S. Sutcliffe

FIG. 158. Peter Volo, a Modern Trotter

Owned by Dr. Edwards

of which were presumably slain and used as food. Subsequently the wild horse seems to have practically disappeared from Europe. But it is possible that some of these horses may have survived in the eastern mountainous regions and have become the ancestors of the heavy European draft breeds.

At any rate, from some such source as this a large Flemish type of horse arose. During the Medieval Ages, when "knighthood was in flower," these heavy horses were introduced into England from Flanders. Their blood appears to have been commingled with that of the pre-Roman horses, and the result was a strong, sturdy horse known as the Great, or Black, horse, the latter name arising from its color. This breed of horse was especially prized because of its ability to carry with ease a knight burdened with his heavy armor. By subsequent breeding this horse's color has been largely changed to bay or brown

Ashby

FIG. 159. Broker's Tip, Winner of the Kentucky Derby, 1933

Owned by E. R. Bradley

and its size has been increased. These are the ancestors of the modern *shire* breed, which is the chief agricultural horse

Fig. 160. Laddie 2nd, a Shire Stallion

of England. They are active, powerful horses, with heavy bodies and large hairy legs.

Another heavy draft horse common both to western Europe and to America is the Belgian. It is regarded as a direct descendant of the early Flemish horses and has been developed for the purpose of drawing heavy loads. The temperament of these horses is rather sluggish as contrasted with the more active shires, and they have little or no long hair on their legs. Their color is usually a roan or a chestnut brown, and it is not uncommon for a well-bred male to weigh from two thousand to twenty-five hundred pounds. They are used almost exclusively as farm horses and for heavy draft work. There are many more breeds of domesticated horses, but the ones given are sufficient to show what careful mating and selection have done.

Cattle have shown a remarkable response to breeding and selection. Of all the animals domesticated by man, cattle are believed to have been among the very earliest.

Fig. 161. A Belgian Stallion

FIG. 162. The Gaur, a Native of the Hills of India

Tame cattle are known to have been kept by the Egyptians thirty-five hundred years before the Christian Era began, and it is believed that domesticated cattle date back to Neolithic times. As in the case of the horse, early exact records as to how the first domesticated cattle were tamed and from what branch or branches of the wild taurine stock they originally came are lacking. Cattle became the servant of man before he appreciated the immense value that reliable historical data would be to subsequent generations.

FIG. 163. The Yak, the Hardiest of All Wild Cattle

Unlike the horse, however, if all representatives of domesticated cattle should become extinct, it is certain that man in a very short time could develop a passably good animal from the wild species that still exist. This would be true for either

FIG. 164. The Water Buffalo of India

the Eastern Hemisphere or the Western. Asia still has the gaur, a monstrous wild ox inhabiting the hills and inaccessible highlands of India and eastward. In the forbidding highlands of Tibet the hardiest of all wild cattle, the yak, still flourishes. Besides other species of wild cattle, Asia also has a native buffalo. In India and Burma these animals have been domesticated and are highly esteemed for their milk. They are an indispensable draft animal in the lowland rice flats, and because of their love for water are called water buffalo. On warm days, as a relief from heat and insect pests, they will completely submerge themselves, leaving only their nostrils protruding above the water. They are self-willed animals, and when they take a notion to submerge, if not promptly disengaged from the crude plow, are likely to drag implement as well as protesting driver along with them into the water.

Africa has its Galla ox, and North America has its bison. Of these latter animals literally millions inhabited the Western plains when the first pioneer settlers arrived. Almost any of these wild bovine races could be domesticated and doubtless would be if man had not already received highly developed animals as a legacy from the breeders of the distant past.

FIG. 165. The Texas Longhorn, a Semiwild Breed of Feral Cattle that ranged on the Plains of Southwestern United States

The cattle of both this country and Europe are believed to have been developed from primitive races of wild cattle that inhabited Europe in very early times. Some authorities think they originated farther east, perhaps in Asia, and were brought westward by man. If the former view is correct, our larger breeds of cattle probably originated from the great wild ox of Europe, *Bos primigenus*. The remains of this animal are found in the brick clays and peat bogs of Europe, and from their skulls some specimens are judged to have had a horn spread of four feet. Some of the skulls are pierced by flint arrowheads, which indicates that they were associated with the men of the Old Stone Age. It is believed that some of the direct descendants of these wild cattle have been preserved to the present day in the hunting parks of certain estates in England. They are, generally speaking, of a dirty white color. Records of their existence go back to the early part of the thirteenth century.

Contemporaneous with these large wild European cattle were one or more smaller races which inhabited the same regions, especially in the west. While there are no historical records to support the theory, according to Davenport it seems quite probable, on the basis of structural resemblances, that

FIG. 166. Maxwelton Monarch, a Grand Champion Shorthorn Bull

Owned by John Alexander and Sons. (Courtesy of American Shorthorn Breeders Association)

these smaller races were the ancestors of our Jerseys and their near relatives, the Guernseys. So through these lines the connection of our modern western European and American breeds of cattle with the wild races of Europe is thought to be fairly well established.

From some such beginnings as these practically all our modern American breeds of cattle have come. Within each strain great improvement has been made and will continue to be made in the future. Space will not permit mention of the more than a dozen different breeds now fostered and pedigreed, but the discussion of one or two will make clear what can be accomplished through breeding and selection.

The shorthorn cattle are one of the most famous of the large beefy types. Their exact origin is not known, but they probably rose from a commingling of the blood of the large European *Bos primigenus* with that of cattle brought in by the Roman and Norman invasions. Later, males were introduced from Holland to add their genetic qualities to the breed. From this medley of bloods, then, the foundation was laid for one of the finest animals whose records have ever graced a pedigree book. Because of hybridization the strain must have been at the outset an extremely variable and unstable type; but from about 1780, the shorthorns have been bred and selected with marked care and persistence. By 1875 there had been developed a male, Comet, which commanded a price of $5000. This was the highest price ever paid for a shorthorn up to that time.

FIG. 167. Melba 15th of Darbalara, a Milking Shorthorn

This cow produced 32,522 pounds of milk and 1614 pounds of butter fat in 365 days — a world's record for all breeds. (Courtesy of American Shorthorn Breeders Association)

The shorthorns have been developed in two directions: one especially for its large size and beef production; the other for a general-purpose animal yielding a generous supply of milk as well as beef. Their color ranges from white to red and roan. With their short horns and generally desirable body contour they are one of the most valuable breeds in existence. Females usually weigh from fourteen hundred to two thousand pounds and males anywhere from eighteen hundred to twenty-five hundred pounds. In modern times a single animal has sold for as much as $23,750.

The milk strain probably had its beginning in a line of shorthorns developed in England by Thomas Bates. Heavy butter-fat production is a hereditary trait. Selection for this quality has been emphasized to the point where the best shorthorn cows will produce more than a thousand pounds of butter a year.

During the last fifty or sixty years a polled, or

FIG. 168. The Polled Shorthorn

A breed derived from the shorthorn. (Courtesy of the American Shorthorn Breeders Association)

hornless, type of shorthorns has been produced. The breed seems to have originated by mating a shorthorn male with hybrid polled females of the shorthorn type. After the horns were eliminated, persistent effort by selection was made to bring the polled animal back to the true shorthorn character.

FIG. 169. A Typical Jersey Cow

Sociable Sybil 478362, grand champion of the 1923 National Dairy Exposition and winner of seventeen other grand championships. (Courtesy of the American Jersey Cattle Club)

Such transformations as these show that within the range of possibility it is feasible by careful, intelligent hybridization and selection to realize almost any type of animal desired.

Genetics has played its part in perfecting the high-class Jersey of the prized dairy herds. The Jersey may have come from one of the smaller species of wild cattle, as Davenport suggests. Be that as it may, the forerunners of the modern Jersey seem to have flourished on the island of Jersey, in the English Channel, just about the time the United States began as a nation. At that time an old record of the island speaks of the introduction of a breed of cattle "to ornament the grounds of the nobility and gentry and to provide rich milk." This strain of cattle was so desirable that a strict law forbidding the importation of cattle from France was passed and rigidly enforced for over seventy-five years. This wise law kept the breed pure, and a rigid selection and breeding have developed one of the finest strains of dairy animals in all the world.

About a hundred years ago a score card for the ideal Jersey was formulated, setting forth standards for body shape, type of head and neck, contour of udder, and other characters. Animals which did not approach this ideal type were readily consigned to the butcher, and breeding was practiced from the finest animals obtainable, with butter-fat production as the primary object. From this island the Jersey has been introduced into England, Germany, Australia, and America.

Wherever the breed has gone its superior butter-fat production has readily brought it forth as one of the leading dairy types.

The Jersey cows are very mild and of a gentle disposition when kindly treated; if mistreated they become wild and nervous. The males are nervous and irritable and should be handled, as Plumb says, "with much care and discretion."

FIG. 170. Oxford Lassy's Design

The grand champion Jersey bull at the 1931 National Dairy Exposition. (Courtesy of the American Jersey Cattle Club)

The quantity of milk yielded by this superb breed is rather low, but its quality is unsurpassed and has been continuously improved. The first American Jersey tests were made in 1853, when Flora, a cow owned by Thomas Malley of Massachusetts, yielded fourteen pounds and eight ounces of butter fat in seven days. About thirty-two years later (in 1885) Princess 2nd is reported to have produced forty-six pounds and twelve and one-half ounces of butter fat in a single week. Thus in three decades breeders had achieved the accomplishment of more than trebling the amount of butter fat which an animal of this breed could produce.

Pigs, like many other domesticated animals, have readily responded to breeding and selection. Wild hogs were formerly abundant in the Americas, Europe, Asia, and Africa. South America still has its peccary, which is a kind of hog, but it is peculiar in that its stomach is more like that of the ruminant than like that of the domestic pig. Africa has its ugly, almost unsightly wart hogs, and both Asia and Europe still have representatives of the true wild boar, from which the domesticated hog has been derived.

The European species formerly ranged all over Europe, North Africa, and Western Asia. It is now extinct in practically all its primitive haunts except in some of the forests of

Germany, where the boar hunt is still a favorite form of sport. The blood of this wild European species has been employed freely in developing the larger domestic breeds of English hog.

FIG. 171. A Wild Boar

Note the long nose and the rangy body. (Courtesy of the Animal Production Subsection, Iowa Agricultural Experiment Station)

The Indian wild boar is very similar and is probably closely related to the European species. It grows a little larger and may reach a height of forty inches. Several smaller species are to be found in different parts of Asia, perhaps a half-dozen being found in the Malay Peninsula alone. One little pygmy hog found in the Himalayan foothills stands only ten or twelve inches at the shoulder.

The wild boar prefers low, marshy land. This animal has no sweat glands to regulate its body temperature, so on hot days it seeks the water and mud as a protection against the excessive heat of the sun. This same characteristic carries over to the domestic hog, which, on warm days, finds its greatest delight in submerging itself in some pool or mud hole. The wild boar, with its canine teeth developed into long tusks, is a formidable animal in combat. These teeth are dangerous weapons, and with its ripping tactics in full play it is one of the most dangerous animals of the wild. Even the "razorbacks" of the southeastern United States, not true wild hogs but only degenerate forms of those escaped from domestication, are redoubtable fighters.

Domesticated hogs were known in China before they were known in Europe and, if their history were positively known it seems probable that much of the blood of the domesticated hog came from Asiatic sources. Either the wild forms found there and in Europe, or else their ancestors, must have formed

FIG. 172. A Typical Lard-Type Poland-China Boar

Courtesy of the American Poland-China Record Association

the original stock from which the modern breeds have descended.

The wild types are inclined to be rangy animals with long noses and thin bodies, but the domestic breeder has constantly selected and bred for size; that is, for flesh and lard. The result of man's selection has been to shorten the snout, to reduce the length of the leg, and to increase both the length of the body and its breadth. Perhaps in no other animal is the modifying influence of man's knowledge of heredity more marked than it is in the domestic hog. A good-sized "razorback" may weigh from a hundred and twenty-five to a hundred and fifty pounds; a well-bred hog may reach the extreme weight of a thousand pounds or more. It is not uncommon for fully-grown market hogs to average from three hundred to four hundred pounds. Of the half-dozen or more types that have been developed by hog-breeders, meat production and lard

FIG. 173. Titan Queen, a Duroc-Jersey weighing 1123 Pounds

Courtesy of the Department of Animal Husbandry, Colorado Agricultural College

yield have been the controlling objective in almost every case.

The principles operative in breeding and selecting. In a former chapter it was found that desirable changes, or mutations, in both plants and animals occur. What causes them to appear is not positively known, but they arrive nevertheless. The mule-footed hog, the white-eyed banana fly, the double petunia, and the Shirley poppy are outstanding examples of mutation. It is probable that most of the qualities desirable in domestic animals have risen in the same manner. For instance, the heavy milk-producing strain of the shorthorn, the high butter-fat yield of the Jersey, the speed of a Hambletonian 10, the heavy egg-laying ability of the Leghorn hen, the prized flavor of the Delicious apple, must have had their inception in changes that embodied greater degrees of excellence than were possessed by other representatives of the same strain.

Once these desirable differences arose the problem of the breeder became purely one of combining and segregating. By hybridizing and selecting he sought to get as many genes of value in one animal or plant as he possibly could. To the degree to which he could aggregate the good genes and eliminate the poor and undesirable ones, just to that degree was he able to develop a superior type. That was precisely what the early breeder of domestic strains was doing when he was closely inbreeding, as he often did, then selecting the individuals that approached his ideal and discarding the others. Before the basic laws of heredity were known the results came more slowly. Nevertheless the achievements even then were marked because these early breeders were really following in a stumbling, halting way the principles of heredity without knowing it. Since the Mendelian laws have been discovered and applied, progress has been much more marked.

In many cases advancement requires much time and effort. In some organisms, as cattle and horses, but one offspring is produced at a time. Moreover, the young requires two or three years before it, in turn, can become a parent. It is for this reason that thoroughbreds often command such a high price. Richard Fairfax, a famous Hereford bull, brought $50,000 — apparently a stupendous sum; yet, considering the amount of careful mating, the exact pedigree records kept, and the time required to bring all these desirable traits together in one animal, he was probably worth the money.

An outstanding example of how breeders proceed to realize a desirable animal has been recently exemplified by Professor H. D. Goodale of the Massachusetts Agricultural Experiment Station. He started with a flock of Rhode Island Red chickens. Of the original group some began to lay earlier as pullets than others, some laid more eggs during the winter months, when the price of eggs is higher, and some had a heavier annual egg production than others. By selecting the earliest layers, the most prolific winter layers, and the heaviest annual layers, and blending their blood, one after the other, then continuing to select and mate to enhance these characteristics, surprising results were achieved. In a period of seven years (1913 to 1920) Professor Goodale was able on the average to clip off fifty-five days from the age of the pullets when they first began to lay. The winter egg production was raised from thirty-six to sixty-seven eggs per bird. The monthly egg yield per hen for the most prolific months (March or April) was raised from seventeen eggs to over twenty-one.

Plants and animals are not yet perfect by any means. Improvement is in progress on a broad front — such improvements as will add much of beauty and excellence and will be worth literally millions of dollars in the end. The method employed is that of the geneticist; the course followed is that of careful selection, hybridization, and then selection again.

QUESTIONS FOR STUDY

1. In what ways may the laws of heredity be applied to the improvement of plants and animals?

2. What evidences can you point out that show application of the principles of genetics to the improvement of plants and animals?

3. What are the purposes of animal-breeding as applied to cattle and hogs?

4. In a dry region, such as that found in the western part of the United States, what characteristics must be developed in food crops?

5. What did Luther Burbank do, and of what value was it to man?

6. The fact that plants and animals may be modified shows the possibility of change. What bearing does this fact have upon the theory of evolution?

7. Trace the development of the modern horse from the wild breeds.

8. What Mendelian laws are involved in the breeding and selection of plants and animals?

9. If all domestic cattle were destroyed how might one proceed to develop new domesticated breeds?

10. What laws of genetics were involved in the development of Marquis wheat?

REFERENCES

DAVENPORT, E. Domesticated Plants and Animals, pp. 3–34, 205–300.

DE CANDOLLE, ALPHONSE. Origin of Cultivated Plants.

DOWNING, ELLIOT R. Elementary Eugenics, pp. 1–13.

GOODALE, H. D. "Changes in Egg Production in the Station Flock," *Bulletin No. 211*, Massachusetts Agricultural Experiment Station.

PLUMB, CHARLES S. Types and Breeds of Farm Animals, pp. 1–251, 323–353, 683–798.

SANDERS, ALVIN H. Shorthorn Cattle.

WHITSON, JOHN ROBERT, and WILLIAMS, H. S. Luther Burbank, His Methods and Discoveries and Their Practical Application, Vol. VII, pp. 267–294.

"Marquis Wheat," *Farmers Bulletin No. 732*, United States Department of Agriculture.

National Geographic Magazine, Vol. 35, pp. 185–280; Vol. 44, pp. 455–566; Vol. 48, pp. 591–710.

UNIT VI

The Knowledge of Man's Origin and of the Laws of Heredity gives Promise of an Improved Society

THE discovery that mankind had a humble origin and gradually evolved to its present estate was a revelation of great importance. It helped man to see that his own development, as well as that of other living things, was a gradual, progressive process. Mendel's work also made a great contribution to human understanding. For from these two discoveries and other similar ones it became apparent that men, other animals, and plants, all follow the same biological laws. This is true both of their origin and of their development.

Turning to heredity, this biological kinship of man holds special promise for the future. For we find that by nature adults as well as children differ enormously. There are superior and inferior men, just as there are thoroughbred and nondescript domestic animals. Moreover, mental ability and social worth run in families. So do their opposites. Society can either increase heavily from the bottom and become inferior or it can encourage reproduction from the top and become increasingly superior. What shall be done in the case depends largely upon the dissemination of biological principles and the response which the average American citizen makes to this knowledge.

CHAPTER XII · Man, although related to Other Animals, holds a Distinctive Place in Nature

Introduction. Man is in many ways a remarkable being. It is surprising what mental acumen and ingenuity he manifests in successfully adjusting himself to the natural forces of his environment. He builds a telescope and communes with the stars, or he fashions a microscope and interviews the otherwise invisible bacteria. His emotional life, too, far surpasses that of the beast. If not more intense, it is of a much higher quality and of a more prolonged constancy. The sacred writer in contemplating the estate of man ascribes to him a place a "little lower than the angels" and perceives him to be "crowned with glory and honor."

Yet, despite man's high position in the universe, he seems to have come from a very humble ancestry. He bears in his body the marks of what scientists consider to be unmistakable evidences of his animal kinship. His developing embryo recapitulates the long story of his ascent. His blood reveals chemical affinities with that of lower animals, and the earth yields fossils that add their testimony as to his lowly origin. The story of man's evolution from the lower animals, as divulged by biological and geological evidences, is one of the most interesting and illuminating in the whole field of nature. We shall have time only to sketch it.

Anatomy gives evidence of man's origin from lower animals. Without special study it is difficult to comprehend how highly organized and complex the human body is; yet in all its aspects it strongly resembles that of the lower animals, bone for bone, muscle for muscle, and nerve for nerve. Moreover, the higher one ascends in the scale the more like man do the animal bodies become.

Man has a certain relation to the frog. Both have internal

skeletons. The central axis of the body in each case is a backbone with a long tunnel through it to house the spinal cord. Both have eyes, ears, and a bony box to protect the brain. Two pairs of limbs protrude from the trunk of the frog. These are matched in man by organs surprisingly alike in structure, but the anterior ones are put to an entirely different use. The internal anatomy bears a close resemblance. Each has a stomach, intestines, a pair of kidneys, and a heart. The liver of each stores glycogen, the pancreas secretes trypsin and lipase, the kidneys throw off urea, and the lungs supply oxygen. Blood is forced through the body of each by the pulsations of the heart. Thus it may be seen that man is surprisingly like the frog. In these same particulars he is like the cow, the horse, and the dog but very much more like the ape.

When man and the apes are compared, their spinal columns are both found to be articulate throughout the greater part of their length instead of being stiff and boardlike halfway up from their posterior end, as in the frog. The cranium of each is large and thoroughly ossified to shelter and protect the well-developed and highly specialized brain. The frog's brain is smooth, with no convolutions; but the brain of man and that of the ape are, so far as form is concerned, almost exact counterparts, down to the very convolutions. The frog can perceive lights and shadows and colors to perhaps a limited degree; he pays attention to moving objects alone. Man and the apes can perceive not only lights and shadows but a whole gamut of colors and shades of color as well. Some of the higher apes are almost as adept at recognizing and matching colors as is man himself. Frogs, with their crude, baglike ears, can hear sounds only; but apes can almost rival man in the variety of pitch and tones which they can appreciate through their long well-coiled cochlea. Frogs use their legs for jumping and swimming; but many apes walk with a semi-upright stride and employ their hands to grasp and manipulate objects much as man does. In many of these respects the chimpanzee and the gorilla are more like man than they are like their cousins, the monkeys.

The lungs of a frog are mere pouches, and the body receives quite as much oxygen through its skin as it does through

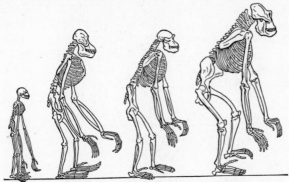

FIG. 174. Comparison of the Skeletons of Higher Apes
After T. H. Huxley

the pulmonary sacs. But the lungs of the higher apes are large, highly vesicled, and can hardly be told from those of man. The frog's heart has but three partly perfect chambers, and the impure blood mixes with the pure. The gorilla's heart, with its four well-formed chambers, can scarcely be distinguished from the human heart except in size. The higher apes are also subject to many of the same diseases as mankind.

In a previous chapter it was also found that blood tests yield strong evidence of man's primate ancestry. To emphasize this fact again, human blood reacts to the serum of other mammals sensitized to human blood in such a way as to give a 100 per cent precipitate. Serum sensitized to man gives, with the chimpanzee's blood, about 90 per cent precipitate; with the blood of the lower monkeys, about 25 per cent to 33 per cent; and with that of the lower mammals, much less.

Farther down the phylogenetic line the precipitate drops out entirely. All of which is taken by scientists to mean that in the chemical properties of the blood also the kinship to man becomes closer as one ascends the mammalian line.

Wiedersheim, the great German anatomist, enumerated as many as one hundred and eighty vestigial organs in man. These are practically all duplicated in the apes. Man possesses

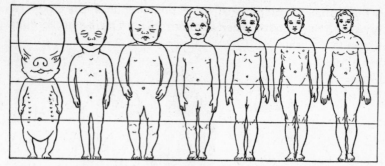

FIG. 175. Changes in Proportions during Prenatal and Postnatal Growth

After Stratz[1]

remnants of twitching muscles to move the scalp and ears. These muscles are also vestigial in the orang-utan and the chimpanzee. The inner corner of the eye in man presents rudiments of a third eyelid which is well developed and functions as a nictitating membrane in the bird. Apes have vestiges of this organ too, and in the Negroes and primitive Australians the fold is larger than it is in the whites. The appendix, so troublesome to man, is likewise found in the ape, although he seems not to suffer so much inconvenience from its presence. If space permitted, this comparison might be extended throughout almost the whole list of rudimentary organs. The instances given, however, are sufficient to show the likeness.

The embryogeny of man and the ape indicates an ancestral relationship. In an earlier paragraph (see page 224) certain recapitulations in man's embryogeny have already been cited. Among them was the appearance of a tail, gill arches, and, in most instances, a fine, silky covering of hair over the body. It will be of interest at this juncture to note other instances of what are taken to be a repetition of ancestral inheritance.

Adult man has his eyes close together, which permits of parallel vision. In mammals below the primates the eyes are usually placed at the sides of the head to permit of sight in

[1] From Sir Henry Morris's *Human Anatomy.* By permission, Yale University Press, publishers.

both backward and forward directions. The embryos of all primates have the eyes much farther apart than at birth. The human fetus, when nine weeks old, has a width between the eyes which measures more than half the width of the face. This gradually narrows until in the white adult the eyes are less than one fourth the width of the face apart. The eyes of the primitive Tasmanians and Hottentots are relatively farther apart than in the more advanced human races.

The human nose becomes higher and narrower in development from the fetus to adulthood. The young child, between the fetus and the adult, has a nose of intermediate height and

Fig. 176. A Month-Old Baby supporting itself by its Own Grasp [1]

width, the change in proportion being more pronounced in the white man than in the Negro. The adult ape has a nose similar in form to that of the early human embryo.

The relative length of arm and leg changes decidedly during the development of the human embryo and after birth. During the early fetal period the arm is much longer. The proportion changes until about a month after birth, when the arm and the leg are about equal. From then on throughout life the leg normally exceeds the arm in length. Relatively long arms and short legs are characteristics of the adult-ape group (see Fig. 174).

Monkeys and some of the apes lead an arboreal life, swing-

[1] Courtesy of *Journal of Heredity*.

ing from limb to limb among the trees. In these pursuits inwardly curved feet with prehensile great toes are a decided

Fig. 177. A Half-grown Gorilla
Note the inturned feet and the widely divergent great toes

advantage. At birth the child's foot is turned inward, and the big toe diverges markedly from the other toes and is very mobile. The whole aspect of the foot is that of an organ to be pressed against the trunks of trees in climbing. The human infant also possesses a surprising degree of grasping power in its hands and arms. A few weeks after birth it may hang from a support placed in its hands for a minute or two without falling.

Other instances of ancestral traits in the human embryo might be cited. These are sufficient, however, to make the principle clear. Scientists are able to account for these characters on no other grounds than that of recapitulation.

Geological evidences of man's evolution. The earth's crust also yields evidences of man's ascent. The story is told by the fossil remains of prehistoric man that have been, and are being, unearthed in various lands. This record is difficult to complete; but, so far as it goes, it is one of the most convincing of all the proofs of man's origin. The difficulty in getting at the facts from this angle arises largely through man's custom of interring the dead. Most animal fossils are derived from strata that have been water-deposited. The animal died when grazing in the lowlands or while seeking water. Subsequently its bones were buried by the sediment carried down from above and preserved. Millions of years afterwards these remains are unearthed either by man or by natural forces and form a regular compendium of knowledge as to the course of life on the earth. With prehistoric men it was different. Perhaps all of them, except the most primitive and animal-like, either cremated their dead or had some sort of burial custom. When they buried the dead they almost never interred in low, swampy ground. On the contrary, they often buried inside the natural caves in which they lived or in the grottoes that served as shelters. Subsequently the bones might be exposed to marauding animals or to excessive moisture, together with oxygen, which would readily promote decay. If interment was made in the earth, it is likely that the looser soils, which would yield more easily to the digging operations of their crude implements, were chosen as burial sites. This soil character would favor decay because of the ready access of both water and oxygen. Because of these conditions the bones of many men, contemporaneous with animals whose remains are found, have very likely been destroyed by the obliterating ravages of time. Moreover, the earth was not so densely populated with early men as it was with other animals.

For these and other reasons the paleontological evidence of man's origin, as we have said, is incomplete. Doubtless time and searching will reveal much more than is now known. However, regardless of the sketchiness of the record, paleontology discloses much that must always be considered in working out the origin and antiquity of man.

The Java man. In the order of estimated chronological sequence the Java man, *Pithecanthropus erectus*, stands first. The fragmentary remains of this ancient apelike man were found near Trinil, on the island of Java, in 1891 by Dr. Eugene Dubois. The remains include the left femur, two teeth, and the skullcap. These parts were scattered over some twenty yards of space and were not all found at the same time.

The teeth are of human type, and the straightness of the thigh bone indicates that its possessor walked almost as erect as present man. The skull has been the center of much dis-

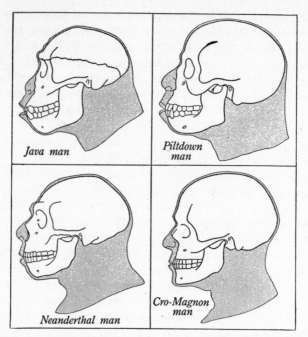

FIG. 178. Comparison of the Skull and Face of Different Types of Man[1]

cussion and controversy. Estimates of the cranial capacity have varied from 710 cubic centimeters to 1060 cubic centimeters, with an average of about 900 cubic centimeters. The cranial capacity of modern man ranges from about 930 cubic centimeters to 2000 cubic centimeters in extreme cases. Some of the apes now living have a cranial capacity of about 600 cubic centimeters.

In 1896 Dr. Dubois summed up the opinions of nineteen men who had studied the Pithecanthropus fossils and found that five of them judged the remains to be those of an ape; seven judged them to be human; and seven, including Dr. Dubois himself, considered them to be intermediate between the ape and man — a sort of "missing link."

[1] From G. A. Baitsell's *The Evolution of Earth and Man.* By permission, Yale University Press, publishers.

On the basis of the measurements of the fragments found and what is known of bodily proportions in man and apes, Dr. McGregor has completed the skull and face of this man in line drawings. The jaw is protruding and powerful to facilitate the grinding of the raw flesh, barks, nuts, and other coarse foods which he ate. The nose is flat, and the bony ridges over the eyes are prominent. The fore part of the skull is low and receding; this characteristic indicates a limited development in those associational areas of the brain which mark modern man. The geological strata in which the remains of Pithecanthropus were found indicate that he lived in the upper Pliocene approximately five hundred thousand years ago.

The Piltdown man. While Darwin sat in his study at Down puzzling over the problem of human origins, the fossils of the Piltdown man were quietly reposing in a gravel pit not thirty miles away. Workmen in excavating gravel for road repair unearthed the cranial remains of this man. The bones were shattered by the picks of the workmen and might have been lost had it not been for the keen mind and sharp eye of Mr. Charles Dawson, a local lawyer and antiquarian who sometimes visited this pit to look for crude implements that could occasionally be found there.

On one of these visits a workman showed him a piece of the cranial bone which he had picked up. Dawson at once recognized the possible value of this fragment and instituted a careful search for other remains. As a result of the most painstaking effort, portions of the cranium, a part of the jaw with teeth in place, a canine tooth, and parts of the nasal bones were uncovered.

The outline of the reconstructed head (see Fig. 178) indicates that this man also had heavy, powerful jaws and a receding chin. His nose was better formed than that of the Java man and the bones over the eyes were not so prominent. The forehead showed some height; but, compared with modern man, the Piltdown man was in truth a lowbrow. The jaws and the chin were decidedly apelike in character, but the cranium was much more strongly human. The cranial capacity is estimated by some to have been about thirteen hundred cubic centimeters as compared with twelve hundred and fifty cubic centimeters for the native Australian of today.

The Piltdown man doubtless had a simple primitive language and fashioned the crude flint hunting implements with which his bones were found associated. The charred remains in the caves that he inhabited show that he had discovered the art of making a fire. He has very appropriately been named *Eoanthropus* ("dawn man") *dawsoni*. There is a wide difference of opinion as to when the Piltdown man lived. Lull places it at from three hundred thousand to four hundred thousand years ago; other authorities believe that his period was from one hundred thousand to two hundred and fifty thousand years ago.

The Neanderthal man. Remains of this man were first found in 1848 near the Strait of Gibraltar. He was well distributed, however, and since that time skeletons have been discovered in Spain, France, Germany, Belgium, Austria, and different parts of Asia.

The number of specimens at hand have furnished rather complete and accurate information as to the height and other general body proportions of this man. He was of low stature, hardly exceeding five feet three inches for the male and less for the female. The head was thrust forward ape-fashion, the thigh bones were curved, and other features of the skeleton indicated that the Neanderthal man did not walk erectly but proceeded along with a sort of bent, shuffling gait. He was clumsy and loose-jointed, but a being of great muscular power.

The cranial capacity was very large, being about sixteen hundred cubic centimeters as compared with an average of approximately fourteen hundred cubic centimeters for modern man. The brow was still relatively low, however, and the development of the brain areas that control the higher intellectual and emotional functions was not great (see Fig. 178). That the Neanderthal being was a true man there can be no doubt.

The most famous skeleton of this race was discovered at the mouth of a cave in the Neanderthal gorge in Germany.

Partly for this reason he has been named *Homo* (man) *Neanderthalensis* (of Neanderthal).

The Neanderthal men were skilled workers in flints, fashioning such implements as spearheads, borers, planing tools, and scrapers. They lived mostly in natural caves and were not over-particular in the niceties of their housekeeping habits. The floors of the caves are strewn with the bones of animals, where they were evidently thrown after the flesh had been gnawed off. They ate such tidbits as the flesh of the mammoth, the woolly rhinoceros, the giant deer, and the cave bear. They must have had a taste for bone marrow too, because the large bones are often found split so as to have exposed this delicacy. The pelts of the animals slain seem to have been fashioned into crude clothing, and it is believed that the fat was often employed to make torches. The dead bodies of these cave-dwellers were apparently reverently buried and were often surrounded with the survivors' most carefully wrought objects and with a supply of food. All of which would indicate a settled belief in immortality.

This species is thought to have migrated into Europe probably from Asia. It is believed to have lived about fifty thousand years ago.

The Cro-Magnon man. The Neanderthals seem to have occupied the most of southern Europe for thousands of years. At length, however, weakened by the extreme cold of one of the glacial periods, they gave way to a much superior people from the east. This race is supposed to have migrated from Asia. The remains of this later man appear to have been much less abundant than those of his predecessor. Nevertheless, parts of skeletons have been found in Wales, in the Pyrenees, and near the rocky grotto of Cro-Magnon in France. If we include in this race, as is usually done, the Grimaldi skeletons which have been disinterred, then the remains representing the Cro-Magnon race are much more abundant. They were well represented across southern Europe and into Asia. A sepulchral grotto in the Pyrenees contained the remains of no less than seventeen skeletons and was discovered by a laborer. Its contents would have yielded abundant ma-

terial for an accurate study of this fine race had they not been
irretrievably lost.   The mayor of the neighboring village, not

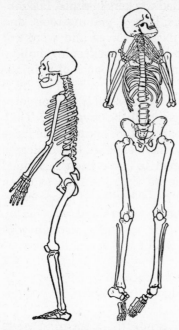

wishing to desecrate the dead,
ordered the remains to be taken
out of the cave and to be buried in the town cemetery.   It
seems that before the anthropological significance of this
important find was realized, the
place of interment had been
forgotten.   Consequently this
most invaluable group of bones
was permanently lost.

The most authentic records
of this race come from a group
of five skeletons discovered by
workmen in 1868.   They re-
posed in a grotto near the natural grotto of Cro-Magnon,
France.   It is from this source
that they received their name.

The Cro-Magnon was a su-
perior man.   It is an open question as to whether he did not
equal or even surpass modern
man in physical vigor, brain

FIG. 179. The Skeleton of Neanderthal (Left) compared with that of
Cro-Magnon. (*Homo sapiens*)[1]

power, and imagination.   At least he was a true *Homo sapiens*.
The men were over six feet in height, and the women probably were taller than those of the present time.   The Cro-
Magnon head was large, and the features were well formed
(see Fig. 178).   The chin was massive, the jaws shortened,
the nose prominent, and the forehead high and shapely.   The
cranial capacity has been estimated to have been almost
sixteen hundred cubic centimeters, which is in excess of the
average for the white race today.

This superb race is thought to have possessed a religious

[1] After Boule and Verneau, from G. A. Baitsell's *The Evolution of Earth
and Man.* By permission, Yale University Press, publishers.

FIG. 180. Bison modeled in Clay

The work of Cro-Magnon man, found in the cavern of Tuc d'Audoubert [1]

sense, and the fact that their remains were interred with implements of industry and warfare, together with food, indicates a belief in immortality. They probably had learned to make and to use the bow and arrow, which gave them a decided advantage over their more primitive competitors, the Neanderthals. The advent of the Cro-Magnons marks the beginning of art in Europe, according to Dr. Osborn. Figures sculptured in clay have been found in their caves, and the walls were often adorned with animal drawings of surprisingly artistic beauty.

The Cro-Magnons are believed to have entered Europe not less than twenty-five thousand years ago. After thriving for ten or fifteen thousand years they declined, leaving no descendants, with the possible exception of a few people now living in France, Brittany, and perhaps the Pyrenees.

There seem to have been several lines of human ascent from the main stem. It was formerly believed that the Java, Piltdown, Neanderthal, and Cro-Magnon men, together with

[1] From H. F. Osborn's *Men of the Old Stone Age*. By permission, Charles Scribner's Sons, publishers.

many other prehistoric types omitted in this discussion, represented what was a possible single line of ascent in man. It is

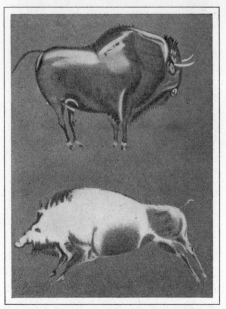

FIG. 181. Paintings made by the Cro-Magnon Man on the Roof of the Altamira Cave in Northern Spain

A bison above and a galloping boar below [1]

now regarded as probable, on the contrary, that these various races were all offshoots from the main human stem. As time passed, these early, more or less divergent branches became extinct, whereas the main stem persisted and continued to develop. The subsequent offshoots became widely separated geographically and, according to one authority, finally differentiated into the four distinct races found on the globe today. Some of these types, such as the native Australians, the Tasmanians, and the African Bushmen, have never evolved much above the Neanderthal level of intelligence and culture. Others continued to evolve biologically and socially into the more dominant and progressive races of modern times.

There are four subspecies of modern man. Of the four subspecies of modern man, the Australians are natives of Australia and Hindustan. They are tall, long-limbed, and are almost chocolate in color. They vary widely in intelligence, some of them being quite teachable, while others represent the lowest plane of human mentality. The Negroid race is found in Africa south of the Sahara and in some of the adjacent islands, especially Madagascar. There are many subdivisions of this

[1] From Thomson's *Outline of Science*. George Newnes, Ltd., publishers.

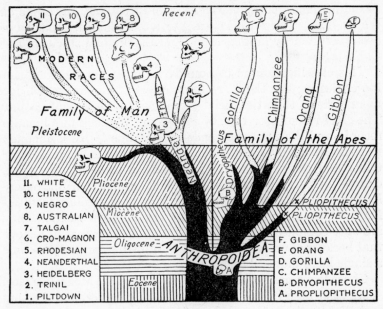

FIG. 182. A Suggested Genealogical Tree of Man and the Anthropoid Apes[1]

group, but for the most part they are characterized by flattened noses, woolly hair, and black, or chocolate-colored, skin. The Mongolians have a high degree of intelligence; but as a group they yield place, in enterprise and aggressiveness, to the fourth subspecies, the Caucasians. The Caucasians are of three types. The descendants of the Mediterranean branch include the Greeks, Italians, Spaniards, and Englishmen. They are dark-complexioned and have brown eyes and long heads. This group entered Europe along the Mediterranean. The Alpine type, which migrated into Europe by way of the Black Sea, took up its abode in central Europe. The descendants of this division include some of the Greeks, Turks, Armenians, North Italians, South Germans, some French and Swiss, and perhaps other peoples of those regions. The Nordic, or blond, type entered northern Europe by way of Russia. The modern descendants of this branch are believed to be the Danes,

2

[1] By courtesy of *Science*.

Scandinavians, North Germans, some Russians, and some of the English. The white representatives in the Americas have been recruited by immigration from the original stocks in the Old World.

This classification as to boundaries must not be examined too critically; for there are few corners of human knowledge more difficult than the racial groupings and origins of people. The chief reasons for this, of course, are the lack of adequate records and the nomadic tendencies of man. He is inclined to roam about over the earth and either to subjugate other tribes and compel them to amalgamate with his own kind, or else to marry peacefully into their groups and ultimately lose his own identity among them. Moreover, modern facilities for transportation are not making the problem easier. They are only complicating it. Indeed, there are those who believe that because of ease of travel the world will eventually become a single large mongrel race.

The native American stock was derived from the Mongolian race. A study of man in the Americas has disclosed the interesting fact that all the nations belong to the same type. The Mayas of Central America, the Incas of Peru, the Aztecs of Mexico, the Indians of both continents, and the Eskimos and Aleuts of Alaska all bear marks of kinship. The evidence is preponderant that these different groups are but the more recent branches of one of the main racial stems. This is true whether the conclusion rests on bodily measurements or on the great similarity of tribal customs, religious beliefs, languages, tools, and legends.

It is believed, then, that the ancestors of the native Americans were Mongolians and that they originally came from Asia. Their immigration into America probably was the result of a hunting expedition to find new sources of game, or was the outcome of journeys inspired by their own adventurous curiosity. At any rate, it seems reasonable to suppose that they crossed from Asia to Alaska — either by a land bridge that has since subsided or in canoes of their own construction. From Alaska they journeyed southward, multiplying and eventually becoming differentiated into the distinct tribes represented today and in the past.

Just how long the native Americans have been here is difficult to say. The oldest human remains in North America do not seem to antedate the last glacial period. From this circumstance their age is estimated at not more than twenty-five thousand years. South America furnishes no certain evidence that man has been there more than five thousand years.

The place of man's origin is uncertain. The question of man's original birthplace has been an interesting one. It has also been one of much discussion and controversy. The preponderance of evidence, however, seems to point to the conclusion that he arose somewhere in the region of Asia.

The exact stages by which man evolved are obviously a mystery sealed in the pages of the far-distant past. Yet geological evidences furnish material from which, as Burlingame and his associates state, "a shrewd guess" may be made. Asia at one time had a far more humid climate than at present, and its expanse was covered largely by great tropical forests. Within the branches of these trees vast hordes of lemur-like animals found food and shelter. The lemurs are a very old stock akin to the monkeys. Later the continental area was pushed up by geological forces and became decidedly drier. Consequently the trees gradually retreated in favor of grassland and desert. With them went most of the lemurs. Their descendants are found today inhabiting the remaining forest regions of Asia and Africa as gibbons, orang-utans, chimpanzees, and gorillas (see Fig. 174).

For some reason one branch of the lemur-like crowd retreated from the trees, so to speak, and took to life upon the ground. This change of habitat was conducive to an upright posture, which permitted its possessor to sight food more easily and to detect the approach of an enemy more readily. Moreover, with the arms and hands relieved from the work of climbing, they came to be used for grasping food and hurling weapons. As time passed the hands became refined in movement and dexterity. Eventually they developed into highly specialized organs to act as servants for a growing and expanding brain. The power of speech was achieved, and in some such way early man evolved into the ancestors of our present races.

It is believed by some that with careful systematic search the sand-drifted desert areas of Asia may yield valuable fossils which will assist materially in unraveling this anthropological story.

There are other evidences that man evolved in Asia, which was a vast area with a great variety of conditions and climate. Moreover, it possessed passageways to Europe, Africa, North America, and perhaps lesser lands, through which nomadic groups might proceed in their migratory movements to more remote parts of the earth. Great waves of restless barbarians are known to have swept out of Asia into other lands, especially Europe. It has already been mentioned that the Americas probably derived their human stocks from this source. It is possible also that other movements into Africa and Australia took place. The most primitive African Pygmies and Australian blacks are believed by some to be but the outer remaining fringes of these early migrations. Asia seems likewise to have been the land in which most of our plants and animals were originally domesticated. The domesticated common fowl, cattle, goats, and camels, together with some breeds of horses, cats, and dogs, are known to have come from Asia. Lastly, southwestern Asia is the seat of the oldest civilizations known to man. The spade of the archæologist is annually uncovering additional evidences of these "venerable abiding places of mankind." The black inscription stones disinterred give mute testimony of cultures that thrived five thousand or six thousand years before the Christian Era and which antedate Archbishop Ussher's chronology by a thousand years or more.

Man's intelligence is believed to have developed gradually. There are those who answer the question as to the origin of man's intelligence easily and confidently. Their solution is *special creation*. What could be more simple and assuring? If this point of view could be established scientifically, it doubtless would readily be embraced by most men. But it cannot. All the evidence summed up in this chapter seems to point with reasonable certainty to the fact that man derived his body from the lower animals. There may have been a great Architect behind the scenes guiding the process,—a belief that is shared by many scientists, — but that man's ancestry runs back to other mammals is a view which they feel can scarcely be questioned. But how, then, is the superior mind which man possesses to be explained. Is it of animal origin too? Or is it not something paramount to and far beyond the mentality of even the highest ape? There can be no question that in his rationalizing, generalizing powers man is far superior to the whole simian group. He knows how to abstract principles from concrete facts and to profit by these principles in experience. There is slight evidence that apes can do this, even to the smallest degree.

On the other hand, that animals do have minds to guide them in their activities is patent. Sometimes the apparent intelligence of a friendly dog or a well-trained horse is most astonishing. Apes do feats involving a degree of mentality that is most surprising. So when we come to look at the question of mind as a whole it also, like the body, seems to have manifested its first faint flickers far down in the animal kingdom. As the body advanced the intelligence grew in power and effectiveness. Finally, in man mental development reached a point where his mind has become the most baffling aspect of his being and, at the same time, the most hopeful.

Animal intelligence can easily be misinterpreted. We have all seen trained horses and "educated" dogs whose feats and tricks astonished us. They would apparently add and subtract numbers, recognize the significance of signals, execute commands, and go and get articles for which they were sent, with a readiness and exactitude that was indeed surprising. Under such circumstances, however, one should beware of the conclusions he draws as to animal intelligence. In the first place, "educated" animals of this character have been brought to a high degree of precision in their performances only after long and patient hours of training. Moreover, in this process all the appropriate thinking ability and ingenuity of the man in charge have been brought to bear upon the problem. So, in reality, the "education" of the animal is the end product of a process in which the human mind has been quite as much involved as has the mind of the animal itself. In most cases there is little evidence that the animal alone could ever reach the height of performance which it is able to attain under man's tutelage. In the second place, the performing animal is always under the direction of a trainer, who is thinking much more than the animal itself while the feat is being accomplished. The result achieved, then, is unquestionably as much a product of the trainer's mind as it is of the mentality of the animal. In the third place, the significance of the acts which animals perform is being interpreted by us who have human minds. A very grave danger of misjudgment lies at this point. When we see the bee visiting flower after flower and storing in the hive the honey produced from the nectar, we are prone to attribute to this insect a degree of providence and foresight all out of proportion to that which it actually possesses. The bee is actuated by instinct and not by motives of economy. What it does is the only way it could react to the situation. So, for these reasons, when we are judging the intelligence of animals we should always be on our guard against fallacies.

Animals do have certain levels of intelligence. To the careful observer, however, it is perfectly evident that animals many times react to situations with a certain degree of intelligence. What they do is often the product of blind instinct. But in many cases, especially among the simian group, there are unmistakable evidences of mental reactions unfettered by any hard-and-fast inherited patterns of action. What they do is appropriate to the occasion, and it varies with the conditions of the situation.

Professor Whitman placed the eggs of a passenger pigeon a few inches from the nest. The bird acted a little uneasy, felt under her body with her beak as if searching for something that was not there, but made no attempt to recover the eggs. In a short time she flew away, abandoning the nest. Certainly the incubating activities of a passenger pigeon are largely the product of instinct. There was little evidence here of action appropriate to the occasion.

Professor Lloyd Morgan raised a couple of moor hens. This bird, which is known in America as the coot or mud hen, not only possesses remarkable diving powers, but the adult seems to enjoy exercising them on the least pretext. Professor Morgan relates that his young coots swam readily, but they would not dive. Suddenly one day a barking puppy came rushing down the bank of the pool in which one of the young birds was swimming. Immediately it dived and then presently rose just enough to bring its eyes above the water line, so it could watch the dog. The diving was instinctive, of course; yet, since the bird would not dive before, there was a suggestion of suiting the action to the occasion — that is, a choice of action had been made.

Mammals have a better-developed brain and consequently manifest more intelligence, but oftentimes their reaction seems extremely unintelligent. A cow whose calf has been removed is restless and agitated. Yet the stuffed skin of a calf is a great comfort to her. Cases have been reported in which the mother, in affectionately licking such a dummy calf, ripped the seams open with her tongue. She then proceeded to eat, with perfect contentment, the hay which formed the stuffing material inside the skin.

When an animal such as a dog does learn to do purposeful things, unless he is taught specifically by man to do them the learning process is almost invariably one either of trial and error or else of accidental association. As has been said, "He muddles through." A dog may learn to open a large box to get food by pressing a lever. Attracted by the scent of the food in the box, he paws at the box, making many errors wholly immaterial to his success. Finally, his paw falls on the lever and the latch is released, presenting the food. A few errors are made the second time the box is presented, but the appropriate connection between paw and lock follows much more quickly. In a short time the relation of the two becomes fixed in the brain somehow, and upon presentation the box is opened immediately. This result is not the product of instinct in any sense. The act has been learned. But the learning is on a very low mental level.

A certain bear in a zoölogical garden would sit up on his haunches and wave his paw for peanuts whenever a boy or girl approached with a paper bag in hand. Such behavior, at first glance, appeared highly intelligent. However, this learning act had evidently been purely adventitious. The eagerness to evoke peanuts had doubtless been accompanied one day by a waving of the paw. The prize was secured. In some way the connection between "waving paw" and "peanuts" had then become registered in the bear's brain cells. This "cute" action was further intensified and fixed by the ready response which the delighted boys and girls made to his ludicrous appeal. Doubtless there is no rational action involved here. The association was a purely accidental one.

Monkeys and apes are the most intelligent of all animals below man. In comparatively recent years apes have been experimentally studied in respect to their habits and intelligence. Professor Köhler of Germany and Dr. and Mrs. Yerkes in America, besides others, have added much to our knowledge of their behavior.

Monkeys and apes have been found to differ markedly from the lower mammals in respect to their restless natures and curious dispositions. Except when resting they are always on the go, looking into things and examining in the most curious manner everything that comes within their reach. As Dr. Thorndike truthfully said: "Watch a monkey and you cannot enumerate the things he does, cannot discover the stimuli to which he reacts, cannot conceive the *raison d'être* of his pursuits. Everything appeals to him. He likes to be active for the sake of activity." The inquisitive proclivities of these simian cousins doubtless account for much of their intelligence. Like the child, they learn much because they are always prying into things.

These animals are also highly emotional and expressive. In speaking of the chimpanzee, which in some respects is most like man, Dr. Köhler gives us the following account:

The chimpanzee's register of emotional expression is so much greater than that of average human beings, because his whole body is agitated and not merely his facial muscles. He jumps up and down

Quietude Sadness Laughter

Weeping Anger Excitement

FIG. 183. Emotional Expressions of a Young Chimpanzee[1]

both in joyful anticipation and in impatient annoyance and anger; and in extreme despair — which develops under very slight provocation — flings himself on his back and rolls wildly to and fro. He also swings and waves his arms about above his head in a fantastic manner, which may not be unknown among non-European races, as a sign of disappointment and dejection. . . . During the leisurely contemplation of any objects which give particular pleasure (for example, little human children), the whole face, and especially the outer corners of the mouth, are formed into an expression that resembles our "smile." . . .

Chimpanzees understand "between themselves," not only the expression of *subjective moods* and emotional states, but also of definite desires and urges, whether directed towards another of the same species, or towards other creatures or objects. . . . Thus, one chimpanzee who wishes to be accompanied by another, gives the latter a nudge, or pulls his hand, looking at him and making the movements of "walking" in the direction desired. One who wishes to receive bananas from another, imitates the movement of snatching or grasping, accompanied by intensely pleading glances and pouts.

[1] From a photograph by Mrs. Kohts in Yerkes and Yerkes' *The Great Apes*. Yale University Press, publishers. By permission, Mrs. Kohts.

The summoning of another animal from a considerable distance is often accompanied by a beckoning very human in character.[1]

Apes likewise exhibit a wide degree of individual differences. Some are happy and vivacious in disposition; others are morose and gloomy. Some are alert mentally; others are dull and stupid. Professors Yerkes and Learned relate the following of two chimpanzees:

Chim is sanguine, venturesome, trustful, friendly, and energetic, whereas Panzee was distrustful, retiring, lethargic. His behavior usually suggested unusual intelligence; hers stupidity. In their relations to people the animals exhibited their usual diversity. Chim would go willingly to almost anyone who seemed friendly. Panzee's reaction was difficult to predict. Sometimes she would meet advances more than half way. Occasionally she would seek out a stranger. Both animals appreciated kindness. Panzee's mode of expressing appreciation was a gentle pat on her attendant's shoulder. ... Chim in a few instances exhibited his friendly spirit toward human companions by bringing objects to them. A case in point is the careful plucking of some blossoms one day in the New Hampshire pasture and the presentation of them to a lady attendant.[2]

Chimpanzees likewise, on occasion, manifest a surprising degree of ingenuity. Dr. Köhler experimented with Sultan, a chimpanzee. He suspended a banana on a string high above Sultan's reach and placed several large wooden boxes inside the inclosure. After futile attempts to get the fruit Sultan finally devised the plan of stacking the boxes one on another to raise himself up high enough to get the prize. The boxes, however, were piled up in a rickety and uneven fashion. Other members of the group readily duplicated this feat.

One of the reactions requiring the greatest display of intelligence was called forth when Köhler placed a banana outside the cage far beyond the chimpanzee's reach just to see what he would do. He describes the experiment thus:

[1] Wolfgang Köhler, *The Mentality of Apes* (second revised edition; translated by Ella Winters). By permission, Harcourt, Brace and Company, Inc., publishers.

[2] Robert M. Yerkes and Blanche W. Learned, *Chimpanzee Intelligence and Its Vocal Expressions*. By permission, The Williams & Wilkins Company, publishers.

This time Sultan is the object of experiment. His sticks are two hollow, but firm, bamboo rods, such as the animals often use for pulling along fruit. The one is so much smaller than the other, that it can be pushed in at either end quite easily. Beyond the bars lies the objective, just so far away that the animal cannot reach it with either rod. They are about the same length. Nevertheless, he takes great pains to try to reach it with one stick or the other, even pushing his right shoulder through the bars. When everything proves futile, Sultan commits a "bad error," or, more clearly, a great stupidity, such as he made sometimes on other occasions. He pulls a box from the back of the room towards the bars; true, he pushes it away again at once as it is useless, or rather, actually in the way. Immediately afterwards, he does something which, although practically useless, must be counted among the "good errors": he pushes one of the sticks out as far as it will go, then takes the second, and with it pokes the

FIG. 184. A Chimpanzee using a Pyramid of Boxes to enable him to reach Suspended Bananas[1]

first one cautiously towards the objective, pushing it carefully from the nearer end and thus slowly urging it towards the fruit. This does not always succeed, but if he has got pretty close in this way, he takes even greater precaution; he pushes very gently, watches the movements of the stick that is lying on the ground, and actually touches the objective with its tip. Thus, all of a sudden, for the first time, the contact "animal-objective" has been established, and Sultan visibly feels (we humans can sympathize) a certain satisfaction in having even so much power over the fruit that he can touch and slightly move it by pushing the stick. The proceeding is repeated; when the animal has pushed the stick on the ground so far out that he cannot possibly get it back by himself, it is given back to him. But although, in trying to steer it cautiously, he puts

[1] From Wolfgang Köhler's *The Mentality of Apes*. By permission, Harcourt, Brace and Company, Inc., publishers.

the stick in his hand exactly to the cut (i.e. the opening) of the stick
on the ground, and although one might think that doing so would

suggest the possibility of
pushing one stick into the
other, there is no indication whatever of such a
practically valuable solution. Finally, the observer
gives the animal some help
by putting one finger in the
opening of one stick under
the animal's nose (without
pointing to the other stick
at all). This has no effect;
Sultan, as before, pushes
one stick with the other
towards the objective, and
as this pseudo-solution does
not satisfy him any longer,
he abandons his efforts altogether, and does not even
pick up the sticks when

FIG. 185. The Chimpanzee Sultan constructing a Serviceable Instrument by joining Two
Sticks[1]

they are both again thrown through the bars to him. The experiment
has lasted over an hour, and is stopped for the present. . . . As we
intend to take it up again after a while, Sultan is left in possession of
the sticks; the keeper is left there to watch him.

The keeper, recalling Köhler, reported as follows:

Sultan first of all squats indifferently on the box, which has been
left standing a little back from the railings; then he gets up, picks
up the two sticks, sits down again on the box and plays carelessly
with them. While doing this, it happens that he finds himself holding
one rod in either hand in such a way that they lie in a straight line;
he pushes the thinner one a little way into the opening of the thicker,
jumps up and is already on the run towards the railings, to which he
has up to now half turned his back, and begins to draw a banana
towards him with the double stick.

When the stick, because of its loose construction, fell apart,
Sultan immediately connected the parts again with "great
assurance" and proceeded to pull in the fruit. Köhler continues:

[1] From Wolfgang Köhler's *The Mentality of Apes*. By permission, Harcourt, Brace and Company, Inc., publishers.

The proceeding seems to please him [Sultan] immensely; he is very lively, pulls all the fruit, one after the other, towards the railings, without taking time to eat it, and when I disconnect the double-stick he puts it together again at once, and draws any distant objects whatever towards the bars.[1]

The next time the experiment was tried the chimpanzee attempted to reach the bananas with the single stick once more, but, failing in this, within a few seconds he coupled up the sticks again and proceeded to get his fruit. Sultan evidently had learned something and had remembered it.

On another occasion this same animal was given two sticks, one of which was hollow, and the other, a thin hard one, too wide to slip into the first. After some manipulation he bit splinters from either side of the wide thin stick at one end until it would telescope into the other. Then, with this self-constructed implement, he proceeded to obtain the distant objects desired. Although Sultan bit the opposite end of the stick he was working on to some degree, he scarcely splintered it. He confined the splintering activities to the sides of the end which he was fashioning into a wedge.

This feat certainly showed a high degree of intelligence for an ape. It was on the order of that shown by a savage when he cuts down a tree and chips its trunk into a crude canoe to be used in crossing a stream.

So, then, it may be seen that among animals there is a growing intelligence, from that of the insect, which works wholly from instinct, on up to the ape and man. Mind, as well as body, seems to have evolved slowly; indeed, the development of one seems to have paralleled the other. It may be that at each step of progress the mind had to wait for a better brain in which to function. Doubtless the evolution of speech, too, had much to do with the matter of an increased mental capacity.

However, one should not connect apes and men on the intellectual side without remembering at the same time that between the living representatives of these two groups there

[1] Wolfgang Köhler, *The Mentality of Apes.* By permission, Harcourt, Brace and Company, Inc., publishers.

is a wide degree of difference. It may be a matter of degree only, but, such as it is, the distance between the two certainly exists.

It should be kept in mind that when Sultan modeled one stick so as to fit into the other, it was an especially brilliant achievement for any ape, even one as intelligent as Sultan. Köhler, in discussing this feat, remarks: "I could not guarantee that each repetition of the experiment would turn out so well. Sultan evidently had a specially bright day."

Moreover, the particular instances of chimpanzee intelligence enumerated must not be taken as the usual level upon which their minds work. On the contrary, the feats depicted are the high lights of their mental achievement. Anyone who has ever watched a cage of these animals knows this to be a fact. The aimless wandering around, the inane things that they do, and the perfectly idiotic reactions to many situations in which they find themselves certainly show that for most of the time, at least, their minds operate on a much lower plane. The simian stunts reported by these observers must be regarded in ape land as being comparable with the rare flights of the human mind when it writes a great poem, invents an intricate machine, or makes an important scientific discovery.

Wherein do the chief mental differences between man and the apes lie? Apes experience intense emotions, express sympathy, have memory, and devise simple implements. Wherein, then, does the principal difference between them and man lie?

Chimpanzees and gorillas do achieve some wonderful things. However, their minds at their best work on what Dr. Romanes has called the "perceptual inference" level. That is, they think about objects and do things with them only when the objects themselves are present. Never, or at least to a negligible degree, do they ever seem to abstract items of information and weld them together into general ideas. As for using these general ideas to profit in experience and to guide future conduct, they seem to have no conception whatever. A simple case will illustrate what is meant.

Both an ape and a man from Professor Yerkes' laboratory

in New England might visit an orange grove in Florida. Both of these primates could see the oranges, smell them, lift them, feel them, and taste them. In fact, so far as we can see, both might acquire the same sense perceptions regarding the nature of this fruit; but when the two returned to New England, from that point on (so far as man can judge) there would be a vast difference between them. The ape apparently has no idea of an orange when removed from its presence. On the other hand, man can remember the smell, feel, color, size, and taste of the orange. These sense percepts have been abstracted from the object itself and may be welded into a general concept or idea of this fruit. Moreover, this general idea, together with its parts, may be employed by the man's mind to think about this object. He might write a short essay describing the orange, telling of his sensations regarding it, explaining its culture, and portraying the kind of tree upon which it grew. The man could do this, even though the closest orange might be a thousand miles away in Florida. But the ape could do nothing of the kind. Even if he possessed language and a knowledge of written symbols, he probably could not do it. For he has formed no abstract idea of the orange with which to work. He has drawn off no mental product from the orange itself. He apparently can think of an *orange* only when the orange itself is present. In this respect, then, there is a wide gap even between the wisest ape and the most ordinary man.

But we may carry the difference farther. The man, either as a result of his own study or by acquiring abstract ideas from other sources, may have learned that the orange is rich in a certain vitamin; moreover, that in the absence of this particular vitamin the human body, as well as that of the ape, will develop bleeding gums, the teeth will become loose, growth will cease, and certain death will ultimately follow. Now the man may use these ideas to guide him in his future conduct to avoid disaster. He will procure oranges or similar fruit as a part of his diet to prevent scurvy in his family. But in the absence of any of these general ideas the ape would be utterly helpless in such a situation. If oranges were present, he might eat them and prevent serious bodily derangements; but in so doing he would not be following intelligent ideas in any sense: he would be acting in response to appetite alone.

This ability to abstract ideas from concrete objects and to use them readily in thought processes puts a wide gulf between the ape and man. The former is concerned from the day of his birth to the end of life with the physical and the concrete. Perceptual inferences are the mountain peaks of his experience. If he deals with ideas at all, it certainly is in a very elementary fashion. Man can think on a low plane too, using perceptual inferences. But he can rise leagues above this plane. He may converse by the hour with a friend about things and objects thousands of miles away. Or he can read a book about Africa with keen interest and delight, forming vivid mental pictures and, in his imagination, building up scenes which he has never experienced. The elements which he is manipulating in such thinking are pure concepts. He handles these ideas, when trained to do so, with as much ease and facility as a player moves "men" about on a board in solving chess problems.

The ability to do abstract thinking carries with it great implications. It raises questions of values. Of two possible lines of action man is induced to ask which is better. This leads to considerations of conduct. It enables man to picture the universe and to realize its vastness and grandeur, together with his own comparative insignificance. Such reflections evoke a spirit of reverence and awaken a religious response. It induces man to consider the rights and relations of other men, and out of this regard come ideas of morals and ethics, of justice and of government. In short, the power to think abstractly with facility has placed man on a pinnacle high above the other animals. In so far as this attainment can make him so, he is an entirely different kind of creature.

QUESTIONS FOR STUDY

1. What evidence is there that man is related to the lower animals?

2. How may the probable relationship of man to apelike individuals be shown?

3. Trace the geological evidence which indicates the evolution of man.

4. What ancient man is regarded as being most nearly related to modern man? Give evidence to support your answer.

5. Show the probable relationship of man to the other primates.

6. Name and briefly characterize the subspecies of man now found on the earth.

7. Where do you believe man first appeared? Give evidence to support your answer.

8. How does man differ mentally from all other animals?

9. What evidence is there that animal intelligence is limited?

10. What did the work of Köhler reveal regarding the intelligence of apes?

11. Point out the fallacies likely to be involved when a very high degree of intelligence is ascribed to a trained animal.

REFERENCES

BAITSELL, G. A., and others. The Evolution of Earth and Man, pp. 148–227.
BEAN, R. B. The Races of Man.
BURLINGAME, L. L., and others. General Biology (second edition), pp. 527–536.
KÖHLER, WOLFGANG. The Mentality of Apes.
MASON, FRANCES, and others. Creation by Evolution, pp. 293–310.
MILLER, G. S. "The Controversy Over Human Missing Links," *Annual Report of the Smithsonian Institution*, 1928, pp. 413–447.
OSBORN, H. F. Men of the Old Stone Age, pp. 49–455.
SHIMER, H. W. Evolution and Man, pp. 139–187.
THOMSON, J. A. The Outline of Science, Vol. I, pp. 155–242.
WELLS, H. G., HUXLEY, J. S., and WELLS, G. P. The Science of Life, Vol. IV, pp. 1147–1269.
YERKES, R. M., and YERKES, ADA W. The Great Apes.

Chapter XIII · Heredity is a Potent Factor in determining what the Individual may Become

Introduction. For some unexplained reason the average man seems loath to recognize the force of heredity in human affairs. He may apply the most rigid genetic principles in the growing of plants and other animals and, at the same time, disregard these principles altogether when it comes to his own kind. A progressive stock-grower will purchase no herd animal which is not accompanied by a long pedigree tracing its ancestry in detail. Any breeder who would release a number of scrubs among his breeding animals would have his sanity questioned at once. Especially would this be true if the scrubs were given to multiplying twice as fast as the thoroughbreds. The final degeneration of the herd would be predicted with certainty.

Strange to say, however, when it comes to man no such fears seem to exist. The defectives of the race often are allowed to reproduce faster than the superior individuals, without any concern whatever. In their altruistic moments some people would even go so far as to let down the immigration bars completely, trusting to luck and to blind faith in their own virtues for the assimilation of every immigrant who might enter the country.

Moreover, in educating boys and girls little attention is given to native ability and individual differences. In many cases the public schools provide identical subject matter for the whole group, use the same methods of instruction, and then assume that the pupils will all acquire desirable habits of study and graduate with proper ideals of industry and correct conceptions of citizenship. From the public schools the parents send their sons and daughters to college. They influence them to study for the professions of law, medicine, engineering,

408

teaching, or for business without any regard for their particular aptitudes and limitations. If the students apply themselves well, follow directions, and remain in school until graduation, the sanguine parents confidently expect them to become potential leaders in their respective fields. Whether or not these young people have the temperament, the physical vigor, or the native intelligence to succeed in the vocations selected is a consideration that receives little attention. The result is often tragic. No one would for a moment expect to take from the pasture a heavy draft colt and to transform him into a race horse by proper grooming and training. Neither would he expect to change a heavy beef heifer into a high-grade butter-fat-producing cow by the diet and environment of the dairy barn. Yet, under our system of education and lack of hereditary knowledge, children are expected to make these often impossible transformations. In order to act with reasonable intelligence, then, toward boys and girls, parents and teachers ought to know something about heredity as it affects mankind.

It is difficult to study human heredity. Despite its great importance in the affairs of men the course of human heredity is very difficult to study and to unravel. The people of civilized races are, for the most part, hybrids of the most complex character. The blond has married the brunette, the tall has mated with the short, and the Englishman has espoused a French wife. Because of the heavy immigration to the United States and the diverse countries from which the parental stocks have come, the blood of this country is perhaps more mixed than that of any other country. The length of human life also makes man's heredity difficult to determine. Fruit flies may be placed in a milk bottle with a piece of stale banana, and in approximately two weeks a new generation with its hereditary characters can be produced; but the time elapsing between human generations is a score or more years.

Flies, plants, and animals other than man can be manipulated and thoroughly controlled in their matings. Man cannot. To do so would violate all established ideas of sentiment and morality. Consequently what we learn of man's heredity must be gained largely from observation.

Finally, endocrine secretions have been found to be involved in man's physiological processes in a most important and often

FIG. 186. A White Blaze in Human Hair[1]

abstruse manner. In certain instances heredity is thought by some to determine largely the functional efficiency of certain of these ductless glands. Just how far environment and ancestry may coöperate in the physiological processes carried on by others of these glands, however, is not definitely known. For these reasons, as well as others, progress in ferreting out and organizing human genetic knowledge has been comparatively slow. It will continue to be so in the future.

Nevertheless, within the past thirty years much progress has been made in this field. Some knowledge of over a hundred physical traits has been attained and much information regarding the more difficult and elusive mental and emotional characters has been determined.

So far, however, the discoveries in this field have pointed to but one conclusion: that human heredity, in general, follows the same laws and principles of transmission that it does in the other higher animals. In this respect there is no fundamental difference between man and the horse, for instance. This fact not only gives a strong hint as to man's evolutionary origin but it enables man to gain much knowledge of himself in this field by studying other animals.

The course of heredity has been found to be complicated. As has been stated, in many cases the course of human heredity, as in all other higher animals, is very complicated. In a few instances, however, it has been found to be comparatively simple. For example, certain people have a white lock of hair. If the surrounding hair is black, the white blaze stands out

[1] From Paul Popenoe's *The Child's Heredity*. By permission, The Williams & Wilkins Company, publishers.

in sharp contrast. This characteristic is usually passed on as a simple dominant. If one of the parents is pure for this character and the other normal, all the children will have the white lock. Its appearance in succeeding generations will depend upon whom the children marry.

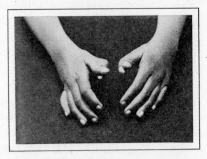

FIG. 187. A Polydactylous Individual

Note the extra digit on each thumb

Brachydactylia (Greek *brachy* + *dactylos*, "short-fingered" or "short-toed") is another character that is inherited in the same manner. This defect consists of an omission of one of the joints in the fingers, making them short, stubby, and unattractive. Polydactyly, or the production of an extra toe or finger, is a more common defect. Its presence is not easily detected, however, because of resort to surgery in infancy to remove the extra members. In the same category may be placed syndactyly, in which two or more fingers or toes are connected by an integument. This condition is often called *webfoot*.

Many human characters are recessive. Straight hair, for example, is recessive to curly hair, and the effect of the blue-eyed gene is also covered and hidden when a darker-eye-producing gene is present. Color blindness is recessive to the power to distinguish red and green. In this particular case transmission is the more readily traced from parent to offspring because it is connected with sex. The gene determining sex and the one for either normal color discrimination or color blindness lie in the same chromosome. Since the female has two sex chromosomes and the male but one, a single defective gene in the former would produce only a carrier, but a single one in the latter would cause color blindness. It is for this reason chiefly that there are about three times as many color-blind men as there are color-blind women.

Many other human characters, by contrast, are more complicated in the manner of their expression. The color of the

Negro's skin furnishes an example. Everyone has observed that there is a very wide range in the degrees of color presented by mulattoes. Davenport believes that color in the African Negro is due to two pairs of hereditary factors working together, and that the color of the hybrid depends both upon what particular genes for color happen to be present and upon the number of them.

Longevity, or length of life, is known to be inherited. Alexander Graham Bell, inventor of the telephone, became interested in the question as to why some people live longer than others. He personally studied the life records of the famous Hyde family. Of these there were sixteen hundred and six males and thirteen hundred and fifty-two females. Bell found that of this group the fathers and mothers who lived to be eighty or past had children whose average age, including infant deaths and all others, was fifty-two years. On the other hand, Hyde mothers and fathers who died under sixty had boys and girls whose average age was but thirty-two and eight-tenths years. Dr. Raymond Pearl, in commenting on the report, makes the grim but facetious remark that, "however the matter is taken, a careful selection of one's parents in respect to longevity is the most reliable form of personal life insurance." The manner in which longevity, or the lack of it, is transmitted, however, is not definitely known.

Stature and weight are recognized as being hereditary. But since environment also plays a large part in determining the growth and development of the individual, the exact manner of transmission has not been ascertained. Tall parents are inclined to have families whose average height is greater than the average of the general population. Among the offspring, nevertheless, there may be some of average height or even shorter. This would not be the case if the parents were pure for this character; but since they are almost always hybrids, it is to be expected that the short individuals would segregate out occasionally. The same relation holds for parents below average weight.

Some individuals are so far undersized as to be classed as dwarfs. Of these there are two types: those that are well proportioned as to length of body and limbs, and those with relatively large heads and bodies but with short, stubby, and often twisted arms and legs. The first type are gracefully built and, aside from height, may not be unattractive; the second class are ill proportioned, grotesque, and may be even hideous in appearance. The relative part that heredity and endocrine-gland disturbance may play in dwarfism has not been fully determined, but that both types of the defect are transmitted

Fig. 188. One Type of Dwarfism. (Plate in *Vererbungslehre*, Engelmann[1])

seems certain. Cretinism, to be discussed in a later chapter, is a dwarfing of both body and mind due to an improper functioning of the thyroid gland. This defect runs in families and is thought to be inherited.

Mental as well as physical traits are inherited. Not only are physical traits inherited, but mental ones as well. In working out the *measure* of transmission of physical and mental characters the coefficient of correlation has been employed. It is a numerical expression of the degree of likeness or unlikeness between two groups of individuals. For example, if tall parents always produced equally tall children, and short parents proportionally short children, then there must be a perfect cause-and-effect relation between the stature of the parents and that of the children. In that case the coefficient of correlation would be said to be $+1$ or just 1. If, on the other hand, tall parents produced proportionally short children, and short parents produced tall children, then the correlation would be -1. This would mean that tall parents cause short offspring and that

[1] From Burlingame and others' *General Biology*. By permission, Henry Holt and Company, publishers.

short parents cause tall offspring. Finally, if tall parents were just as likely to have children of short or intermediate stature as they were to have tall offspring, and the same relation held for short parents, then the coefficient of correlation would be 0. This would mean that there was no cause-and-effect relation whatever between the heights of parents and children.

Karl Pearson, a noted statistician, found the coefficient of correlation between brothers for the following traits to be:

Eye color 52
Stature 51
Cephalic index [1] 49
Hair color 59

For a long list of physical traits the average coefficient comes out about .52. When parents and children are compared for similar characters the result is approximately .49. Roughly speaking, then, we say that the coefficient of correlation between the children or between the parents and the children in respect to physical traits is .50. Now these physical characters are not the product of the environment. There might be some question as to the effect of food supply on stature, let us say, but certainly eye color, hair color, blonds, and brunettes are not the product of the environment. Children have these characters because the genes which produce them were received from the parents.

Pearson also compared siblings with respect to certain mental traits. His results are presented in the following table:

	Brothers	Sisters	Brothers and Sisters
Vivacity	.47	.43	.49
Assertiveness	.53	.44	.52
Introspection	.59	.47	.63
Popularity	.50	.57	.49
Conscientiousness	.59	.64	.63
Temper	.51	.49	.51
Ability	.40	.47	.44
Handwriting	.53	.56	.48
Average	.52	.51	.52

[1] Cephalic index is the ratio of the width of the skull to its length.

These results are not so accurate as the results of physical measurements. Psychologists do not yet know how to measure mental abilities with the same degree of exactness that they can measure physical dimensions. Nevertheless, that the average coefficients of correlation for these mental and personality traits are almost identical with those for physical traits is highly significant. In other words, Pearson's results furnish the strongest evidence that mental traits are inherited in exactly the same manner and to approximately the same degree as are physical characteristics.

Inheritance is an important factor in intellectual achievements. Many people will agree that physical characteristics are transmitted, but they are reluctant to accept the idea that mental traits are also transmitted. Their opposition to this conclusion grows largely out of their supposed regard for the individual. If it be admitted that there is a wide innate difference between individuals in regard to their mental powers and capacities, then but one conclusion can be reached: that some individuals genetically have a very great advantage in life, and that others are seriously handicapped by nature. To avoid the discomfort of such a conclusion, then, these people seek another way out. They refuse to recognize the force of heredity in determining human character and turn to the environment as the dominant factor; in short, they become environmentalists. The environmentalist may recognize heredity in mankind, but he sees in it only a minor force. In his opinion the big factor is the individual's surroundings. Give the child a proper environment and desirable social advantages, and if he tries to make the best of them he will succeed, the environmentalist claims, regardless of his heredity.

Now no reasonable man would decry the importance of a good environment. Man is the product of two forces — nature and nurture. The same individual can certainly make much more of himself under the influence of good schools and cultural surroundings than he can if he resides in the slums. But there is much evidence to show that heredity is a very potent factor in life, if not the dominant one. One may take two balls, one of rubber, the other of wood, and throw them

on the floor with equal force. The rubber ball will bounce
several times higher than the wooden one. The difference

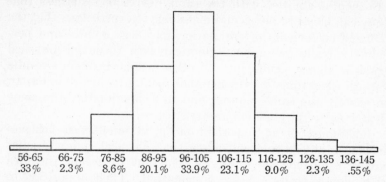

56-65	66-75	76-85	86-95	96-105	106-115	116-125	126-135	136-145
.33%	2.3%	8.6%	20.1%	33.9%	23.1%	9.0%	2.3%	.55%

Fig. 189. Graph showing the Results of Terman's Measurement of
Intelligence[1]

here is not one of the environment, but of the stuff composing
the balls. So it is with people. Given the same cultural influences, some individuals can rise intellectually much higher
than others. As to the wide spread of intelligence among
unselected people there is ample evidence.

Intelligence is distributed over a wide range. On the basis
of their "intelligence quotient," or I.Q.'s, Terman classified
905 unselected California school children. The average intelligence is indicated by an I.Q. of about 100. The highest-
grade feeble-minded have an I.Q. of from 65 to 70, and their
number, compared to the whole group, is small. The merely
dull are found at the 80 to 85 mark, and the percentage of
their number is greater. The highest percentage as to numbers is found in the average grade of intelligence. It may be
noted also that there is no sharp line of division between any
two classes, but a gradual transition from one grade to the
other. Above the median line the change is gradual and continuous, with the number decreasing until the brightest children are reached. These have an intelligence quotient of from
140 to 145. Individuals with I.Q.'s of 140 and above have

[1] From L. M. Terman's *The Measurement of Intelligence.* By permission,
Houghton Mifflin Company, publishers.

been arbitrarily designated as geniuses by psychologists. An occasional one may reach 190 or 200. These are the potential leaders of society. A later study by Terman has shown that the percentage of geniuses in the general population is very low. On the average there are but four or five in a thousand. The very lowest in the scale of intelligence (not represented in the graph because they do not attend school) drop to an I.Q. of 10 or below. These represent hopeless idiots.

So, then, the range of human intelligence on the basis of I.Q. may extend from practically 0 to something like 200, or from stark idiocy to men and women of profound and brilliant intellects. These natural differences, too, are not acquired. They are inherited. Moreover, the degree of intellectual capacity with which one is endowed determines largely what he can do successfully in life; it should also be recognized as a voice shouting from the very "house tops" to educators and parents, telling them that not every boy and girl can profit by the same kind of training.

Intelligence runs in families. Another indication that native intellectual capacity is inherited is afforded by the fact that ability tends to run in families. Frederick Adams Woods made a study of American eminence. The measure of eminence which he adopted was the appearance of the person's name in the American biographical dictionaries. At the time the study was made the list contained about thirty-five hundred names.

On the basis of careful estimates Dr. Woods determined that if eminence were a matter of chance alone,— that is, if it were the result of the environment into which one might be thrown or of an especially advantageous opportunity that might come his way,— then one of these eminent people out of every five hundred should be related to another eminent person in the list. As a matter of fact, however, the thirty-five hundred people whose names were listed in the dictionaries were found to be related to each other not as 1 to 500, but as 1 to 5. If the more celebrated were taken, the ratio increased so that about one out of every three had a relative in the list. If the forty-six names included in the Hall of

Fame at the time the study was made were selected, it was found that more than half of them had at least one other eminent relative on the roll.

These eminent Americans, then, are from five hundred to a thousand times as much related to other distinguished people as is the ordinary undistinguished man or woman; or, as Popenoe puts it, 1 per cent of these eminent people is as likely to produce a child that will be eminent as is the whole other 99 per cent of the ordinary American people.

The different social and occupational groups in the United States differ widely as to the number of bright children they produce. It is commonly recognized that it takes a better mind to become a doctor or a lawyer, for instance, than it does a day laborer. In other words, professional people as a class are superior intellectually to the unskilled laboring class. If this be true, then the children of professional people, taken as a group, should by heredity be more intelligent than the offspring of common workingmen. Terman has made one of the most thorough studies of the parentage of children with high I.Q.'s yet undertaken. Of almost a thousand California children with intelligence quotients of 140 or above, he found almost 50 per cent of them had fathers engaged in the professions, such as law, medicine, teaching, business administration, and banking. More than 35 per cent of them were from the semiprofessional and mercantile classes. A little over 10 per cent were from the skilled labor-classes, such as auto-mechanics and steam engineers, and but 5 per cent of them were from the unskilled and semi-skilled laboring classes.

This does not mean, of course, that there are no laborers at all who have good minds. There certainly are. And there are some individuals attempting to do professional work whose native mental equipment is far inferior to that of some laborers. But as a group the mentality of the latter is much below that of the former. The mental status of the parents tends strongly to be transmitted to the children, too, as the percentages just quoted show.

Mental inferiority tends to be transmitted to the offspring. Not only does mental superiority tend to run in families, but its opposite, mental inferiority, does also. Studies, such as those of the Jukes, Nams, Zeros, Ishmaels, Hickories, Piney Folk, and the Kallikaks, prove this fact conclusively. Moreover, this is exactly what one would expect to find if his inherent mental powers are a matter of heredity rather than of environment.

As an example we may take the Kallikaks. "Kallikak," of course, is a fictitious name for a real family, just as are the other names in the preceding paragraph. Miss E. S. Kite made the study of the Kallikak family under the direction of Dr. Henry Goddard, formerly head of the well-known Vineland, New Jersey, home for the feeble-minded. In that institution, while he was directing it, there lived a feeble-minded girl by the name of Deborah. Her ancestry was traced back to the period of the Revolutionary War. While the colonial troops were stationed at .Philadelphia, one of the soldiers, designated as Martin Kallikak, Sr., took advantage of a feeble-minded girl. From that illegitimate union a son, Martin Kallikak, Jr., was born. He was feeble-minded and became the father of the line from which Deborah came several generations later.

At the close of his period of army service Martin Kallikak, Sr., received an honorable discharge and returned home. Here he married into a very superior family and became the father of a second line of reputable and highly respected people. The succeeding generations of the legitimate line married into the best families of the state. Nothing but good citizens were found. They became doctors, judges, lawyers, educators, traders, and landowners. They have become scattered over the United States and are prominent members of the communities wherever they reside.

In sharp contrast to this branch of the family stand the descendants of Martin Kallikak, Jr., the early, illegitimate son. Including Martin himself, 480 have been traced. Of these, 143 were definitely feeble-minded, and but 46 could be designated as normal. Among the descendants there were 36 illegitimate children, 33 were sexually immoral, mostly prostitutes, 24 were confirmed alcoholics, 3 were epileptics, 82 died in infancy, 3 were criminals, and 8 kept houses of ill fame.

Other mental defects are also inherited. In addition to feeble-mindedness, there are many other mental defects that may be transmitted from parent to offspring. Among these are a tendency to become insane, epileptic, and criminally inclined. To be sure, not all insanity, epilepsy, and delinquency is inherited. Many defects are produced by injury, disease, drugs, and other unfavorable forces of the environment. No one knows just what percentage of the general population in the United States are the victims of a bad heredity; but mental tests administered to over a million and a half American soldiers during the World War, faulty though these tests may have been, together with the intelligence scores of millions of school children, give unmistakable evidence that a very large number of children and adults are seriously handicapped in this respect. Dr. Paul Popenoe estimates that there are at least six million people in the United States whose intelligence quotient places them in the mentally defective class.

There is evidence that much crime may be due to bad heredity. No one knows accurately to what extent crime may be attributed to an unfavorable heredity. Estimates vary widely. Some place it as low as 20 per cent, and others as high as 80 or 90 per cent. The forces of heredity and environment are so complex and interwoven in their interaction that it is impossible at present to separate the one from the other in the individual. Doubtless some people are bad because of an evil heredity; others are criminal because of a vicious environment; and in still another class of delinquents the cause may be attributed to the effect of both a poor heredity and an unfavorable environment. Nevertheless mental tests administered to criminals indicate strongly that many of them have a low mentality.

Miss Augusta F. Bronner of the Juvenile Court of Chicago, after a study of more than five hundred cases of delinquent boys and girls selected at random in the detention home, found less than 10 per cent of them to be subnormal. This figure is a very low one. Much of the petty-law violation of these juvenile offenders may have been the result of a bad environment rather than of an undesirable heredity.

Mental tests administered to two hundred and fifteen boys in the Whittier, California, industrial school showed 53 per cent of them to be either feeble-minded or border-line cases. In the same group only 20 per cent of the inmates were normal or superior in ability.

Dr. Walter S. Fernald, formerly of the Massachusetts School for the Feeble-Minded, estimates on the basis of his experience that "at least 25 per cent of the population in our penal institutions are mentally defective."

Again, we repeat, it is impossible in the present state of our knowledge to designate any definite percentage of crime as being due to the specific influence of heredity. The percentage of defectives in penal institutions is doubtless higher than it is in the general population; but the true percentage, as in most other cases, must lie somewhere in between the upper and lower extremes of the estimates. However, if this intermediate view be taken, then the cost to society of crime which may be attributed to a vicious heredity must be very great.

Estimates as to the cost of crime in the United States are enormous. It is probably as difficult to arrive at a reliable estimate as to the cost of crime in the United States as it is to say what part of law violation is due to the influences of heredity. Crime raises its head in so many forms and, like an octopus, its tentacles ramify into so many phases of human activity that it is very hard to get at the problem at all. Law-enforcement officers must be employed, courts of justice must be set up, jails must be built, penitentiaries must be established, insane asylums and custodial homes must be provided. These are only a beginning; for damage to property, stolen goods, broken homes, contributions to charity, and loss of man power must all be counted in. When we do this the total crime bill mounts up to enormous proportions.

Edward H. Smith, writing in *Business*, has estimated the total annual crime bill of the United States at $10,000,000,000. This figure has also been indorsed by the chairman of the National Surety Company, a firm which has as one of its chief functions to furnish bonds for trust officers. Mark O. Prentiss,

FIG. 190. The Cost of Crime is Twelve Times the Cost of the Upkeep of the
Army and Navy[1]

who organized the National Crime Commission in 1925, places his estimate at the staggering sum of approximately $13,000,000,000. This stupendous sum is too great for the average imagination to comprehend. But if a man were to start to count out silver dollars at the rate of one hundred a minute, it would take him two hundred and fifty years to pay the annual crime bill as estimated by Prentiss. This amount would support all the public schools of the country for four years. It would build two $10,000,000 universities in each state of the Union and endow each with $125,000,000 to pay the expenses of operation.

If Mr. Prentiss is right in his estimate, and even 25 per cent of the cost of our crime can be laid at the door of an evil heredity, then it would pay the American people to begin to give serious attention to the kind of people with which they are replenishing the nation. We ought to give at least as much

[1] From M. F. Guyer's *Being Well-Born*. By permission, The Bobbs-Merrill Company, publishers.

consideration to this important question as we do to the kind of live stock we raise. Much is heard these days about the cost of popular education, and the total bill is often criticized in the strongest terms. If the American people would take rational measures as to the composition of its population, much of its bad blood could in time be eliminated. We might even reach the point where we could save enough on our present annual crime bill to run the public schools and also to reduce taxes to the extent of more than a billion dollars.

The people of lower mentality are increasing the faster. If we are to eliminate the cost of crime due to a bad heredity, we should be reducing the birth rate of the defective and intellectually inferior and increasing the contribution of the superior. But the figures show that we are doing the very opposite. Within the past fifty years the lower levels of society have reached a point where they are increasing almost twice as fast as the upper levels. The unskilled laborers are producing on the average three or four children per family, whereas the semi-professional and professional groups are averaging only about two children to each family.

In the past, nations such as Babylon, Egypt, Greece, and Rome have had their periods of ascendancy, a golden age of supremacy, and then their decline. Will the European and American civilizations, in the centuries to come, have a similar experience? While many factors were doubtless involved in the decline of these ancient nations, there is much evidence to support the theory that the degeneracy of the human stock itself was one of the major causes. The upper classes prospered, became rich and dominant. In their luxury and self-indulgence they not only degenerated in most cases but failed to produce sufficient potential leadership to maintain themselves. In the end they fell before the onslaughts of strong, virile, unspoiled barbaric invaders — people who were not softened by the enervating influences of a voluptuous life and who had a high rate of reproduction. Does such decadence await modern nations, including the United States? We do not know, but at the present we seem to be headed in that direction.

2

Eugenic principles should be applied to human society. In the light of this racial trend is there anything that can be done about the matter? There is; and to do this is perhaps one of the fundamental social responsibilities of our day. Those who believe in the theory of evolution say that such human superiority as we have is the product of long ages of progressive change and development. It has taken, they say, the forces of natural selection millions of years to produce a man equal to the tasks of civilization and culture. Now that same man threatens to commit racial suicide in a few generations. As fast as the "cream" rises to the top it is skimmed off and largely lost through the differential birth rate. If the superior elements of society are to maintain themselves, then it behooves them to be alive to the situation and to see what can be done to rectify the present trend.

The remedy for this evil lies, it is maintained, in the application of eugenic measures, in working out the laws of heredity as they apply to man and then following them. In short, the perpetuity of civilization seems to depend upon what kind of eugenic policy man adopts. Galton, the father of the movement, defined eugenics as follows: "Eugenics is the study of all of the agencies under social control which may improve or impair the inborn qualities of future generations of man, either physically or mentally." What, then, are some of the things man might do to promote a sound eugenic policy?

The public, in general, needs to be educated as to the perilous trend of the race. The majority of our citizens are unaware that a eugenic problem really exists. They are conscious of mounting taxes, of overflowing jails, penitentiaries, and asylums for the insane, and of a demand for increased charity; but they do not know that one of the major causes of the situation is an overproduction coming from the dregs of society.

It is believed by many that all hereditarily mental defectives who are left outside detention institutions or who are turned loose after incarceration should be sterilized. This would apply to the periodically insane, the feeble-minded, and the habitual criminal. Under methods of modern surgery now employed the sterilized individual is perfectly normal in all sexual urges and impulses. He simply has been rendered infertile and cannot reproduce tainted offspring. California has sterilized approximately nine thousand persons during the past twenty-four years (most of them during the last ten years), and the results are very satisfactory. It is gratifying to know, too, that six out of every seven who are sterilized are themselves pleased with the results. Those who are not satisfied seem to be mostly the mildly insane, or the periodically insane who are self-deluded to the point of believing that they are not mentally impaired. Many of the sterilized higher-grade defectives actually marry those of their own station when released from the detention homes. Relieved, now, from the dangers of reproduction and the responsibility of supporting a family, they are often able to engage in menial tasks to the point of becoming self-supporting.

Defectives who require constant care that cannot be given in the home, or who cannot compete with normal-minded people to the extent of making a living, should be segregated from the rest of society. In custodial homes they should be given simple tasks, medical care, and be made as comfortable as their unfortunate condition will permit. A humane society should do nothing less. But these dysgenics should not be permitted to pass their tragic defects on to generations still unborn. The cost of segregation would be rather heavy at first, perhaps, but it would tend to become lighter with each generation. Under our present policy we are drifting farther and farther into a situation from which it will be increasingly difficult to retrace our steps.

Lastly, eugenic knowledge should be so widely and thoroughly disseminated as to influence young people in the selection of their mates. In its last stages love is "blind" and perhaps ought to be. But it certainly should not be in the early stages of attraction. Proper eugenic education should cause a young person at the beginning of such experiences to stop and ask himself what the final consequences may be? Marriage into a family having feeble-mindedness, insanity, or criminality scattered back through previous generations is likely to show the same defect in the next generation. Especially is this true if the individual from outside the family should also be a carrier of the defect. Most of these dysgenic traits are recessive; but, like all recessives, they are sure to crop out any time that two genes for any particular character are brought together.

Hereditary differences should always have weight in deciding one's vocation. In closing this chapter let us consider briefly the influence that should be accorded native ability in choosing a vocation. It is trite to say that the world is full of square pegs trying to fit into round holes. A man with just enough mentality to make a good, semi-skilled laborer cannot become a successful surgeon. The converse is probably true that one whose mental powers would enable him to become a skilled surgeon would not make a happy automechanic. All of us are contented and happy when working at something that places a reasonable demand upon our mental ability. For this reason occupations and vocations should, in general, be chosen in the light of the individual's capacity and the demands to be made upon him by any particular line of work. Effort, enthusiasm, interest, and training (these may depend in part upon hereditary endowment too) are all important factors in vocational success; but they cannot wholly compensate for intelligence when one presumes to take on the heavier responsibilities of life. One should attempt to do only what his native capacity makes possible. He should not attempt the impossible. If he does, failure, disappointment, and often tragedy await him in the end. With the public schools setting up testing programs and establishing vocational-guidance agencies in many quarters, it is possible for each young person and his parents to get much valuable information to help them in these problems. Mental and other personality tests are not wholly reliable, to be sure, but they are highly indicative of one's native capacities and aptitudes. So if the information available from these sources were used, it would, beyond any reasonable doubt, often result in much better service to society and a happier individual. The admonition of Socrates to "know thyself" is more pertinent in these days than it ever has been before.

QUESTIONS FOR STUDY

1. Are the laws of heredity as applicable to man as to other animals? If so, what evidence can you present to support this fact?

2. What is brachydactylia, and how is it transmitted?

3. Make a list of pairs of contrasted human characteristics which are inherited.

4. Explain what is meant by the "coefficient of correlation."

5. How is the intelligence of children related to parental occupational groups?

6. On the basis of heredity does every child have an equal chance to succeed? Explain your answer.

7. To what degree are eminent people related?

8. What proportion of the normal-minded population is believed to be carriers of genes that determine mental defectiveness?

9. What are some of the important causes of crime?

10. Why should the United States government restrict immigration and subject those who do enter to rigid mental and physical examinations?

11. If it continues, what effect will the differential birth rate now acting have upon the probable future of the United States?

12. Some states require certificates of health before marriage. Is this desirable? Why?

13. Should young people contemplating marriage know something of the heredity of the family into which they are marrying?

14. What, in your judgment, may be done to reduce the racial contribution of the physically and mentally unfit?

REFERENCES

BURLINGAME, L. L., and others. General Biology (second edition), pp. 537–562.

GUYER, M. F. Being Well-Born, pp. 11–180.

HOLMES, S. J. The Trend of the Race.

HUNTINGTON, E., and WHITNEY, L. F. The Builders of America.

POPENOE, PAUL. The Child's Heredity.

POPENOE, PAUL, and JOHNSON, R. H. Applied Eugenics (revised edition), pp. 1–18, 80–279.

TERMAN, L. M. Genetic Studies of Genius, Vols. I and III.

TERMAN, L. M. The Intelligence of School Children.

WALTER, H. E. Genetics, pp. 293–331.

UNIT VII

Man has discovered Much as to the Nature and Control of Disease

DOUBTLESS the most interesting pages of the history of human biology are those connected with disease. Ancient and primitive man conceived disease to be caused by evil spirits and possession by demons. When these superstitions became untenable he then came to regard illness as the result of bad humors produced in the body. But within the last seventy-five years his ideas as to the cause and control of disease have completely changed again.

He now knows that disease may be caused by many different agents. A diet deficient in vitamins may produce crooked limbs, an impaired mind, and even death. Proper functioning of the ductless glands also has been found to be essential to a healthy body and a sound mind. If these glands speed up their rate of secretion or go on a strike, dire consequences may follow. Finally, microörganisms may infect the body to torture and rack it with pain.

Not only has man learned much about the cause of disease, but he has made equally rapid strides in learning how to prevent and control it. His successes in this field are perhaps the outstanding human achievements of the past fifty years. In a former age these advances would have been pronounced miraculous. The account of them and what they mean to the race reads almost like magic.

CHAPTER XIV · The Discovery and Application of Scientific Facts regarding Food have contributed much to Man's Physical Fitness and Welfare

Introduction. Little is definitely known about the food habits of primitive man. By studying the barbarians inhabiting the remote corners of the earth today and by examining the cave dwellings of extinct races some facts have been gathered. That knowledge, however, is sketchy and incomplete. It is probable that early man ate food when he could get it and that the periods between meals must often have been long and irregular. When, after periods of enforced fasting, he did get food, it is probable that he gorged himself much as wild animals do. Then, after satiating his appetite, he would retire to his den or cave to rest and sleep. Even among the ancient Hellenes, Milo of Crotona, who was six times victor in wrestling at the Olympic games, is boasted to have carried a four-year-old cow around the stadium on his shoulders and then to have eaten the whole carcass in a single day. Milo's fame probably exceeded his actual accomplishments; but the literary incident may be taken to indicate that at that period men were often inordinate eaters.

Indeed, after we span the whole period of the Dark Ages and emerge into medieval life, the appetite, perverted and unrestrained, seems often to have been the guide to eating habits rather than the needs of the body. Mrs. Glasse, author of an English cookbook published in 1788, in the preface reproaches a contemporary for using six pounds of butter to fry twelve eggs. She follows with the comment that "everybody knows, who understands cooking, that half a pound is enough." The same authority gives a menu for a three-course January dinner which in the light of modern dietetics seems almost incredible.

431

January (a Dinner)[1]

First Course

Leg of Lamb
Chicken and Veal Pie
Tongue

Chestnut Soup
Petit Patties
Cod's Head
Raisolds

Boiled Chickens
Roast Beef
Scotch Collops

Second Course

Marinated Smelts
Roast Sweetbreads
Almond Tort
Roast Turkey

Vermicelli Soup
Tartlets
Stands of Jellies
Maids of Honour

Woodcocks
Mince Pies
Larks
Lobsters

Third Course

Artichoke Bottoms
Custards
Scalloped Oysters
Morels

Dutch Beef, Scraped
Cut Pastry
Potted Chars
Rabbit Fricaseed

Macaroni
Black Caps
Stewed Celery

In these days of dietetic knowledge one stands aghast as he contemplates such a menu. Certainly giants, gastronomically speaking, must have lived in those days, and one cannot escape the conclusion that even they must have lived to eat rather than have eaten to live.

Scientific knowledge as applied to foods has been attained slowly. Like every other segment of scientific knowledge, our understanding of foods and their function in the body has grown up gradually and laboriously. The means by which this knowledge was fostered has been slow, patient, and painstaking investigation and experimentation. Formerly the disappearance of food in the body was attributed to "insensible perspiration" and "heat," without any real understanding of what either term meant.

In 1614 a university professor by the name of Sanctorius devised a chair suspended from a steelyard to weigh himself before and after meals in order that he might determine the amount of this "insensible perspiration"; for, said he, "he only who knows how much and when the body does more or

---

[1] C. H. La Wall, Philadelphia College of Pharmacy and Science, *Popular Science Talks*, Vol. VI, p. 25.

less insensibly perspire, will be able to discern when or what is to be added or taken away either for the recovery or preservation of health."[1] Painstaking as Sanctorius's efforts were, however, they availed little; for men in those days knew practically nothing about chemistry, and chemistry is the very foundation of dietetics.

Robert Boyle, by incasing a mouse in a glass vial, demonstrated the dependence of animals upon the air they breathe to sustain life. Stimulated by the work of others who preceded him, Lavoisier, the brilliant young French chemist, shut a sparrow up in a small chamber. After the bird died he concluded from his observations that the oxygen had

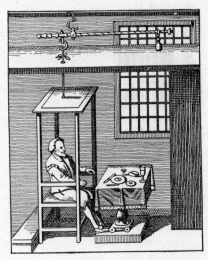

Fig. 191. The Device arranged by Sanctorius to determine the Amount of "Insensible Perspiration"

disappeared from the inclosed air and that carbon dioxide, which he absorbed with limewater, had appeared in its stead. Then he and Laplace put a guinea pig into a chamber. For a period of ten hours they measured the carbon dioxide formed by its respiration. This they found to be equal to the weight of the same gas produced by burning 3.33 grams of carbon in a closed vessel. Next they put the guinea pig for ten hours in a chamber containing a known weight of ice and determined the amount of ice melted by the heat of the animal's body. The amount of heat required to melt the weight of ice that disappeared was found to be almost exactly the quantity liberated when 3.33 grams of carbon were burned. The inference readily drawn, disregarding the slight fall of temperature in the pig's body, was that the carbon dioxide exhaled by the animal

[1] Sanctorius, *De Medicina Statica Aphorismi*, Venice, 1614. Translation by John Guiney, M.D., London, 1737.

was evolved by oxidizing in its body the equivalent of 3.33 grams of carbon. Subsequent work with improved and elaborately designed calorimeters showed that the body uses food to get energy from it, as a steam engine uses fuel.

FIG. 192. A Water Calorimeter of 1823[1]

We have already found that the energy which the body releases is derived originally from the carbohydrate food components. From time immemorial it has also been observed that living things grow. The human infant at birth weighs, on the average, about six or seven pounds. When he reaches adulthood he may weigh one hundred and seventy-five pounds. Apparently this increase in volume and weight was from the first rightly conceived to have come from materials appropriated from the food. Scientific study during the past century has revealed that the tissue-building component of food is the proteins. Proteins form the essential building blocks for every living cell, whether for growth or for repair.

The body contains fats too; but these are generally looked upon as a form of reserve energy food built up from the carbohydrates. Except where certain special functions are to be served, the fats will either be reduced to carbohydrates again and be broken down chemically to supply energy, or they will be combined with other chemical constituents to form proteins.

For a time these three classes of foods, together with chemical salts and water, were considered to meet the full requirements of the body. Carbohydrates, fats, and proteins were the only food substances that could be found chemically and be isolated from food materials. Since chemical analysis had come to be

[1] From Mary S. Rose, *The Foundations of Nutrition*. By permission of The Macmillan Company, publishers.

looked upon as the very handmaiden of dietetics and food values, the verdict from the standpoint of many food experts was final. Carbohydrates, fats, and proteins, then, when supplemented with water and minerals were all the body needed.

Atwater, who was the outstanding American investigator in the field up to the year 1900, advised housewives freely as to the purchase of foods. In doing this his eye was ever on the calories and protein content. Aside from palatability, which seems to be largely an individual matter, the energy and tissue-building content of the food appears to have been his only criterion. In a United States bulletin published in 1895 he contrasted the nutritive values of a list of everyday foods on the basis of the amount of proteins, carbohydrates, fats, and energy that could be purchased in the Eastern markets for twenty-five cents. After giving a table of different foods he wrote:

[The figures of the table] tell their story so plainly that they need very little comment. A quarter of a dollar invested in the sirloin of beef at 22 cents per pound pays for one and one-seventh pounds of the meat with three-eighths of a pound of actually nutritive material. This would contain one-sixth of a pound of protein and one-fifth of a pound of fat, and supply 1,120 calories of energy. The same amount of money paid for oysters at the rate of 50 cents per quart brings 2 ounces of actual nutrients, an ounce of protein, and 230 calories of energy. But in buying wheat flour at $7.00 a barrel, the 25 cents pays for 6.25 pounds of nutrients, with eight-tenths of a pound of protein and 11,755 calories of energy. The price of food is not regulated solely by its value for nutriment. Its agreeableness to the palate or to the buyer's fancy makes a large factor of the current demand and market price.[1]

This statement, together with many others made by Atwater during this period, showed clearly the basis upon which food was judged. In Atwater's opinion the twenty-five cents could be much more economically invested in flour than in beef or oysters if the purchaser could bring himself to like bread as well as meat. But the food specialists of that period

[1] W. O. Atwater, "Foods: Nutritive Value and Cost," *Farmers' Bulletin No. 23*, p. 23, United States Department of Agriculture.

were resting too complacently upon their chemical knowledge. Each was like a miner working over his ore to extract the copper and lead while the rare and invisible platinum slipped through his hands unnoticed.

There are other important elements in food. The calorie and protein advocates were overshooting the mark; for while they were making these positive statements as to the nutritive properties of foods and their relative values, trouble for their point of view was brewing. Even as early as 1881 a Swiss named Lunin was experimenting at the University of Basle with mice. He fed one group a diet of milk sugar, milk fat, milk protein, and the minerals found in milk. These substances had been chemically extracted and purified. The other group he fed with straight milk. The milk-fed mice lived and thrived; those fed with the mixture extracted from the milk died within a month. From this result Lunin drew the conclusion that "other substances indispensable for nutrition must be present in milk" in addition to the sugar, fat, mineral salts, and protein products.

Dr. McCollum performs a noted experiment. Many other investigators in the field of foods got similar results when using experimental diets with different animals. Consequently there gradually grew up a conviction that certain food principles still undiscovered were necessary to sustain life. One of the notable experiments was carried on at the University of Wisconsin between the years 1906–1911. Dr. E. V. McCollum, now of Johns Hopkins University, directed the investigation. The purpose was to find out whether diets as nearly chemically alike as possible but from different plants were of equal value when used as food for cattle. The part of the experiment of interest to us involved three groups of young heifers. The animals in each group weighed about three hundred and fifty pounds at the outset and were as nearly alike in size and vigor as could be selected.

One group was fed a diet made up wholly of different parts and products of the wheat plant. It included straw, leaves, and grain. The second group was fed a similar oat diet, and the third group was given a ration made up of all parts of the corn plant. The groups were watered, salted, fed, and exercised in the same way.

For a time all went well, and it began to look as if the carbohydrate, fat, protein, and mineral advocates were being vindicated. However, scientific work requires both patience and persistence, and Dr. McCollum and his assistants had both. It was almost a year before differences began to appear. By a year, or more in some cases, they were easily observable.

The corn-fed animals were sleek-coated and finely nourished in every respect, but the wheat-fed group stood in marked contrast. These animals were gaunt and rough-coated. Their girth also was considerably smaller than the corn-fed cattle, although their weights did not differ greatly. The oat-fed lot stood intermediate between the corn-fed and wheat-fed lots.

These experimental animals were all mated. The record of their offspring is of great interest. The corn-fed mothers produced well-developed calves. They were carried the full period and, when born, were vigorous and able to stand and suck within an hour. They lived and developed normally. In sharp contrast stood the offspring of the wheat-fed mothers. These calves were born from three to four weeks too soon and weighed on the average but forty-six pounds as against from seventy-three to seventy-five pounds for the young of the corn-fed heifers. They were all either dead when born or died within a few hours after birth. The young from the oat-fed mothers averaged seventy-one pounds, but they were born about two weeks early. Of the four one was dead at birth, two were very weak and died within a day or two after birth, while the fourth, although weak, was kept alive.

The experiment was continued through the second year. The results obtained the first year were repeated in all essentials, including the condition of the mothers themselves and the birth and death of the offspring. The milk yield of the different groups also presented sharp contrasts. After the first calf the average daily milk production of the corn-fed group was 24.03 pounds; of the wheat-fed group, 8.04 pounds; and of the oat-fed group, 19.38 pounds. After the birth of the second calf the corn-fed mothers produced an average of

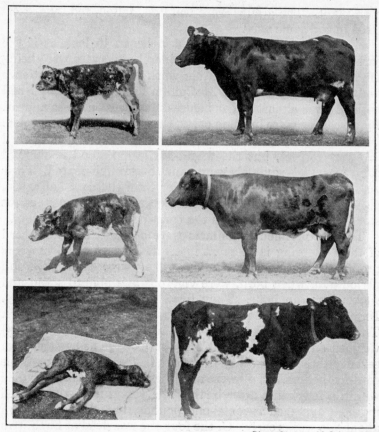

Diemer Photographic Laboratories

FIG. 193. **Different Foods have Different Effects on Animals**

Top, result from corn and corn fodder; middle, result from oats and oat straw; bottom, result from wheat and wheat straw

28.0 pounds per day; the wheat-fed mothers, 16.1 pounds; and the oat-fed mothers, 30.1 pounds.

These results clearly indicated that in some foods there was present a very important element of nutrition which was not present in other foods. This strange principle could not be determined by any chemical skill known at that time. It was revealed only by its physiological effects.

Physicians with patients suffering from what are now recognized as dietary diseases had also discovered, doubtless by trial and error, the marked beneficial effects of certain foods. As early as 1720 Kramer wrote that neither medicine nor surgery would cure scurvy.

But if you can get green vegetables; if you can prepare a sufficient quantity of fresh antiscorbutic juices, if you have oranges, lemons, citrons, or their pulp and juice preserved with whey in cask, so that you can make a lemonade, or rather give to the quantity of 3 or 4 ounces of their juice in whey, you will, without other assistance, cure this dreadful evil.[1]

Funk names this active food principle. Finally, by 1912, different investigators had been able to extract from different foods in impure form an active principle of some kind that would cure birds affected with beriberi. Casimir Funk announced that he had been able to cure polyneuritic pigeons in a few hours by the use of certain crystals which he had prepared from wheat bran. Other workers had been able to connect different diets with the cure of other diseases, just as Kramer had done. Finally Funk, recognizing the extreme importance of these unnamed principles in connection with foods, coined the term *vitamine* to designate them. This was afterwards, for reasons which do not concern us here, shortened to *vitamin*.

Here, then, was opened up a whole new field in food science — the realm of vitamins. By this discovery a new standard, in addition to calories and proteins, was set up by which to judge the value of food products. From then on, dietitians must not only consider the energy and tissue-building content of meals but must also know their growth-promoting vitamin content as well.

This was a remarkable achievement for science. For now future generations could be spared the pitiful sight of thousands of diseased, twisted, and malformed children. Bleareyed, palsied-bodied, bowlegged boys and girls on the threshold of a miserable existence could be made strong and normal.

[1] H. C. Sherman and S. L. Smith, *The Vitamins.*

2

As time went on it became evident that there was a whole group of these vitamins, each performing its beneficent service to both man and beast. By usage they came to be designated after letters of the alphabet. In discussing them we shall take them up in their alphabetical order and shall emphasize only the leading facts that are connected with each.

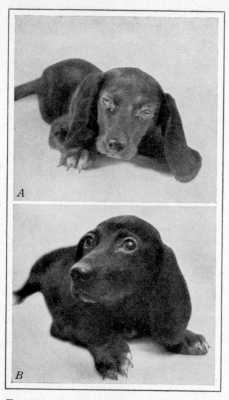

FIG. 194. *A*, a Dog with a Well-Developed Case of Xerophthalmia; *B*, the Same Dog a Short Time after Vitamin A has been added to the Diet

Vitamin A. Vitamin A is essentially a vitamin for promoting growth. Young rats when fed a diet devoid of this substance grow normally for a few weeks until the vitamin already in the body is exhausted. Growth then slackens and in a short time usually ceases altogether. Weight is often lost rapidly, and the animal presents a condition of general decline which sooner or later results in death.

Children and other animals when fed with a diet deficient in this vitamin often develop well-marked symptoms of such disease. The tear glands apparently go into a resting stage, and the normal supply of water to the eye dries up. As a result the periodic washing and wetting of the eyeball by the blinking of the lids is prevented. The tissues of the eye, becoming dry and unwashed, now fall an easy prey to bacterial infection. In response to these conditions white blood corpuscles tend to accumulate in the eye. Finally the eye becomes inflamed and full of pus. The exudate thickens and dries, and the lids are often stuck together in such a manner as to be opened with difficulty. The condition is known to the medical profession as xerophthalmia, or dry eye. Night blindness, or the inability to see well under any condition but bright light, is also a symptom of this dietary deficiency. The epithelial tissues of the body are often affected also. It has been found that the nasal passage, the salivary glands, the larynx, and the bronchial tubes as well as certain parts of the urinary tract may present the characteristic desiccated and inflamed symptoms. In general it seems that a lack of vitamin A in the diet decreases the general resistance of the body and makes it susceptible to bacterial invasion at many points.

Vitamin A is found most abundantly in butter fat, cod-liver oil, egg yolk, liver, leafy vegetables, and yellow-pigmented vegetables, such as carrots and sweet potatoes. One of the most interesting recent developments in connection with this vitamin is its relation to these yellow pigments. It has been found that the animal body after eating such plants can readily transform these pigments into vitamin A.

During the World War many children in less favored parts of Europe, where the diet was poor, developed the symptoms of this disease to a marked degree. In 1904 Japan suffered a severe food shortage. Among the poorer classes large numbers of infants and children developed severe cases of xerophthalmia. For many years there had been a "grandmother remedy" in Japan for the treatment of this disease. It consisted in feeding chicken livers or eel fat to the sufferers. Its efficacy, of course, arose from the fact that these dietary articles are rich in vitamin-A content.

Vitamin B. Vitamin B is often called the antineuritic vitamin. The lack of this principle causes the dreadful disease beriberi, in which the nervous system is seriously impaired and paralysis often follows. Beriberi has been known for centuries and was especially prevalent in the Orient, where the diet was largely confined to polished rice. In certain years during the last century almost a third of the men in the

Japanese navy would fall victims to this plague. A young Dutch colonial official from the East Indies relates his observations regarding the disease. He states:

I lived at Buitenzorg in the neighborhood of the so-called beriberi hospital, and daily I saw hundreds of the poor sufferers of that mysterious disease passing my home in batches. It was a pitiful sight. Natives, Chinamen and a few white men dragged themselves along with their swollen legs. They had to take a daily walk, as exercise in the fresh air, if possible, was in those days the panacea for this disease, wherefore no other cure was found. Many of the numerous cases in the crowded hospital were not even able to walk; they died slowly within the precincts of the hospital, and more than once it happened that some of the patients having their daily exercise collapsed and died on the road of heart failure.[1]

To Dr. C. Eijkman, a Dutch physician, we are mainly indebted for our knowledge as to the cause and remedy for this disease. In 1897 he succeeded in producing all the characteristic symptoms of beriberi in chickens by feeding them a diet of polished rice. Birds fed upon whole rice remained well and hearty. When whole, or unhulled, rice was substituted for the polished rice early enough, in the case of the afflicted fowls, they promptly recovered their normal health and vigor. This necessary food principle contained in the hulls of the rice is what has come to be known as vitamin B.

Beriberi, as has been stated, is characterized by degenerative changes in the nervous system. Extensive neuritis may develop, often accompanied by a dropsical condition. Liquid is effused into cavities in the linings of the abdomen and thorax. The heart becomes affected, is often enlarged, and death frequently comes from its failure. The symptoms vary considerably, and in some cases a "dry" form of the disease develops. In this type the muscles lose their power to function, become shriveled, and are extremely painful to the touch.

Vitamin B is very abundant in many common articles of food. It is found in whole wheat, unpolished rice, corn, oats, peas, beans, and the liver and kidneys of animals. Green

---

[1] Julius Steiglitz and others, *Chemistry in Medicine*.

vegetables, such as spinach, leaves of beets, and lettuce, are especially rich in this food principle. Vitamin B is very stable and is not easily destroyed by ordinary cooking temperatures.

Beriberi was almost completely eradicated from the Philippine Scouts in 1910 by Dr. Chamberlin of the United States Army Medical Corps. Until that year the cases among the scouts had ranged annually from one hundred and fifteen to six hundred and eighteen in a force of five thousand men. By substituting in the ration 16 ounces of unpolished rice and 1.6 ounce of dry beans for 20 ounces of polished rice, the patients dropped to fifty the first year, and in three

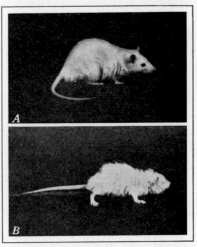

FIG. 195. Two Rats of the Same Litter. *A*, Rat given Balanced Diet; *B*, Diet Deficient in Vitamin B

years the plague was completely wiped out. Today the hospital at Buitenzorg is closed — closed for the want of patients.

Vitamin C. Scurvy has been known for centuries. Ever since captives have been taken in war and subsequently herded together in prison camps, this scourge has been present. The cheap, limited, and monotonous diets of these detention camps have furnished the very conditions from which this disease results.

The affliction develops gradually. The victim loses weight, becomes pale, weak, and anæmic, and suffers from shortened breath. The gums become swollen and bleed profusely, and the teeth loosen. The ankles and lower limbs become swollen and sensitive to the touch. Dark areas appear on the skin. The blood vessels rupture easily, spilling their contents out among the tissues. In its final stages death comes to relieve the sufferer.

In 1536 Jacques Cartier spent the winter in the Gulf of St. Lawrence, where he lost many of his crew from scurvy.

Those who lived to return home did so largely because they followed the advice of some Indians who directed them to drink an infusion prepared by soaking pine needles in water. The pine leaves contained vitamin C. Scurvy was so prevalent among the early French forces during the long, severe winters in Canada that they frequently debated the wisdom of abandoning the settlements altogether.

Attention was sharply focused upon scurvy during the sixteenth century. This was peculiarly the period of long sea voyages of exploration in sailing vessels. Oftentimes many months would elapse during which little or no green-vegetable food was available. Among such crews scurvy was a dreaded affliction. A few navigators and doctors, however, learned empirically how to control the disease. In the eighteenth century Lind, the English naval hygienist, advocated the use of citrus-fruit juices and whey. Captain Cook boasted that he sailed the seas for three years without having a single sailor fall victim to scurvy. He did it by providing liberal portions of "sweet wort" and "sauerkraut."

So the knowledge that scurvy can be both prevented and cured by the use of raw foods gradually developed. The effective principle in accomplishing this we now recognize as vitamin C.

Vitamin C is irregularly distributed in food materials. Because of its wide use the potato is perhaps the most valuable source of this food adjunct. Fresh raw tomatoes are more abundantly supplied with vitamin C than potatoes; as are also oranges and lemons. Bananas and apples are good sources too. The root vegetables rutabagas and turnips and the leafy vegetables cabbage, spinach, and lettuce may be used very effectively to prevent scurvy.

Vitamin C is very easily destroyed by excessive heat when exposed to the oxygen of the air. It is doubtless for this reason that raw fruits and green vegetables have been emphasized as having peculiar antiscorbutic properties. This also accounts for the fact that commercially canned products of this kind, prepared in sealed containers, are more desirable than those preserved by the open-kettle methods of the home.

Vitamin D. We have all seen unfortunate, dwarfed, hunch-backed men and women walking along the streets and have thought how unhappy their condition must be. It is gratifying, however, to note that these disabled people belong mostly to the adult generation. Today we very seldom see a young child so afflicted. This change can be attributed to the progress of science, in that rickets have been found to be caused by an insufficient amount of vitamin D in the diet, and that it is now known what particular kinds of food are rich in this principle.

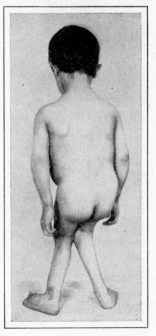

FIG. 196. A Typical Case of Pronounced Rickets

Note the enlargements at the joints and the curvature of the long bones. Courtesy of James L. Gamble, M.D., Harvard Medical School

The case for rickets as a diet-deficiency disease was closed when investigators were able to produce the malady experimentally in animals. By subjecting dogs to faulty diets all sorts of dwarfed, deformed dogs were produced. Moreover, it was found that both dogs and children when in the incipient stages of rickets could be cured by eating food containing antirachitic properties. Both adults and children are susceptible to this deformative disease when improperly fed. It is more marked in children, of course, because their poorly calcified growing bones yield more readily to its disturbing and often distorting effects. At the outset the child is usually constipated, restless, and irritable and does not sleep well. The muscles are lax, and the ligaments and tendons elongate. Because the muscles of the intestines and abdomen are weakened, the condition known as "pot-belly" is likely to follow. As the disease progresses deformities of the bones begin to appear. One of the first symptoms is the "rickety rosary."

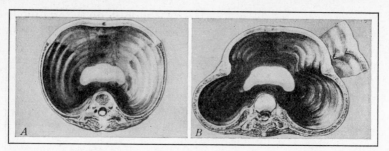

FIG. 197. *A*, Section of the Thorax showing the Normal Condition; *B*, the
Rachitic Rosary

Courtesy of James L. Gamble, M.D., Harvard Medical School

This is a line of beadlike knots which develop down either side
of the breast at the points where the cartilage and bone of the
ribs join. The walls of the chest are drawn inward and become compressed from the sides, leading to the well-known
"pigeon breast" development. Curvatures of all sorts may
occur in the spine. The bones of the arms and legs, especially
those of the legs since they support the weight of the body, may
become crooked and deformed. Owing to unnatural bone
growths the cranial walls may be modified also, producing
a peculiar square-headed appearance. Extreme nervousness
manifests itself in advanced cases, and anæmia frequently develops. These conditions, if they themselves do not cause
death, weaken the body to the point where it falls an easy
prey to invading bacteria of all kinds.

Vitamin D is abundant in cod-liver oil, and the diet of
every expectant mother and young child should include limited
quantities of this valuable substance. This vitamin is also
found to some extent in the flesh of fish and other marine
animals. Foodstuffs in general are very deficient in this
vitamin. Outside the sources mentioned, it is found in appreciable quantities in no other common foods except in butter
fat and egg yolk, which contain limited amounts of it. From
these statements it may be seen that before the cause of
rickets was discovered, fish-eating peoples were about the only
ones who were likely to be free from this disease. Other

peoples were, as a rule, getting the minimum amount of vitamin D. For this reason the malady was very widespread.

Almost all children in industrial districts of the British Isles suffer to some extent from this food deficiency. Rickets to some degree is prevalent in many parts of Europe. In some areas of the United States where the poorer and less intelligent parents live from 50 to 80 per cent of the children may be afflicted in varying degrees of severity.

Sunlight also has been found to be very beneficial in preventing rickets. The peoples of the

FIG. 198. Effect of Cod-Liver Oil on Egg Production

Courtesy of the Minnesota Agricultural Experiment Station

tropics seldom have this disease because of the abundance of sunlight. Peoples of the arctic regions, where the sunlight is reduced to the minimum, are not often the victims of rickets either, but this is because of the large amount of fish, whale, walrus, and seal flesh which they eat. Rickets, then, is preëminently a temperate-zone disease. But people who are freely subjected to the influence of sunlight seldom are afflicted. This is because the sun's rays act upon a sort of fat in the body called ergosterol to produce vitamin D. Hence sunlight has come to be looked upon as being equally important with food in the prevention of rickets. Adults should seek as much sunlight as possible and every young child should be subjected to regular sunshine baths.

An interesting report has recently been made from the University of Minnesota on the effects of feeding cod-liver oil to laying hens. At the latitude of St. Paul the winters are long and cold, which requires that poultry be confined to houses a large part of the time. Under these conditions of limited sunshine it was found that the fowls suffered a decrease in egg production due to a deficiency of vitamin D in their diet. By the addition of 2 per cent of cod-liver oil to the diet the six-months winter egg production was raised from eighty-two to one hundred and five per bird. The cost per dozen eggs was also cut from 7.3 cents to 5.3 cents.

Vitamin E. Two investigators, Evans and Bishop, found in their work on vitamins that rats when fed a diet including all other known food elements except vitamin E suffered a peculiar disability. The reproductive functions degenerated, and the rats became sterile. In the males the germ cells were destroyed, and eventually the whole sperm-cell-producing testis was irretrievably impaired. The result in the females was different. Their reproductive cells and tissues apparently were not destroyed, but the young embryos themselves, after starting to form, were arrested in their development. A few days later fetal death occurred.

Because of the rather complete degeneration of the testes of the male it is evident that he could not be restored to a condition of fertility by a corrected diet. Afflicted females, however, can be brought back to a complete state of reproductive power through proper diet.

Lettuce leaves and the germ of the wheat kernel have been found to be especially rich in this vitamin. The vitamin E content of oil from the wheat germ has been found so potent that the administration of but $\frac{3}{10,000}$ grams a day keeps the male rat perfectly fertile.

Vitamin G. For almost two centuries a peculiar disease called pellagra has been known. It has invaded every known part of the world to a greater or less degree. In 1917–1918 there were two hundred thousand cases in the United States, the majority of them in the Southern states. In 1915 more than ten thousand people in the United States died of pellagra.

The onset of the disease is gradual; but if it is not checked, it is very deadly in the end. The symptoms manifest themselves first as a soreness of the mouth and tongue. Later the digestive tract becomes generally disturbed, including diarrhea. The skin takes on a peculiar bronzed appearance, which is limited to the exposed portions of the arms, face, and neck. If the victim goes barefooted the bronzing will appear on the legs and feet as well. The discolored areas are especially susceptible to infection. In its advanced stages mental disturbances also develop.

This affliction is especially prevalent in the poorer rural districts where the dietary is limited and likely to consist largely of the cheaper, starchy foods. For years before the vitamin itself was discovered certain individuals engaged in health work suspected that faulty diet was connected with pellagra. Foremost among these was Dr. Joseph Goldberger of the United States Health Service. At several orphanages in certain of the Southern states pellagra was especially prevalent. The Health Service secured permission to experiment with the diet of these children. At each of the institutions a liberal allowance of fresh milk and lean meat was added, and the children were encouraged to partake heartily of these foods. Previously these items had been served only sparingly and irregularly. All other conditions continued as they had been. In a reasonable length of time the disease completely disappeared. In one of the institutions after the experiment a relapse was made to the old diet. Within a period of from three and one-half to nine and one-half months 40 per cent of the inmates were again victims of the disease. Other investigations, exacting in character, followed which established beyond doubt the connection between pellagra and a hitherto unknown vitamin. This essential food principle (in the United States) has come to be recognized as vitamin G. The letter is the initial letter of Dr. Goldberger's name. He died shortly after proving the existence of this necessary food adjunct. Thousands of lives have been saved in the poorer charitable institutions in this country and among the lower classes by including in the diet articles of food containing this vitamin. Milk, lean meats, especially liver and kidney tissue, as well as yeast have been found to be very rich in vitamin G.

Scientific knowledge is necessary in the selection of food. It has been widely proclaimed that the appetite is a safe guide in the choice of diet. The work in vitamins has completely exploded this ancient idea. It is true that a rather wide and free use of various kinds of food will include appreciable amounts of most of the vitamins. But the hunchbacked victims of rickets, the paralyzed beriberi individuals, and the bronze-skinned pellagrins of other years show, in no uncertain terms, that appetite cannot be relied upon entirely in the choice of food.

Vitamins have much to do with resistance to disease too. In many diets these elements may not be lacking to the degree that specific diseases are produced, yet they may be deficient. The result, then, is not to produce visible evidence of faulty feeding. The effects are much more insidious. Organs and tissues of the body are weakened, natural barriers against bacterial invasion are broken down, and the resistance of the whole system is reduced. In this condition the body becomes a ready prey to all kinds of microörganisms that fasten upon it with the least exposure. Doubtless thousands of children, and older people as well, have suffered and died from specific diseases, when the real cause was not the disease itself but the lowering of resistance through the deficiency of diet that made the visible disease possible. Taste, then, supplemented by scientific knowledge is the only safe guide in the choice of food.

QUESTIONS FOR STUDY

1. Trace the historical development which has led to our present ideas concerning the value of foods.

2. What are our modern ideas of foods?

3. By whom were the vitamins named, and what function do they serve in the body?

4. Make a list of the vitamins, with the foods in which each is found.

5. What disease is caused by the lack of vitamin A in the diet? What treatment is given to cure this disease?

6. Give the method employed by Dr. McCollum to show the value of the vitamins in the diet.

7. What is the cause of beriberi, and how may it be cured?

8. What is the effect of rickets on growing children? What treatment is given to them?

9. As a source of vitamins which are better, cooked or uncooked foods? Why?

10. What relationship is believed to exist between the vitamin supply and the resistance of the body to disease?

11. Why is it important that orange juice or some nonacid fruit juice be included in the diet of the infant?

12. How is it supposed that vitamins are produced in such vegetables as the carrot?

13. What is the influence of direct sunlight upon the human body?

14. Why are sun lamps so beneficial to an individual where it is not possible for him to receive the benefit of direct sunlight?

REFERENCES

LA WALL, C. H. "The Romance of Cookery," *Popular Science Talks*, Vol. 6, pp. 5–31.

McCOLLUM, E. V., and SIMMONDS, N. Food, Nutrition, and Health.

ROSE, MARY SWARTZ. The Foundations of Nutrition, pp. 1–25, 199–299.

SHERMAN, H. C., and SMITH, S. L. The Vitamins.

STEIGLITZ, JULIUS, and others. Chemistry in Medicine, pp. 112–190.

CHAPTER XV · The Discovery of Hormones and an Understanding of their Function have added greatly to Man's Health and Efficiency

Introduction. It has long been suspected that chemical substances produced in the body exert an important influence upon man's behavior. This idea bore fruit in the belief that one could enlarge his own virtues by eating certain organs of his enemy. For instance, certain primitive combatants ate the hearts of their slain foes to increase their own courage.

Other ancients believed that diseases of various kinds were produced by a lack of certain chemicals manufactured by different organs. Hence the logical conclusion followed that to cure these diseases one should eat the organs of animals which supplied these deficiencies. Celsus and Dioscorides thereupon prescribed wolf's liver for hepatic troubles, hare's brains for nervous disorders, and fox lungs for respiratory disturbances.

In this day of scientific discovery man has become entirely free from such superstitious and mythical connections. He reads the historical account of them and smiles. He wonders how man could have been so simple; then he picks up a modern book on the chemistry of life, reads, and ceases to smile. For, while all ideas as to the connection between heart tissue and bravery and between hare's brains and nervous diseases have been repudiated, man does find that the human body, as well as the bodies of most other animals, is a veritable chemical laboratory, and that the products of this laboratory frequently act upon the body in what appears to be a most magical way.

For instance, it had been noticed for centuries that the castration of domestic animals oftentimes produced a very marked alteration both in their physical characteristics and

452

in their behavior, especially toward the opposite sex. The result was thought, of course, to be related in some way to the removal of the sex organs. Just how, no one knew. It was not until about the middle of the last century that biologists began to offer an explanation of the true relationship. From studies thus initiated, other investigations followed that have resulted in the most marvelous discoveries regarding glands and their internal secretions.

The hormones regulate certain bodily activities. The secretions of the endocrine glands differ from the ordinary secretions of the body in that they are absorbed directly into the blood stream from the gland tissue that produces them. We have found that the liver and pancreas are provided with a special duct to carry the bile and pancreatic juice into the intestine. The saliva flows through well-formed tubes to reach the mouth. Many other products of the body follow a similar course. But these internal secretions are absorbed directly into the blood circulation through the thin capillary walls. By this provision they are not brought into contact with other body fluids that might destroy them. This also makes it possible for them to be caught up by the speeding blood and to be carried with the greatest promptness to the points where they are to be used. This journey between the place of origin and the organ whose action they are to influence requires in most cases but a fraction of a minute. Because the glands which produce these substances have no ducts they are often spoken of as *endocrine* glands.

The internal secretion itself has come to be recognized as a *hormone*. Starling proposed the term, which, considering its derivation, is peculiarly appropriate. The word itself comes from the Greek *hormao,* "I arouse" (or excite). Just how hormones work may be seen by considering one called secretin. Formerly it was a mystery as to how activity in the pancreas was initiated. The individual would eat a meal, and immediately after the first semi-digested food began to pass from the stomach into the intestine the pancreas would go into action. In this way a liberal supply of pancreatic juice would be poured into the food in process of digestion at exactly the right moment. But how did the pancreas "know" when to begin secreting? That was the question. It responded with unfailing regularity and apparently with almost human intelligence.

For a time it was believed that some sort of reflex nervous mechanism must exist between the alimentary tract and the pancreas; that when food reached a certain point in its passage an unknown reflex nerve was touched off whose impulse immediately aroused the gland to secrete. In 1902, however, Bayliss and Starling discovered that an entirely different mechanism existed. By careful work they ascertained that when the acidulated food is poured through the pylorus into the small intestine, the acid has a direct effect upon certain glands in the intestinal wall. It stimulates them to liberate minute quantities of the substance *secretin*. Secretin is a hormone. Now, caught up by the blood, it is rushed to the pancreas. This organ responds immediately by pouring forth the requisite amounts of digestive juices.

Hormones have the distinction of being the most powerful physiological agents known to man. Adrenine, for example, which we shall study in more detail later, is produced by glands which in man rest upon the kidneys. This hormone is so potent that 1 part added to 300,000,000 parts of water produces a visible biological effect. This solution would be so dilute that it is hard to comprehend. If one ounce of adrenine crystals were added to water to make the dilution indicated, the solution would just about fill 235 tank cars such as are used to transport gasoline by rail. Dr. Abel, of Johns Hopkins, whose work in this field has been recognized by his election to the presidency of the American Association for the Advancement of Science, has isolated a substance from the pituitary gland in the brain so powerful that one ounce of it would have to be distributed through 29,375 such tank cars before its dilution would reach the point where its effect could not be detected.

The exact way in which these chemical messengers act to accomplish their results in the body is in many cases obscure or incompletely understood. However, Dr. Kamm, who in 1929 was voted the annual $1000 prize by the American

Association for the Advancement of Science for his work in this particular field, tells us that hormones are catalysts.

Altogether the human body is known to secrete at least seven recognized endocrine substances, and the evidence for eight others is fairly convincing. In the space allotted for our discussion we can take up the most commonly known ones only.

Thyroxine regulates the rate of metabolism. The thyroid gland is a two-lobed organ that fits around the front of the windpipe, just below the larynx. At first it was sentimentally regarded by some as being a means adopted by the Creator to round out the contour of the neck and make it beautiful. Later careful research disclosed it to be one of the most important of the internal

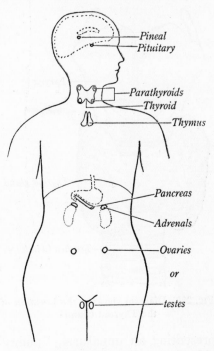

Fig. 199. Diagram showing the Location in the Human Body of the Principal Endocrine Glands

glands. It secretes a substance called thyroxine. Thyroxine seems to regulate the "speed of living" by governing the rate of metabolism.

A good supply of food and exercise speed up the rate of respiration. Hence the amount of oxygen consumed under such circumstances varies widely with the individual and with the degree of exertion. But under the same conditions the general rate of metabolism among people who are quiet and who have not eaten food for twelve hours is surprisingly constant. So consistent is it that a metabolism test, which may be made

2

at almost any first-rate hospital these days, furnishes a very reliable indication of the rate of action of the thyroid gland.

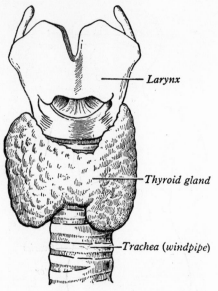

FIG. 200. Diagram showing the Location of the Thyroid Gland[1]

Underactivity of the thyroid. When the thyroid becomes depressed functionally, the amount of thyroxine produced is lowered. This is followed by well-defined characteristic symptoms. The patient is likely to be eating about the normal amount, but his food is not utilized at the proper rate. Consequently (except in extreme cases, when the rate of metabolism is insufficient to support life at all and death ensues) the victim of the disease is almost certain to deposit the surplus as fat, as a result oftentimes presenting an unnatural, "pudgy" appearance. It is probable that many so-called normal people who are inclined to become fleshy do not produce enough thyroxine. When this happens during childhood the cartilage hardens into bone very slowly, often producing a bad case of bowleg, the skin and hair are improperly nourished, and the muscles remain weak and limp.

Another symptom, especially if the deficiency of thyroxine be pronounced, is a lowering of efficiency in the nervous system and a sluggishness in the patient's mental reaction. In extreme cases the result is feeble-mindedness or even complete imbecility. There are some human strains known in which the thyroid gland is believed to be hereditarily defective or wholly

[1] From Burlingame and others' *General Biology* (second edition). By permission, Henry Holt and Company, publishers.

nonfunctional. This condition is known as cretinism, and the sufferers themselves are called cretins. This defect prevails to a marked degree in certain valleys of Switzerland, the Tyrol, and the Pyrenees. To some degree it is doubtless found in every country of the globe.

Congenital cretinism usually develops in early childhood. The skeleton shows arrested growth, particularly in the long bones of the arms and legs. The head also is generally misshapen. The tongue often protrudes, and the doughy skin hangs in folds over the abdomen and joints. The hair is coarse and sparse. The full-grown cretin rarely exceeds four feet in height and is often less than three feet. The

Fig. 201. A Cretin, before and after Treatment with Thyroxine

Courtesy of Edward C. Kendall,
Mayo Foundation

most intelligent of these unfortunates rarely exceed the mind of a child three or four years old. The cretin's voice is harsh and unpleasant, and as he walks his low, pudgy, misformed body goes waddling along in a most uncouth manner.

Before 1891 the helpless victims of this disease lived in imbecility until death. That year, however, Dr. Murray reported that he had relieved patients suffering from thyroxine deficiency by injecting under the skin a thyroid extract taken from the gland of an animal. Immediately the practice was adopted by other physicians with astounding results. A treatment had been found for this dreaded disease which gave relief in both hereditary and nonhereditary cases. Practice

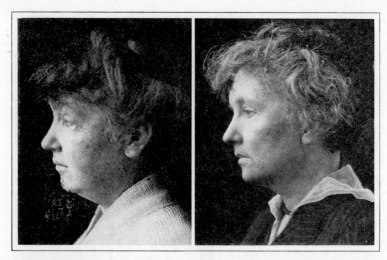

Fig. 202. Left, Myxœdema, a Disease due to Improper Functioning of the Thyroid Gland; right, after the Administration of Thyroxine for Thirteen Months

Courtesy of Edward C. Kendall, Mayo Foundation

revealed that the patient could not be cured by thyroid extract, even when the extract is regularly administered. It was found, however, that as long as the dosage was maintained the individual continued to develop and behave in a perfectly normal manner. This first patient treated for thyroid deficiency in 1891 died in 1920. For twenty-nine years she was perfectly normal, but the only thing that stood between her and a living death was her regular dose of thyroxine.

Overactivity of the thyroid. If the thyroid functions too freely, opposite physical symptoms may follow. Not infrequently individuals are found whose thyroid is either overactive or else produces a secretion modified in chemical composition, or both. Such a condition produces its characteristic symptoms, which vary in degree according to the deficiency or composition of the thyroxine thrown into the blood stream. The level of basic metabolism is raised. The food products are burned rapidly; and although the sufferer is generally a heavy eater he tends to become weak and emaciated. The whole nervous system, so to speak, is thrown into "high." The patient, contrary to expectation, nevertheless is not more efficient mentally. He becomes nervous and is inclined to be irritable. In advanced cases his whole health becomes seriously undermined and his mental condition approaches complete derangement.

The iodine supply seems to be definitely related to the effective functional processes of the thyroid. Chemical analyses show that iodine is a necessary component of thyroxine. Normally each molecule contains four atoms of iodine, and this element is believed to be the most active physiological component of the hormone. Under normal conditions the thyroid gland itself accumulates heavy stores of iodine. No other vertebrate tissue compares even remotely with this gland in respect to its iodine content. In the human body iodine comprises on the average about .2 per cent of the dry weight of the organ. An insufficient supply of iodine-containing food, then, may lead to a diseased condition known as *goiter*.

There are different forms of goiter. Overactivity of the thyroid, with the characteristic symptoms already portrayed, produces a form of goiter known as exophthalmic goiter. It is recognized by protruding eyeballs, a much quickened and irregular heart action, high temperature, nervousness, insomnia, and, in advanced cases of severe character, hallucinations and death. Relief is usually sought through surgery in which a part of the gland is removed to cut down the amount of secretion.

The second form of goiter is called simple goiter and is the result of an insufficient supply of thyroxine. It is often characterized by an enlarged thyroid. Mild cases show only a slight enlargement of that region of the neck, but in pronounced cases the gland may become the size of a pint cup or larger. Its weight produces a pouch in the skin which hangs suspended from the throat in a most unsightly manner. In extreme cases it may weigh several pounds and hang out over the chest or even down over the abdomen. The whole enlargement seems to be an attempt on the part of nature to increase the size of the gland in order to produce more thyroxine. This unfortunately cannot be done, since the supply of iodine is inadequate. In other instances of thyroid deficiency the gland itself is not enlarged, but the chemical composition of the thyroxine secreted seems to be altered. In such cases a molecule of the

FIG. 203. Marked Enlargement of the Thyroid Gland[1]

secretion instead of containing four molecules of iodine contains a smaller number. This modified form of thyroxine is a more potent stimulator of metabolic processes than the normal product. Consequently this condition may present symptoms very similar to those produced by an overactivity of the thyroid. Simple goiter when not too pronounced may be relieved by taking iodine in connection with the food. The doses, however, are very minute and should always be taken under the direction of a competent physician. More pronounced cases can be relieved only by the use of thyroxine.

The iodine content of food is believed to have come principally from the sea. Each liter of sea water has been estimated to contain on the average about .05 milligram of iodine. All plants and animals living in the sea, so far as our knowledge goes, contain this element. For this reason people living along the seashore, or those living inland who eat sea foods, are not likely to be afflicted with simple goiter. On the other hand, people far inland or in mountainous districts have a higher goiter rate. Particularly is this true in older countries where prolonged cropping and removal of the plant products have depleted the soil of its original iodine content. The inhabitants of such districts should always be alert to the possibility of iodine deficiency in their food.

The parathyroids are believed to influence the amount of calcium in the blood. The parathyroid glands are small and are closely associated with the thyroids. They usually are four in number, and the upper pair is frequently embedded in the

[1] From *Public Health Bulletin No. 192*, Public Health Service, United States Treasury Department.

thyroid tissue. Formerly death promptly followed the complete removal of the thyroid glands and was thought to be due to the immediate effects of thyroxine deprivation. Though anyone deprived of all thyroid secretion will die in time, yet the rather quick death that followed the removal of the glands was not due primarily to the loss of the thyroids: it was caused rather by the unwitting removal of the parathyroids along with the thyroid gland.

The parathyroid secretion is believed to regulate the amount of calcium in the blood. If the parathyroids of a dog be removed, the calcium content of the blood serum falls in a pronounced manner. Intermittent muscular spasms quickly follow, and unless relief measures are taken the animal dies in a few days. If a solution of calcium chloride be injected into the dog, it will recover from the muscular disturbance promptly and for a period will appear perfectly normal.

In man cases both of insufficient and of excessive secretion of the parathyroid hormone are known. An extract of animal hormone has been prepared to supply the first condition. The second may be relieved by the surgical removal of some of the parathyroid tissue to cut down the amount of the secretion.

The pituitary gland supplies hormones important to the physiological well-being of the body. Attached to the base of the brain near the middle, like a cherry suspended from a tree, is a minute gland which secretes hormones whose physiological functions have proved most baffling to students of endocrinology. In man the organ itself weighs only about .6 gram, or about the weight of "ten small grains of wheat." The gland is a sort of composite structure made up of several diminutive lobes, or parts. That a glandular bulb of such tiny proportions could secrete minute quantities of substances so elusive chemically and yet so potent in influencing the functions of the body is one of the intriguing wonders of human physiology. Dr. Abel and his coworkers at Johns Hopkins report the extraction, from the posterior lobe of this gland, of a hormone in the form of a chemical salt so powerful that when diluted in the proportions 1:15,000,000,000 it produced well-marked contraction of the uterine muscles in the female guinea pig. This dilution would mean one gram in something like twenty railroad tank-cars of water. If there are any physiological T.N.T.'s this extract would certainly qualify for the honor.

FIG. 204. Extremes in the Size of the Human Body

The claims of investigators as to their findings relative to the secretions of the pituitary gland are very diverse and often actually conflicting. Its physiological effect is apparently interrelated with that of both the thyroid and the adrenals in a most puzzling manner, but the investigation of this gland and its secretions is being pursued relentlessly. In time doubtless all the conflicting ideas regarding it and its hormones will be cleared away and its physiological function will be completely understood. For the present, however, we shall have to confine ourselves to a brief survey of what has been fairly well established concerning the pituitary gland.

The front lobe of the pituitary gland secretes a hormone which is known to influence growth to a very marked degree. Giant rats have been produced by the daily injection of this substance. In one instance a rat after almost a year's treatment weighed 596 grams, whereas its healthier mate weighed but 248 grams. The giant in this case was more than twice as large as the untreated rat.

This hormone is especially active in influencing the growth of the skeletal bones. If the secretion is deficient the child becomes a dwarf. He does not necessarily remain an infant, but may develop into a little man perfect in bodily proportions.

Tom Thumb, the dapper little circus man, was probably the victim of an underactive pituitary gland. On the other hand, an overactive gland may result in a giant. The bones grow long, and the person stretches up to huge proportions. Many circus giants belong in this category. The famous Kayanus was 9.2 feet tall. Goliath, the Philistine bully whom David slew with his sling, doubtless owed his prowess to an excess of pituitary secretion.

If the overactivity occurs subsequent to adolescence the result is a misshapen person whose bodily aspect is more like a gorilla than a man. One peculiar manifestation of a deranged pituitary front lobe is known as *acromegaly* (Greek *akron*, "extremity," and *mega*, "large"). As the name implies, after adolescence the person begins slowly and insidiously to grow at the extremities of his body. His hands and feet grow out of all proportion to the rest of his person. The head enlarges, the nose and the lower jaw increase in length, and the tongue may become so extended as to protrude between the lips. The tragedy of the situation lies in the fact that a perfectly comely, even handsome person may be transformed by this trick of nature into a perfectly hideous and repulsive-looking human anomaly.

The anterior lobe also secretes a hormone that stimulates development of the reproductive organs and the subsequent sexual emotion and function. When the extract from this lobe is administered to a very young female rat, the adolescent condition develops promptly. The ovaries mature, Graafian follicles enlarge and rupture to release the eggs, and the whole sex mechanism reaches the state preparatory to mating and fertilization. A corresponding sexual effect is likewise produced in the young male.

Recently Dr. Bengston of Chicago has reported that baldness may be successfully treated by the use of anterior pituitary extract. As Dr. Hoskins says, "A considerable proportion of humanity will no doubt await with interest the confirmation of Bengston's discovery."

The posterior lobe of the pituitary gland, according to Kamm, produces two very important hormones. He has called them the alpha and the beta hormones. The first tends to raise the blood pressure; the second is important in contracting the muscles in child delivery and is used extensively in obstetrics. It is also coming into use to relieve intestinal gas pains following an operation. The effect is to stimulate peristalsis and to move the gas along the intestinal canal. This hormone also controls the utilization of water by the tissues. People whom Kamm facetiously calls the "physiological wets" are very sensitive to the action of this extract. Along with food they utilize much water in the tissues of the body and are almost invariably fleshy. Others whom he designates as the "physiological drys" are much less sensitive to this secretion and tend to be slender and bony. Despite the amount of food they eat, they apparently will not take on flesh. He suggests that the differences of persons with respect to this hormone may be the chief factor in determining whether one is obese or lean.

Secretions from the adrenal glands play an important rôle in the life processes of higher animals. The adrenal (added to the kidney) glands are two small yellowish bodies which in man fit snugly, like a cocked hat, upon the top of either kidney. Each gland weighs about four grams. Their importance may be inferred from the fact that they have the heaviest blood supply of any organs in the whole body. About six times their own weight of blood courses through them every minute. The adrenals were discovered and described by Eustachius, an early Italian anatomist, about the middle of the sixteenth century. It was almost three hundred years, however, before man began to discover the large rôle which these tiny organs play in keeping his physiological machine running smoothly and adjusting itself quickly and effectively to emergencies that may arise.

The adrenals produce two hormones: one by the outer part of the gland, or cortex; the other, by the inner medulla. The hormone from the cortex is called *cortin* and seems to be the more essential to life; for when the cortex is surgically removed from experimental animals, death follows in a short time. Swingle tells us that cats deprived of their adrenal glands by an operation live from but eight to ten days.

Operated cats which had developed conditions of extreme adrenal hormone deficiency were prostrate, unable to stand or move about, and if placed on their feet promptly collapsed. Their temperature had fallen six or seven degrees, and if left untreated they would have died in a few hours. He took such animals and, by injecting beef cortin, was able to revive them so rapidly that within seventy hours they had completely recovered and were running and playing about the laboratory.

The dread Addison's disease, in which the skin becomes discolored and the heart action extremely feeble, and in which a general fatal languor in the end fastens itself upon the whole body, is due to cortin deficiency.

Young male rats when fed cortex extract responded by a much more rapid growth of the sex organs than the controls did. Sometimes in children the adrenals develop tumors which produce an enlargement of the cortical tissue. In such cases the masculine characters are strongly accentuated. Under such conditions the generative organs of boys from even four to six years of age may be of adult size and have complete functional powers. Such sexual precocity, it should be pointed out, may be induced also by tumors in other glands, such as the pituitary. It may be that young boys often condemned for premature sexual behavior may be the victims of diseased glands rather than of a depraved nature.

Young girls under similar circumstances are believed to take on masculine characters. Occasional cases have been known in which attractive young girls have had the painful experience of undergoing what practically amounted to a reversal of sex. Examination has revealed a slow-growing suprarenal tumor which interacted with the sex organs and other endocrine glands so as to produce this distressing effect. In such cases the breasts atrophy, the arms and legs take on the male aspect, hair grows on the face, and the voice becomes masculine. As Hoskins says, "The deep-voiced, coarse-featured, bearded ladies of the circus side-shows are probably victims of this glandular mishap." By surgery, in which the suprarenal tumor was removed, a few of these unfortunates have been restored to their normal, feminine condition.

The adrenal medulla produces a hormone called adrenine or epinephrine. This hormone has the same effects upon the body that stimulation of the sympathetic nervous system has. When administered to patients in proper dosages epinephrine increases the heartbeat and constricts the walls of the arteries and capillaries, resulting in an increased blood pressure; the bronchial muscles relax and those of the stomach and intestines are inhibited. Muscles of the skin contract, producing "gooseflesh" and raising the hair, and the eyeballs tend to protrude and the pupils to dilate. In the fourth chapter it will be remembered that these were largely the conditions found to follow a strong emotional state, such as anger or fear, and that these physical reactions were of great value to the organism in preparing to meet a crisis. It seems probable that the emotional state in some way stimulates the adrenal gland to a more than normal secretional response.

Adrenine is now prepared synthetically, but we still get most of our commercial supply from the glands of animals slaughtered for food in the packing houses. It is employed very extensively in medical and surgical practice. Victims of acute asthmatic attacks are treated with this drug to relax the bronchi. The response is rapid and affords marked relief. It is also applied locally to wounds to stop bleeding and is used in surgery to diminish hemorrhages along the edge of the incision. In certain cases of heart failure and collapse it is especially effective. Epinephrine administered in the operating room has been known to initiate contractions again in a heart that from clinical shock had entirely ceased to beat.

The pancreas secretes a hormone which is necessary for the proper utilization of sugar by the body. The pancreas has long been known as a gland that pours into the duodenum, through a special duct, a copious supply of an important digestive juice. The knowledge that this organ also secretes an indispensable endocrine substance is of much more recent origin. The history of how this discovery was made forms one of the most dramatic episodes in physiological science. The series of incidents connected with the work itself, moreover, reveals in a most vivid manner not only what science is — truth established by incontestable facts — but also portrays in a striking fashion how scientists proceed to discover new truth; in other words, the method employed by the scientist as he works.

For many years prior to 1921 it had been suspected by certain medical investigators that the pancreas was in some abstruse manner connected with the dreaded disease called diabetes. When seized with this malady the body is unable to utilize effectively the sugary part of the food. The blood becomes charged with sugar, and the system, in an attempt to get rid of it, throws it over into the urine in great quantities. If the disease is permitted to take its course, the victim becomes emaciated and suffers from extreme weakness and lassitude; he also becomes anæmic. In advanced cases ulcers may develop, and gangrenous sores are imminent because of the impure condition of the blood. The end comes usually in a deep coma from which the sufferer never awakens.

Until the alleviating principle was discovered children who suffered from diabetes almost never recovered. Physicians knew no relief except to starve the unfortunates by putting them on strict diets as nearly free of sugar-producing foods as possible.

Not only had many previous workers, as has already been stated, tentatively located the hormone regulating sugar metabolism in the pancreas, but some had even gone so far as to suspect that the active principle was connected with certain glands in the pancreas called the islets of Langerhans. But none of them had quite reached the coveted goal of a complete solution as to the cause and relief of diabetes. Finally, in 1921, there lived at London, Ontario, a young doctor by the name of Banting. The remuneration from his practice was insufficient to meet his needs, so he pieced out his income by teaching some classes in the University of Toronto Medical College. One night he had just finished the preparation of a lecture connected with the subject of the pancreas and diabetes when he picked up a medical journal. His eye fell upon an article. In the article was the statement that when patients die of gall stones that block off the flow of pancreatic juice through the duct, post-mortem examination reveals that the digestive-juice-secreting glands themselves are shrunken and atrophied but not the islets of Langerhans. The islets of Langerhans are little groups of isolated cells in the pancreas, named after Langerhans, the anatomist who discovered them. These islets had for several years previous to that time been thought to be connected in some unexplained way with sugar metabolism and diabetes. This statement was the spark that ignited the latent interest of Banting to white heat.

Forthwith he obtained permission from Dr. J. J. R. Macleod of the medical school to procure ten experimental dogs and the services of Charles Best, a physiological chemist, for a period of eight weeks. To begin with, Banting took a dog and tied off the pancreatic duct to stop the flow of juice. The digestive-juice-secreting glands degenerated promptly, leaving the islets of Langerhans well formed and functional. From a second dog, by a surgical operation in which the animal had been carefully anæsthetized, he removed the whole pancreas. By the end of nine days this dog had developed a well-defined case of diabetes. It could not stand alone and could hardly wag its tail it was so weak. A short time later the dog was on the verge of death. Now the pancreas of the first dog was removed and mashed up to form a salt-water solution which was injected into the jugular vein of the diabetic dog. Best stood by ready to test the injected dog's blood for sugar. In a brief period its blood sugar was down. The dog raised its head, presently it sat up, and within an hour it stood up and in a very short time was walking about the laboratory. The experiment had worked. A diabetic dog at the brink of death had been restored by extract from the Langerhans tissue of the pancreas. Other experiments followed with feverish intensity. It was presently found that a diabetic dog could not only be brought back to health by this extract but, so long as regular injections were made, could be kept perfectly well and contented. Other men now joined Banting and his assistant, and improved methods of extracting the hormone from these islets were perfected. Moreover, it was found that generous supplies of *insulin*, as the extract came to be known, could be obtained from the pancreas of cattle slaughtered in the packing houses.

Now the critical question arose, Could insulin, which had been found to restore diabetic dogs to a condition of health, be used to relieve human sufferers? Fortunately Banting found a willing subject ready at hand. A classmate of his in the medical college, Joseph Gilchrist by name, and a fellow soldier in the World War, had fallen victim to a severe case of diabetes. He was in the advanced stages of the malady, and there seemed no hope for him. Gilchrist was ready to try anything that might give a remote promise of help, even if the attempt was attended with considerable risk. When everything was ready the sufferer was fed an ounce of glucose and, at the proper time, the insulin was injected. The first trial did not give results. The experiment was tried again. It worked beautifully. The patient's blood sugar fell promptly; his legs lost their heaviness; his head felt clear; and after dinner this man, who had been temporarily snatched from the very jaws of death, was able to take a short walk. Today Joseph Gilchrist, who in 1921 was a mere living skeleton, is a healthy, energetic, and successful medical consultant. He has a chance to live as long as any other man of his age. Medical science had triumphed again. One of the most insidious and hopeless diseases had been conquered.

Banting's remedy was tried out extensively in a hospital for diabetic soldiers with astounding results. Today all those early experimental patients, except those who have died from other causes, are alive and well. Insulin has been adopted everywhere as the standard treatment for diabetes. Literally hundreds of thousands of lives have been saved by its use. Insulin does not cure the patient in severe cases; but so long as this extract is administered the patient remains perfectly well and active. The possibility of preparing insulin chemically gives fair promise of an abundant supply shortly which will be much less costly. This, if realized, would make it easily available to sufferers who now are able to obtain it only at great financial sacrifice.

Sex hormones are extremely important in the proper development and sexual functioning of the human body. The sex glands (testes in the male and ovaries in the female) have a double function to perform. The best known of these is the production of germ cells which when united in fertilization form the beginning of a new individual. Both the eggs and the sperms pass down specific ducts in their movement from the points where they were generated, before they unite to form the new individual. In this sense, then, these organs are duct-supplied glands and function as such. But it has long been suspected, and is now definitely known, that the glands function in another very important manner. They are also ductless glands and pour into the blood hormones which are very essential in the physical development and sexual behavior of the individual. The physiological action of these hormones from the sex glands is very complex and at some points imperfectly understood. So, for our purpose, we shall confine ourselves to their more simple and elementary functions, which are fairly well known.

The sex hormones are secreted in the reproductive organs by cells mingled with the tissues that produce the germ cells. We have already found that the sex of an individual is determined largely by the combination of sex chromosomes that it receives from the parents. But the development of the internal and external sex organs in the embryo is largely stimulated and induced by sex hormones. Lillie has found that in male and female twin calves it sometimes happens that the blood vessels from the placenta that supply the two embryos become fused, thus mixing the blood supply. When this happens, the male calf develops normal sex organs; but the female, owing to the sex hormone coming over from the male twin, has her sex organs so modified and imperfectly developed that she is sterile when grown and unable to function sexually.

The sex hormones are also very important later in life in effecting those changes connected with the period of adolescence. For instance, the general body form in both sexes, the development of a beard in the male and the deep quality which his voice takes on, are secondary sex characters due to the sex hormones. The eunuchs mentioned in the Bible were emasculated men. In many instances their strikingly feminine characteristics were due to the fact that during growth their bodies had been deprived of the sex hormones.

In the female the menses, together with the physiological reactions accompanying them, are induced by sex hormones. For example, a few drops of a fluid extracted from the ovary of a pig when administered to a newly weaned female rat will bring her to sexual maturity and full menstruation in four days. Normally it would take such a young rat from thirty-six to eighty days to reach this condition. The development of the mammary glands is also induced by sex hormones.

There is a very close and intricate relation between the action of the sex hormones and the organization of the nervous system. Other hormones may have for their function the preservation and welfare of the individual, but the sex hormones are for the specific purpose of perpetuating the race. Their function is to prepare the body for mating and to insure the care and succor of the offspring both before and after birth.

The public should beware of hormone quacks and charlatans. Since the discovery of endocrine secretions and their important offices in the body, no branch of legitimate medicine and surgery has had so many outrages perpetrated upon it. Quacks have arisen who purport to cure all manner of sexual defects for fat fees. The future does give great promise as to the possibilities of endocrine science and glandular therapy. Of a truth the ductless glands seem to be in fact "glands regulating personality." In some instances, as in the adrenal, thyroid, pituitary, and sex gland secretions, much very important progress has already been made. But the public should know that no reputable surgeon ever prescribes as a matter of established practice the substitution of animal sex glands for those of the human body. Only frauds and cheats pretend to do this. Moreover, there are all sorts of "hormone doctors" who claim to be able to cure various and sundry diseases by the use of internal gland extracts. Fortunately most of their dopes are simple deceptions and harmless. Sometimes, however, their dosages are very dangerous, and permanent harm, if not death, may result from their use. Only two hormone substances, thyroxine and adrenine, have been prepared synthetically and are available to the medical profession in their pure chemical form. Practically all other hormones used by physicians have to be extracted from the glands of other animals. Their purity and potency vary, and they must be constantly checked up by using experimental animals before their dosage is safely determined for human use. If that be the case, then it may readily be seen how dangerous such powerful poisons are in the hands of a quack. Hormone extracts should never be taken except when prescribed by well-trained and competent physicians. Moreover, it should be remembered that medical men of standing never disseminate their claims in glaring newspaper advertisements.

QUESTIONS FOR STUDY

1. What is a hormone, and what is its function in the body?

2. Describe the kinds of goiter and give the symptoms of each.

3. What reasons can be given to explain the differences between the secondary sex characteristics of men and women?

4. How is the calcium metabolism of the body regulated, and what gland is responsible for this hormone?

5. Give a brief account of the discovery and use of insulin.

6. What factors influence the growth of an individual?

7. What hormone works principally upon the autonomic nervous system?

8. Explain the probable cause of cretinism.

9. If all or a large portion of the thyroid gland is removed, what must the person do to overcome the effect of the lack of thyroid secretion?

10. In what sections of the United States is goiter most prevalent and what means is taken to remedy this condition?

11. How are metabolism tests run? Tell what is shown by such tests.

12. Where are the various endocrine glands located in the body?

13. How has our knowledge of the secretions of the ductless glands aided in our understanding of bodily activities?

REFERENCES

BURLINGAME, L. L., and others. General Biology (second edition), pp. 172–191.

DARROW, FLOYD M. The Story of Chemistry, pp. 269–278.

DE KRUIF, PAUL H. Men against Death, pp. 59–87.

EAST, EDWARD M. Biology in Human Affairs, pp. 299–304.

GREISHEIMER, DR. ESTHER. "The Endocrine Organs," *The American Journal of Nursing*, Vol. 32, pp. 241–254, 395–400.

HOSKINS, R. G. The Tides of Life. (Endocrine glands in bodily adjustment.)

KIMBER, D. C., and GRAY, C. E. Textbook of Anatomy and Physiology, pp. 333–344.

STEIGLITZ, JULIUS. Chemistry in Medicine: "The Internal Secretions," Dr. R. G. Hoskins; "The Hormones of the Suprarenal Glands," Drs. J. J. Abel and E. M. K. Geiling; "The Story of Thyroxine," Dr. E. C. Kendall; "The Hormones of the Pituitary Secretions," Dr. E. M. K. Geiling; "Hormones of the Sex Glands — What They Mean for Growth and Development," Dr. C. R. Stockard; "Iodine in the Prevention and Treatment of Goitre," Dr. D. Marine; "Insulin to the Rescue of the Diabetic," Dr. J. J. R. Macleod; "The Internal Secretion of the Parathyroid Glands," Dr. J. B. Collip.

Chapter XVI · Science aids Nature in the Fight against Infectious Diseases

Introduction. Today, with a rather extensive knowledge of disease and its causes, it is difficult for one to picture the conditions of the past. Without due consideration he is likely to think that from remote days man has understood the nature of infection and the chief cause of illness. Such, however, would be far from the truth. Our present knowledge of diseases and their causes scarcely antedates the Civil War. Indeed, there are many people still living whose parents never heard of a germ. Before microörganisms were discovered and their causal connection with disease was established, all sorts of mystical and superstitious ideas prevailed.

Primitive man, unacquainted with natural phenomena, was prone to personify everything. The rustle of the leaves in the forest, the play of sunbeams on the ground, or the reverberating roll of thunder, he recognized as a manifestation of some spirit or deity. What could be more natural, then, than for him to place disease in the same category. He saw in sickness and suffering the response of the individual to some malevolent spirit that had taken possession of him. To effect a cure the logical thing to do was to cast out the evil spirit by some sort of incantation.

Hippocrates, who lived about 400 B.C., had little faith in the evil-spirit superstition and, after studying the human body, advanced an entirely different theory of disease. He conceived the essential elements of the body to be phlegm, blood, yellow bile, and black bile. As long as these elements appeared in proper proportions a condition of health followed. Just as soon as they were thrown out of balance, however, the patient became ill and recovery could be effected only by reestablishing a state of equilibrium among these elements.

The Greek ideas, in modified form, influenced medical beliefs up to the sixteenth century. Thomas Sydenham, who lived during the latter half of the seventeenth century and who has been called the "English Hippocrates," asserted that "a disease is nothing more than an effort of nature to restore the health of the patient by the elimination of the *moribific matter*." Although our ideas of "moribific matter" were to undergo marked changes at a later date, Sydenham's announcement is practically in accord with the modern view of disease. Most illness is recognized today as an attempt on the part of the body to eliminate germs and to recover from the injurious effects of their invasion.

The perfection of the compound microscope and of certain micro-staining technics paved the way for the germ theory of disease, which was formulated and established during the latter part of the nineteenth century. Robert Boyle, the English physicist and philosopher, writing in the seventeenth century, in a prophetic moment predicted that the problem of infectious diseases would be solved by him who cleared up the nature of fermentation. Boyle's prediction was fulfilled two hundred years later by the brilliant work of Pasteur. The latter showed that beer and wine "sickness" — the undesirable fermentations that often occurred, resulting in the spoiling of the product — was due to vitiating wild yeasts and bacteria that got into the liquors. Subsequently Pasteur succeeded in proving that the deadly silkworm disease pebrine, which devastated the silkworm hatcheries of France and threatened the very existence of the whole industry, was caused by a bacterium within the eggs, larvæ, and bodies of the adult moths. Aroused by these discoveries, Pasteur in France and Koch in Germany, both working with feverish energy, showed unmistakably that the destructive live-stock disease anthrax was caused by a bacterial infection which pervaded both the tissues and the blood of the afflicted animal. Taking his cue from Pasteur, Joseph Lister, at the noted hospital in Edinburgh, was able to reduce the death rate from amputations from 45 per cent to 15 per cent in a single year by spraying the air, the instruments, the incision, and the hands of the surgeon with a weak solution of carbolic acid while the operation was being performed.

With these achievements the astounding truth had been laid bare. Most human diseases, as well as those of other animals, are caused not by demons or bad humors, but by invading germs that get into the body in one way or another. Once established, this new view was decidedly revolutionary in its effect. It has changed our whole conception as to how diseases are contracted, and it has refashioned our ideas as to how the body naturally wards off disease and as to how it combats pathogenic organisms once they have gained access to the tissues. Moreover, it has completely changed and remade the whole science of both prophylactic (preventive) and therapeutic (curative) medical practice. In addition it has led to the development of a new science, the science of immunology, which offers bright prospects in the future both to the investigator and to the physician who would alleviate human suffering.

A more recent point of view regarding the cause of certain diseases is admirably expressed by Ward of Harvard:

A further development of the bacterial origin of infectious disease was the discovery by Löffler and Frosch in the year 1898 that foot and mouth disease in cattle was caused by an invisible virus, often called a "filterable virus" because the infecting agent can pass through the pores of fine filters which will hold back the ordinary visible bacteria. The nature of a filterable virus is still uncertain, because we can neither see it, nor cultivate it outside the body except in living cells. There is a certain amount of evidence that a filterable virus is a very minute microörganism but this point is not yet settled. Many human and animal diseases are caused by filterable viruses — small-pox, yellow fever, measles, mumps, chicken pox, rabies, infantile paralysis, the common cold, etc.

The body is constantly beset by disease-producing organisms. As has been made clear in a former chapter, bacteria are abundant almost everywhere. The particles of dust in the air which people breathe, the food they eat, and the water they drink carry them in great numbers. There is scarcely anything around ordinary human habitations free from them unless it be deep down in the earth or be cleansed by adequate processes of sterilization. Even the body itself normally harbors microörganisms by the millions. The outer layer of the integument, particularly in the armpits and other areas of the body where moisture, warmth, and food in the form of decomposing skin cells are abundant, supports them in almost unbelievable numbers. It is for this reason (in warm weather especially) that persons must give careful attention to bathing to keep down body odors. The fingers, that come in contact with everything, the membranes of the eye, the alimentary canal, all the respiratory passages, and often the urinary tract as well furnish a suitable abode for a rich bacterial flora.

Fortunately under normal conditions most of these microorganisms are harmless. But almost always lurking among the saprophytic ones are dangerous parasites which enter into the body tissues whenever the least opportunity is afforded and cause serious trouble. Against these devastating invaders the bodies of all animals must vigilantly protect themselves. Happily nature has provided for this protection in very elaborate, interesting, and effective ways. The manner in which the body intrenches itself against infection and the way it fights the invaders, once they get in, reminds us very much of the tactics employed by a besieged army. The besieged barricade themselves against the entrance of the enemy. They erect barbed-wire entanglements, throw up earthworks, and, if opportunity affords, retire behind well-constructed walls. But if these defensive measures prove ineffective, and the enemy gets in, then the besieged army engages the hostile force in a hand-to-hand conflict. If those within are strong enough they conquer the attackers and either kill them or take them prisoner. The body in the presence of infecting agents acts in a quite analogous manner.

The human body barricades itself against germ invasion. The body is well barricaded against the entrance of disease-producing organisms. Covering the whole outer surface of the body is the skin. This serves to protect the tender and sensitive tissues within against mechanical injury and against an excessive loss of water to the surrounding air. It also serves in a remarkable way as a protection against the ingress of dangerous microörganisms. Most of these minute invaders cannot penetrate the unbroken skin. A few, however, are reported to be able to "bore" through the protective integument of the body and thus gain access to the delicate tissues within. The bacilli which cause the black plague are said to have infected guinea pigs when placed on a shaven spot. Tularemia, or rabbit-fever microörganisms, are also said to be capable of entering the body through the unbroken skin. In most of these cases, however, it is probable that infection took place through unobserved microscopic lesions or through the duct of an oil gland or a sweat gland. It may even be that the disease-producing organisms gained entrance through a hair follicle. Almost everyone has suffered the pain and inconvenience of a boil or a carbuncle which appeared to form directly below an unbroken area of the skin. Probably in all such instances the pus-forming bacilli gained entrance through the avenues already mentioned. It is for this reason too that the pus discharge from a subcutaneous abscess should be meticulously taken up with absorbent agents and promptly removed. The surrounding skin also should be frequently sterilized by some effective means, such as washing with soap and clean water to prevent secondary infections through these paths.

The hair in the nose, which strains dust particles out of the air as it enters the nasal passages, and the mucous lining of the body tracts and cavities, all interpose distinct barriers to pathogenic organisms. The mucous membranes, however, are not nearly so effective in this respect as is the skin. It is believed by some, also, that the connective tissue under the skin possesses some antiseptic property. Large quantities of thick mucus exude from certain surfaces, such as the nasal passages, when the membranes are irritated by the presence of dust particles. In this way great numbers of germs are engulfed and expelled from the body. Nevertheless the mucous membranes themselves may be penetrated by microörganisms. Diphtheria bacilli and streptococci have been found capable of settling down on the membranes of the throat and nose and producing local inflammations and infections. Through this avenue the latter germ may enter the blood stream and be carried to other parts of the body.

The gastric juice normally contains about 0.2 per cent of hydrochloric acid. Though this dilute acid certainly does not kill all microörganisms that enter the body through the alimentary canal, it does destroy many of them. This is particularly true of microörganisms that flourish under alkaline conditions. Koch, the great German pathologist, found it necessary to neutralize the gastric juice with sodium carbonate before he could infect guinea pigs with cholera germs introduced through the mouth.

So in these and in many other ways the human body and that of many other animals as well are protected against invasion by pathogenic organisms. But despite these safeguards infections occur. Lesions are produced by breaks in the linings of the body cavities; the skin is rent by injury or disrupted by tiny pimples and abscesses; sometimes the bite or sting of an insect penetrates the protective coverings. When this happens the microörganisms come trooping in like an invading army through a break in the protective wall of a besieged enemy. Now, in military parlance, a crisis exists. There is nothing to do but fight. The attackers must be defeated and routed or else the army within the walls must capitulate. When infection occurs, the body is faced by the same sort of issue. Either it must engage the enemy and defeat them by reaching a state of immunity or it must succumb.

Different microörganisms affect the body in different ways. After microörganisms have gained entrance through the protective coverings of the body they do not all produce injurious effects of the same nature. Their action varies; and upon the specific type of injury which they produce depends the character of response which the body itself makes to the invaders. Upon this also depend diagnosis on the part of physicians and the therapeutic treatment to follow.

Some microörganisms produce exotoxins. Some microörganisms work injury to the body in a peculiar manner. Instead of destroying the tissues on an extensive scale, they diffuse toxic substances into the general circulation. These are conveyed through the body of their host and act as powerful poisons to the vital centers. Such toxins are appropriately called *exotoxins*.

The diphtheria and tetanus (lockjaw) bacteria are perhaps the most common examples of exotoxic microörganisms. For the most part their infection is a local one, produced, respectively, either in the membranes of the throat and nasal passages or at the site of some wound. Having gained access to the tissues the pathogenic organisms multiply within the local area. Their deleterious effect, however, does not come chiefly from injury at these points of infection; but, once established, they soon begin to throw off powerful toxic substances which are caught up by the blood stream and carried to the vital organs, such as the heart, brain, lungs, and kidneys. The toxin of tetanus, formerly believed to be carried along the nerve fibers from the points of injury, is now thought to be carried by the blood stream too. Arriving at the organic centers, the toxins paralyze or destroy the functional powers, and death speedily follows. The dreaded botulinus poison is an exotoxin liberated by bacteria working in decaying food substances, and is one of the most deadly toxins known. Snake venoms also owe their fatal properties to toxins which they contain.

Some microörganisms produce endotoxins. There are other organisms which do not normally liberate their toxic products through the cell wall. Instead their toxins are retained within the bacterial bodies and are freed only when the microörganisms themselves die and decay. These poisons are called *endotoxins*. It is probable too that the toxins are partly derived from the split proteins formed in the decompositional process itself. Cholera spirilla and typhoid-fever bacilli are found in this class. Colon bacilli, which normally inhabit the intestinal tract without apparent injury, may become endotoxic also; for sometimes the individual has his resistance reduced to the point at which these bacteria enter the tissues and become deadly in their effect. Many cases of appendicitis are believed to be caused largely by colon bacilli that have thus become parasitic. Other microörganisms, such as the tuberculosis bacilli, the staphylococci (connected with boils and abscesses), and the streptococci (which are the chief cause of blood-poisoning and septicæmia), are tissue destroyers. In their metabolic processes they attack the tissues of the host, break them down, and destroy them. In the case of tuberculosis the whole lung, as well as the lymph glands, may be practically consumed. The deep, scarlike depressions left by carbuncles and abscesses give mute evidence of the tissue-destroying power which these pus-forming organisms possess. It is very probable too, considering the general discomfort and suffering of the host, that these microörganisms also produce toxic substances of various kinds which are caught up and carried through the body by the general circulation.

Pathogenic microörganisms vary in virulence. The reader has heard of "heavy" attacks of a specific disease and of "light" forms of it. The degree to which a given infection affects the individual depends upon many factors. Some persons are more susceptible than others, as we shall see later; the same person falls a victim to the attacks of germs much more readily on one occasion than on another; but, disregarding all circumstances of time, place, and person, virulence depends largely upon the nature of the infecting microörganism itself. Some strains of the same germ are a great deal more virulent than others. For instance, streptococcus grown in broth may not kill mice when as much as a cubic centimeter of the culture is injected; but, by being passed through the bodies of mice for several generations, the strain may become so potent that $\frac{1}{100,000}$ of a cubic centimeter is fatal. Pasteur found that hydrophobia virus gained in virulence to the maximum when inoculated into rabbits. On the other hand, after being grown for three generations in monkeys its virulence for rabbits decreased almost to the point of extinction. It is possible that sudden outbreaks of an infectious disease whose source is a mystery may sometimes rise from organisms that have become highly virulent by passage through the bodies of common animals.

Susceptibility to disease germs varies greatly. The susceptibility of different races with respect to certain diseases is almost as variable as their physical and mental characteristics. For instance, the white races of Europe and North America are much more resistant to such diseases as smallpox, influenza, and syphilis than are the Negroes, the South Sea Islanders, and the Eskimos. On the other hand, Negroes are much less susceptible to malaria and yellow fever than are the whites. The Mexicans possess a high degree of resistance to typhus fever.

Such differences in respect to susceptibility as do exist probably may be attributed more to the results of natural selection than to original racial variation. The whites have been subjected to germs of smallpox and influenza for centuries. As a result the more susceptible have died of these maladies, and the more resistant have survived. Negroes and other tropical peoples likewise have become selected for warm-climate diseases, such as malaria and yellow fever.

It is even seriously questioned by some immunologists whether the marked racial differences supposed to be found between different races with regard to certain diseases due to natural selection really do exist. That there is a difference between the adults cannot be gainsaid. However, these critics of the theory prefer to believe that the different degrees of immunity have been produced by childhood attacks in the countries where the diseases are found rather than by racially acquired biological differences.

There also appears to be a wide variation between different persons of the same race with respect to their resistance to disease. It is a common observation that when an epidemic sweeps through a population some persons are carried off summarily, others have light cases, and still others escape without the semblance of an attack. Some of the immune persons, of course, may have been made so by unobserved childhood infections. But, aside from this consideration, wide differences seem to exist. Indeed, it seems that even the same person is much more susceptible to a particular disease at one period than at another. Just why this should be so, it is difficult to say.

However, there are certain physical conditions that predispose one to disease. Fatigue, poor nutrition, worry, exposure, and physical weakness due to the ravages of other diseases may all be contributing causes. Pasteur rendered a naturally resistant hen susceptible to anthrax by standing her in cold water until the fowl was thoroughly chilled. Pigeons are normally immune to the germs of this disease; but if they are partly starved just before and after inoculation they readily succumb.

The body is best able to ward off and to cope with germs when it is in good physical condition and can muster all its powers for defense. Any condition that impairs the general health is likely to pave the way for the entrance and growth of microörganisms within the system.

The body combats infection in many ways. As has already been stated, when once disease germs gain entrance to the body, nature has but one recourse, that is, to fight. It does this chiefly in two ways: through the production of counteracting substances called *antibodies* and through the agency of *phagocytes*, which, under proper conditions, attack invading germs and destroy them.

Antitoxins. Diphtheria bacilli sometimes become so virulent and abundant in the throat and nasal passages as to initiate the growth of extensive membranes. These obstructions may mechanically shut off the supply of air and cause the patient to die of suffocation. Far more frequently, however, the deleterious effects of diphtheria arise from the deadly exotoxins which the microörganisms liberate. Normally the body of the sufferer reacts promptly by producing counteracting antibodies called antitoxin. This must not be thought of as cells or corpuscles, but as a chemical substance of some sort. Apparently this antibody does not chemically neutralize the poison either, but in some unexplained way engages it and renders it noneffective and harmless. If this antibody is formed in sufficient quantity, the patient overcomes the disease and recovers. In the same way nature combats the organisms producing tetanus. Recovery from the deadly botulinus toxin and from poisonous snake bites also seems to depend upon the power of the body to produce antibodies in sufficient quantities to counteract these destructive exotoxins.

Lysins. Disease-producing bacteria which do not liberate toxins may be combated in the animal body by the elaboration of *lysins* (Greek *lysis*, "a loosening"). These substances attack the invading microörganisms and destroy them by breaking down the minute cell walls and reducing them to a solution. Pfeiffer and his associates injected cholera spirilla into the peritoneum of a cholera-immune guinea pig. The spirilla were promptly attacked and rapidly dissolved. By taking out small quantities of the peritoneal fluid periodically with a capillary pipette, the bacteria could be observed in all stages of dissolution. When bacteria from the same culture were injected into the peritoneum of a guinea pig which had not been immunized to cholera, however, no such destruction of the pathogenic organisms was observed. Not being immune the pig possessed no lytic antibodies to attack and dissolve the cholera bacilli.

Phagocytes (Greek *phagos + cytos*, "eating cell"). One of the most spectacular and important rôles of all the body cells is taken by the phagocytes. They combine two functions in one: they are both policemen and scavengers for the system. Whenever a city is threatened with riot the police department responds by detailing a larger number of policemen and sending them to the point of trouble. The body reacts in a similar way. Within the blood stream there is normally a considerable number of special white blood corpuscles, or leucocytes, whose function seems to be chiefly protective. They move about through the circulation, and if they encounter an invader of any sort, such as a bacterium or other foreign cell, they promptly engulf, digest, and destroy it. Their action is very similar to that of an amœba when it ingests food particles and consumes them. When an infection of any sort occurs, these leucocytes multiply in great numbers and rapidly accumulate at the point of trouble. Here they congregate to combat the pathogenic organisms. Being amœboid cells they even squeeze through tiny porelike interruptions in the capillary walls and swarm out into the lymph spaces to attack and arrest the destructive action of the enemy.

In addition to these mobile leucocytes in the blood stream, there are believed to be many stationary protective cells which

also guard the welfare of the individual. Certain connective
tissue cells, endothelial cells, spleen-pulp cells, liver cells, and

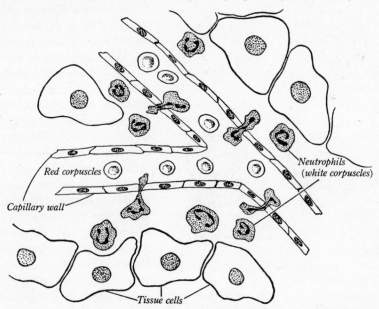

FIG. 205. The White Blood Corpuscles pass through the Capillary Walls

cellular components of the lymph nodes, scattered through
the body, assume the function of phagocytes too. So silently
and effectually do these various kinds of phagocytic cells do
their work that invading microörganisms are doubtless often
arrested in their development and cleared from the system
without the individual's ever knowing that he has been subject
to infection.

But it seems necessary, if phagocytic action is to be sufficiently rapid to produce immunity, for the microörganisms
to be prepared beforehand for the attack of the phagocytes.
For instance, if human leucocytes be washed free of all other
blood constituents and be mixed with certain forms of bacteria,
the white blood cells are practically indifferent toward the
bacteria. They probably would be wholly so if, in washing,
every trace of blood serum could be removed from the leu-

cocytes themselves. But if bacteria from the same culture first be brought in contact with human blood serum and then the leucocytes be mixed with them, a wholly different scene takes place. In this case the leucocytes become energetic phagocytes, engulfing, digesting, and destroying the microörganisms with avidity. The difference manifestly lies in some active chemical component of the serum itself which prepares the bacteria for the attack of the phagocytes. Because this blood principle makes ready the microörganisms for the action of the phagocytes a very vivid term has been applied to it. It is a specific antibody, and is called *opsonin* (Greek *opsoneo*, "I prepare a table" or "I prepare a banquet"). Recovery from disease often seems to hinge upon the potency and the amount of opsonin present in the blood. If this substance is present in sufficient amounts, or if the body when infected can adequately increase the supply, then immunity and recovery appear to be merely a matter of time; but if the "opsonic index," as medical men call it, is not high enough, then entirely different consequences to the patient may follow.

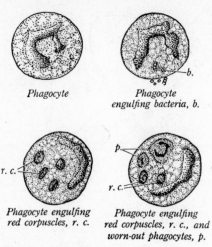

Phagocyte

Phagocyte engulfing bacteria, b.

Phagocyte engulfing red corpuscles, r. c.

Phagocyte engulfing red corpuscles, r. c., and worn-out phagocytes, p.

FIG. 206. White Blood Corpuscles assist in keeping the Body Clean

Certain phagocytes are also believed to act as scavengers in the body. For instance, old worn-out red blood corpuscles and decrepit leucocytes themselves are believed to be engulfed and removed from the circulation by the action of certain large phagocytes. Indeed, Metchnikoff, the great Russian physiologist, who was the leading proponent of the phagocytic doctrine, went so far as to maintain that some of the prominent effects of old age are due to the unrestrained action of phagocytes. For example, he claimed to have shown that the senile degeneration of the nervous system is brought about by phagocytosis.

He further attributed the whitening of the hair, in both men and dogs, to phagocytes that creep in through the root sheaths and devour the pigment.

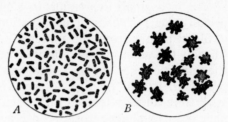

FIG. 207. Agglutination

In serum from individuals who are not immune the distribution of the bacteria throughout the field is homogeneous, as in *A*; in immune serum the bacteria collect in clumps, as in *B*

Agglutinins. Agglutinins are substances found in the blood stream which cause certain forms of microörganisms to group together in clumps. Some investigators are not certain whether this clumping is a more or less accidental reaction to protective substances formed by the system or whether agglutination is a definite protective measure.

Others feel that agglutination is a rapid protective reaction. They believe it to be a preliminary step in phagocytosis. If pneumococci, the germ causing pneumonia, be injected into a rabbit in sufficient quantities, they will in due time appear in large numbers in the blood stream. If, now, an adequate dose of anti-pneumococcus serum containing agglutinin from an immune rabbit be introduced into the circulation, a very rapid agglutination of these microörganisms occurs. Blood taken through an immediate heart puncture shows clumps of these bacteria evenly distributed in the circulation. In from three to five minutes, however, it may not be possible to find any of these germs in the circulating blood. Those who regard agglutinin to be a true antibody hold that this clumping greatly facilitates the concentration of the invading germs in the capillaries and lymph nodes, where the phagocytes attack and destroy them.

This agglutination reaction is used everyday in medical practice to diagnose many germ diseases, such as typhoid fever, pneumonia, cholera, and meningitis. The method may be illustrated by the famous Widal test for typhoid fever. A drop or two of a *laboratory culture* of typhoid bacilli are placed in a small depression in a microscope slide made especially for purposes of observation. Now a small amount of blood serum from the patient is introduced into the culture on the slide and carefully observed. In a short time, if the individual has typhoid, the bacteria will be clumped. The grouping, it will readily be seen, is brought about by the agglutinin already formed in the patient's blood as a preliminary step to phagocytosis. Since agglutinin, like other antibodies, is specific and will clump only the particular type of microörganism for which it was prepared by nature, the test becomes practically conclusive.

The discovery of antibodies and phagocytes has made artificial immunization possible. When one is attacked by a germ disease, if his body is in good condition, and the microörganisms are not too virulent, he is likely to recover of his own accord. That is, his system will produce antibodies in sufficient quantities and rapidly enough to counteract any toxins that may be formed and to prepare the germs for the work of the phagocytes. Oftentimes, however, the patient cannot recover of his own accord. His resistance may be low, the infecting microörganisms may be unusually virulent, or he may be subject to an extremely large number of them at the time of infection. Under such conditions he may not be able to respond rapidly enough to combat the disease successfully. In such cases artificial measures to supplement the patient's protective responses have in certain diseases proved highly efficacious in overcoming the attack. The account of how artificial immunization was discovered and of the circumstances surrounding the discoveries forms one of the most gripping chapters in medical science.

Artificial immunization was discovered by Jenner and Pasteur. Edward Jenner, an English physician, is regarded as the original discoverer of artificial immunization. He practiced in a rural community where dairying was the main industry and where young women did most of the milking. At that time smallpox was a very virulent disease which often swept the country as an epidemic. Jenner was told by some of his patients that dairy maids who had contracted cowpox through their chapped hands while they were milking were immune from smallpox inoculations. Being a man of scientific curiosity and inquiring mind, Jenner decided to investigate. He found that the dairy cows were subject to a pustule-forming disease very similar to human smallpox. Moreover, he ascertained that when the contents of a cowpox pustule were transferred to an abrasion of the skin in a man, the man generally suffered a light attack similar both to cowpox and to smallpox.

FIG. 208. Edward Jenner discovered that Smallpox and Cowpox were Similar Diseases

The first human being to be inoculated was Jenner's own eight-year-old son. The trial took place on May 14, 1796. Virus from the hand of a milkmaid afflicted with cowpox was used in the vaccination. The experiment was successful. The boy contracted cowpox and when, on the first of the following July, he was inoculated with smallpox virus he proved to be immune.

Cowpox is believed today to be but a light form of smallpox in which the virus has been greatly weakened by passing through the body of the cow. This was the beginning of the practice of vaccination for smallpox, which is now almost universally practiced among civilized peoples. The virus used is taken from the pustules of young heifers. It is interesting to note that the word *vaccinate* comes from the Latin word *vacca*, "a cow."

The next great impetus to artificial inoculation was the result of an accidental discovery on the part of Louis Pasteur. We may be very thankful that he had the mental acumen to grasp its significance. He had been working with chicken-cholera cultures. As long as these organisms were transferred to new chicken-broth media every twenty-four hours, they remained perfectly virulent and active. Cholera could be

Ewing Galloway

Fig. 209. The First Vaccination with Cowpox Virus to prevent Smallpox

produced at will by injecting them into the fowls. But one day while Pasteur was preparing to leave for a vacation, certain cultures with which he had been working were overlooked and left standing. Upon his return his curiosity led him to inoculate some chickens with bacteria from these old dry cultures. The fowls in due time showed symptoms of cholera, but to Pasteur's great surprise they all soon recovered. He now tried them with some new culture; but the unexpected phenomenon of resistance occurred, and he could not produce death with the most virulent germs. What had happened? What had weakened the germs that had stood exposed to the air, and what effect had been produced in the fowls which made them immune? Subsequent work showed that oxygen had attenuated (weakened) the bacteria. By using cultures exposed to the air for variable periods, Pasteur could on the average produce at will the death of eight hens, five hens, or one hen out of ten. Finally, when the cultures were let stand as long as the original forgotten ones, no hens died. Best of all, those which recovered from the attacks of the attenuated virus were all immune to subsequent attacks of cholera.

This incident was presided over by a genius, and society has continued to reap the benefit of that chance discovery of cause-and-effect relationship. For Pasteur immediately began with feverish interest to wonder whether other animals, and man as well, could not be made immune to various diseases by similar treatment. First anthrax yielded to his hypothesis; then the dreaded hydrophobia and other diseases followed. Medical science was on the right track now, and all the benign discoveries in immunology which have resulted in the saving of hundreds of thousands of lives date their beginning from the monumental discoveries of Jenner and Pasteur.

In general there are two kinds of acquired immunity. Immunity acquired by artificial means may be said in general to fall into two classes: *active* and *passive*. In active immunity the patient's system itself is induced by special treatment to produce the antibodies necessary to recovery; in passive immunity the body is rendered so by the direct injection of the antibodies themselves. These antibodies are derived from some other source; as, the blood of another animal.

Passively acquired immunity. Passive immunity is best exemplified by the antitoxin treatment for diphtheria. The horse is the best-known source of this antibody. A young sound horse free from all disease is treated with diphtheria toxin. The toxin for injection is produced by growing the bacteria in flasks of bouillon. The bouillon is then filtered free of the germs, and a small amount of it containing the toxin is injected into the horse. At the outset only a fraction of a cubic centimeter of the toxin from the cultures constitutes a dose. The animal shows a slight rise in temperature and a temporary loss of appetite. It soon recovers, however, and a slightly larger dose of the toxin is introduced into its blood with similar symptoms following. This treatment is continued at intervals of a few days for a period of two or three months. By this time the horse is able to receive as much as three hundred cubic centimeters of the toxic bouillon with but slight if any reaction. The horse is now immune by reason of the large amount of antitoxin generated in its blood. Under aseptic conditions a tube is now introduced into the animal's jugular vein and as much as five to eight quarts of blood is extracted. After a period of care the horse may be used again for the same purpose.

After being drawn from the horse the blood is permitted to clot, and the antitoxin is recovered in the clear, amber-colored serum. Its strength is tested by using some animal, such as a rabbit or a guinea pig, and it is then put out in what are known as dosage *units*. When these units are injected into the blood stream of a patient, an immense number of antibodies are released at once to counteract any toxin that may be carried in the circulation. The patient's antitoxic response, now supplemented by the animal serum, is usually sufficient to overcome the deleterious effects of the disease. Finally, phagocytosis steps in to clear the germs from the system, and recovery is accomplished. It must be remembered that such immunity is but temporary. Its limit of effectiveness has been roughly estimated to be about twenty-eight days on the average.

The remarkable effectiveness of the antitoxin treatment for diphtheria may be seen by a comparison of the mortality figures before the treatment was applied and since. Before 1894, when the treatment was perfected, the death rate of those contracting the disease was about 38 per cent. Now it is less than 10 per cent, and if antitoxin be administered promptly the death rate is scarcely more than 1 per cent.

Tetanus antitoxin confers immunity for only about twelve days. On this account wounded soldiers during the late war were inoculated with tetanus antitoxin about every nine days.

Artificial actively acquired immunity. Artificial actively acquired immunity is the result of antibodies produced by the individual's own body in response to vaccines. The object of such treatment is obviously prevention. The vaccines are cultures of the pathogenic germs either killed or attenuated by special treatment of some sort. The agents usually employed in preparing vaccines are high temperatures, exposure to light, or harmful chemicals. In this way the dangerous possibilities of the germs are obviated; nevertheless they still retain a sufficient degree of their pathogenic property to incite the body to produce antibodies in adequate quantities.

Within the past few years a combination of toxin and antitoxin known as the toxin-antitoxin treatment for diphtheria has been devised. The antitoxin is administered along with the toxin in order partly to counteract the latter's injurious effects. Meanwhile the patient's system is responding by producing antitoxin of its own. Such immunity, it is evident, is active and may last for some time, but it is not permanent. This treatment has proved very effective in immunizing the individual when other members of the family are attacked or when an epidemic is imminent.

Prophylaxis against typhoid fever may be taken as an example. This is a disease in which the germs are generally acquired through either food or drink. The most prolific sources are the water supply, ice, milk, and other foods. Vegetables that have come in contact with either polluted water or soil are also likely to be fruitful sources of contagion when eaten raw. The vaccine is prepared by growing the bacillus for two or three weeks in broth. The organisms are then killed by heating, and the culture is finally preserved by the addition of one half of 1 per cent of carbolic acid. The strength of the vaccine depends upon the number of dead organisms it contains, which is determined by special methods of counting. The treatment is usually administered under the skin in from three to five inoculations. When three are made, the number of dead organisms introduced each time are, respectively, 500,000, 1,000,000, and 1,000,000, or a total of 2,500,000.

The dead microörganisms, as is obvious, cannot multiply and produce the disease. However, they do retain sufficient potentiality to induce the body to respond by freeing into the circulation an abundance of typhoid antibodies in the form of agglutinin. These remain in the system for some time. Immunity to typhoid so acquired usually lasts for a period of from one to two years.

The efficacy of the anti-typhoid treatment may be seen by comparative figures from the army. During the Spanish-American War, in 1898, before typhoid vaccine had come into use, almost one third, or about 30 per cent, of the American soldiers contracted typhoid. Of the 1,900,000 soldiers in France during the World War, however, there were but 885 cases of typhoid, or a small fraction of 1 per cent. Once in a while one out of a great number who have been vaccinated will contract typhoid fever; but such a result is so rare that to have typhoid today is almost a reflection on one's intelligence.

Artificially acquired immunity holds great promise for the future. Many diseases are now combated with vaccines and serums. Diphtheria, tetanus, typhoid, anthrax, and hydrophobia have been practically conquered. Serums, with varying degrees of success, have been used in treating pneumonia, spinal meningitis, infantile paralysis, and pus infections. Medical-research laboratories everywhere are working to overcome the difficulties encountered in treating other infectious diseases by this means. Nowhere is the old adage "An ounce of prevention is worth a pound of cure" truer, however, than in the field of infectious diseases. It has proved to be far easier to prevent the disease, either by insuring that the person does not come in contact with the infective agent or by raising his resistance to it by prophylactic inoculation, than it is to cure the patient once the disease has fastened itself upon him.

Each year sees some additional maladies conquered. Others are attacked for the first time. Beasts of burden and other domestic animals have been the recipients of these benefits as well as man. There are no fields of human endeavor that offer greater promise for the relief of the suffering than the study of immunology and its methods.

QUESTIONS FOR STUDY

1. How do the old ideas of disease differ from our present ideas?

2. In what ways has the work of Jenner and Pasteur affected the treatment of disease?

3. Contrast the medical treatment which gives active immunity with that which gives passive immunity.

4. In what ways is the body protected against the entrance of disease-producing microörganisms?

5. What relationship may exist in certain diseases between heredity and susceptibility?

6. How may the body be incited to produce antibodies?

7. Describe some ways by which the body combats disease.

8. Where are leucocytes believed to be produced by the body, and what is their function?

9. Describe the commercial manufacture of smallpox vaccine.

10. Describe the method of immunization against typhoid.

11. What means are taken by health authorities to prevent the spread of contagious diseases?

12. How did vaccination for smallpox and inoculation for typhoid effect the death rate from these diseases during the World War as compared with previous wars?

13. What is the difference between endotoxins and exotoxins, and how is each produced?

14. Of what value has the microscope been to the advancement of medical knowledge?

15. How do races differ with respect to susceptibility in the case of some diseases? What causes have been suggested to account for this difference?

REFERENCES

DE KRUIF, PAUL. Microbe Hunters.

GARRISON, F. H. History of Medicine.

GIESEN, JOHN, and MALUMPHY, THOMAS L. Backgrounds of Biology, pp. 171–198.

JORDAN, EDWARD O., and FALK, I. S. The Newer Knowledge of Bacteriology and Immunology, pp. 702–891.

LUTMAN, B. F. Microbiology, pp. 243–312.

STITT, E. R. Practical Bacteriology, Blood Work and Parasitology, pp. 215–344.

VALLERY-RADOT, RENÉ. The Life of Pasteur. (Translated by Mrs. R. L. Devonshire.)

ZINSSER, HANS. Resistance to Infectious Diseases (fourth edition).

UNIT VIII

Man's Cultural Development moved Slowly at the Outset but has been greatly accelerated by Science and Invention

THE stream of culture is cumulative. At the outset it was an uncertain rivulet. Probably the first thing that marked man's departure from the beast was his ability to pick up a rock with which to crack the shell of a nut, to repel a wild animal, or to bring down a bird for use as food. Gradually intelligence evolved. Man conceived the idea of better tools and weapons than nature furnished, so he began to fashion crude scrapers and axes. The cultural stream was increasing, but it took man a half-million years to pass beyond even the dawn Age of Stone.

Tributary forces entered from the sides. Flint, a stone peculiarly well adapted to primeval needs, was discovered; plants and animals were domesticated; copper and tin were probably accidentally smelted and later purposely combined to make bronze; iron was recovered from its ores. Presently, through these agencies the feeble stream of social development had become a swift current. Then, when power was discovered and employed, the cultural stream became a wide river sweeping along with almost incredible speed. Man now advanced more in a decade than he did in a century of primitive life. It is said that society has moved forward farther since the discovery and industrial use of power than it did in all the ages which preceded that date.

But speed and volume have not proved to be clear gain. With this scientific and mechanical advance have come grave social and economic problems—problems so complicated and pressing as to demand the highest degree of man's wisdom for their successful solution. Indeed, whether man is equal to the task of making this new adjustment remains to be seen.

Chapter XVII · Man's Cultural Development has come Gradually from Very Humble Beginnings

Introduction. Primitive man was a hunter. With no knowledge of husbandry, either plant or animal, he had to depend upon the wilds for his food. When the resources of one region were reduced or depleted, he was compelled to move to a more bountiful one. Because of this his problems were largely those of a physical existence. Until the Cro-Magnons appeared little, if any, art is found, and historical records of any kind were not kept until very much later. Consequently, for our cultural knowledge of savage man, which antedates even the Cro-Magnons by hundreds of thousands of years, we must seek other sources.

As might be expected, the earth itself is the most fruitful and reliable witness as to how man lived and what he did in the remote ages of savagery. Primitive man at the outset adopted handy objects which he found in nature as his first crude implements and weapons; later he learned to work in stone, bronze, iron, and other materials. As he moved about from place to place, died, or was ejected by stronger invaders, he naturally left behind in caves and other dwellings many of these artifacts. Wind-borne particles of weathered rock covered these treasures. They were buried deep in débris and thus preserved. If the spot proved to be a favored one, often later races appeared and established their habitation on the buried remains of the first race. These secondary levels might in time become interred and furnish a floor for still later peoples.

In some parts of the world these successive habitation levels are much deeper than in others. By digging down through them, anthropologists discover a veritable book whose pages they read from back to front. In this objective literature of the past the Old World is especially rich and varied.

Dr. MacCurdy tells us that in the province of Dordogne, in France, practically every level of prehistory may be found in the deposits that have accumulated in four rock shelters.

Anthropology is a relatively new science. Anthropology (the science of man in general, particularly his antiquity) is a relatively new science. A hundred years ago little more was known of man's prehistory than was known of electricity, radiant energy, or genetics. However, once man got to delving into the floors of these old habitations, these concrete tales of the past, his enthusiasm knew no bounds. Objects by the score were turned up by the spades of investigators. Not only did the artifacts themselves begin to mount in number, but man's interpretation as to the cultural significance and meaning of these discoveries grew apace. Soon some sort of classification for this body of accumulating knowledge became imperative. C. J. Thomsen of Denmark, assisted by other scientists, advanced a plan which proposed a relative scale for prehistoric time. His system applies especially to Europe and provides for a sequence of cultures based largely upon the materials which primitive man used in making his implements, weapons, and other objects of adornment or utility. The basis of this classification represents what may be called three ages, or stages of cultural development: the Stone Age, the Bronze Age, and the Iron Age.

Although it is impossible to think consecutively of prehistoric ages and events without a time scale, the reader must not be too exacting as to precise periods or dates. Scientists find it about as difficult and hazardous to make definite time statements as to the events of prehistoric ages as they do to make positive time predictions relative to the events and changes of the future. The chronology of anthropologists is cast in general terms which represent thousands of years and often periods of even much greater duration. It deals largely in estimates based on the facts in hand and not in precise figures. The chart on the following page,[1] based on Thomsen's scheme, will give the reader a relative idea of prehistoric ages.

[1] Adapted from George G. MacCurdy's *Prehistoric Man*. By permission, American Library Association.

Iron Age (from 1000 B.C. to 1 A.D.)	{ La Tène Epoch (from 500 B.C. to 1 A.D.) { Hallstatt Epoch (from 1000 to 500 B.C.)	
Bronze Age (from 3000 to 1000 B.C.)	{ Often divided into four epochs based upon the { type and kind of artifacts produced	
Stone Age (from 1,000,000 to 3000 B.C.)	{ Neolithic Period (from 12,000 to 3000 B.C.) Mesolithic Period (from 20,000 to 12,000 B.C.) Paleolithic Period (from 500,000 to 20,000 B.C.) Eolithic Period (from 1,000,000 to 500,000 B.C.)	{ Upper Paleolithic subdivision Middle Paleolithic subdivision Lower Paleolithic subdivision

The lines of demarcation between these ages and subdivisions are not sharp and well defined. The intervals are, however, very certainly arranged in a correct sequential order. New types of implements would come in slowly, overlapping those of the previous age. These would rise to a place of dominance both as to kinds and number, then gradually give way to more advanced types as the latter came in. The development of human culture has ever been after this fashion. At one time the automobile, a rarity on our public highways, troubled the drivers of the dominant horse-drawn vehicle. Gradually carriages and wagons gave way to the mounting number of automobiles. Then the automobile took the ascendancy; and now a horse-drawn vehicle, especially a carriage, is rapidly coming to be a rarity. A complete change in methods of local transportation has been made. Yet there has been no sharp, well-defined line in the transition: the horse-drawn vehicle has gradually given place to the now dominant automobile. So it is with the different ages of prehistoric man: one type of culture has been gradually superseded by another more advanced type.

The Stone Age was characterized by a preponderance of stone products. The artifacts produced in the Stone Age are not confined to stone exclusively. Wood, bone, horn, ivory, and perhaps other available materials were used as well, but stone was the most abundant natural resource at hand. Because of this advantage it easily outstripped all other substances as the raw material upon which the most primitive men worked. Moreover, stone was the material most resistant to the ravages of use and decay. This was another strong point in its favor. To prehistoric men, without refined tools to assist them, the fashioning of stone must have been a slow, laborious process. When a hammer or a knife was toilsomely shaped to the desired form, it was highly desirable that it should prove to be durable. Indeed, it is this very lasting property of stone, together with its extensive use, that have led scientists to characterize this early age of human culture as the Stone Age. The most that we know of these very primitive people we infer from the stone and other more limited artifacts which investigators have found in their burial grottoes, habitations, and other centers of activity.

Various kinds of stone were used by Paleolithic (Greek *palaios*, "old," and *lithos*, "stone") man; but of all the types employed, flint easily played the dominant rôle. Flint was abundant, was widely distributed, and, from the point of view of the primitive tool and weapon maker, possessed highly desirable characteristics. It is composed of quartz so fine that no distinct crystals can be seen even under the microscope, it has a semi-glassy luster, and its hardness exceeds that of steel. In addition, when it is subjected to either pressure or pounding, it breaks readily, releasing thin, flakelike chips. These qualities enabled the primitive worker to bring the edge of his stone to a smooth or wavy sharp line.

The culture of the Stone Age is divided into three chief periods. The culture of the Stone Age is divided into three chief subdivisions, and if we admit the earliest one the number becomes four. They are the Eolithic (Greek *eos*, "dawn," and *lithos*, "stone"), the Paleolithic (Greek *palaios*, "old," and "stone"), the Mesolithic (Greek *mesos*, "middle," and "stone"), and the Neolithic (Greek *neos*, "new," and "stone").

The Eolithic period is one of uncertainty. The roots of this period go so far back into prehistory that investigators are often at variance and are uncertain as to the conclusions to be drawn from the evidence. According to the topmost portion of the geological table on the following page certain objects studied as possible artifacts go back to the upper Miocene and the Pliocene. Good authorities believe that some of these objects are actually crudely fashioned implements; others consider most of them to be natural, sharp-edged flints that these early men, or manlike creatures, found and learned to use. Close examination makes it difficult to escape the opinion that some of these best-formed

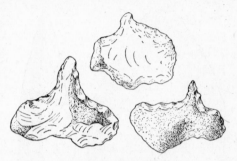

FIG. 210. Pointed Eoliths[1]

objects are artifacts. But objects similar to some of the Eoliths have been discovered in the lower Tertiary epoch, which is believed to have been long before man appeared on the scene. For this reason the whole question is beset with some uncertainty, although it seems probable that artifacts were produced.

Era	Epoch	Period	Millions of Years	Representative Animals
Cenozoic	Quaternary	Holocene (the present) Pleistocene (the world-wide glaciation)	1	Mammoths, mastodons, elephants, primitive man, present man
	Late Tertiary	Pliocene Miocene	7 19	Pithecanthropus (may be lower Pleistocene), three-toed horse, horse, lemurs, elephants, early camels, apes
	Early Tertiary	Oligocene Eocene Paleocene	35 55	Ancestors of the rhinoceros, primitive pigs, hippopotamuses, horses the size of rabbits

The Paleolithic period was of great duration and marked a high development in stone art. The term *Paleolithic* is applied to the stage when man was definitely shaping implements of

[1] From G. G. MacCurdy's *Human Origins*. By permission, D. Appleton and Company, publishers.

2

stone, but as yet had no agriculture, domestic animals, or pottery. In the lower Paleolithic the workmanship was crude,

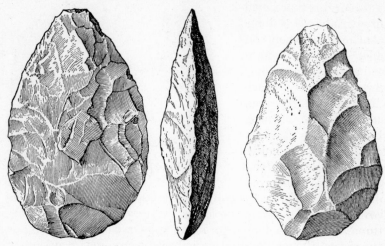

FIG. 211. A Paleolithic Flint Ax[1]   FIG. 212. Cleaver of Quartzite[2]

and the different objects which early man of this period fashioned into tools were limited as to kind. The principal tool was a kind of stone hand-ax. The oldest ones were crudely chipped on both faces and tended to be pointed at one end and rounded at the other end. The edge formed a wavy line at the point where the two roughly formed faces met. Another implement of this period was the cleaver made of heavy stone, which may have been used for cutting. It may also have been employed as a kind of saw and, very likely, as a scraper. Other tools of the lower Paleolithic included hand-scrapers, used in cleaning and scraping surfaces, and points, obviously used as punching instruments to make holes.

Many wild animals lived contemporaneously with the lower Paleolithic peoples, and from these the latter evidently obtained much of their food. Elephants, large horses, and lions

[1] From G. G. MacCurdy's *Human Origins*. By permission, D. Appleton and Company, publishers.   [2] Ibid.

have left their fossil remains scattered about from place to place, as well as the mammoth and the woolly rhinoceros.

Little is known definitely as to the physical characteristics of the men that made these crude stone implements. Such fragmentary evidence as is at hand indicates that the Heidelberg, the Piltdown, and perhaps other men of early type were associated with these objects.

The *middle Paleolithic* covers a long period of time and marks the con-

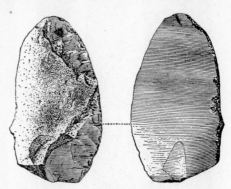

FIG. 213. A Flint Scraper left by Neanderthal Man [1]

tinuation of the lower Paleolithic culture. The implements and artifacts representative of this primitive life are found mostly in natural caves and rock shelters, which indicates that these men were true cave-dwellers.

The race which left this culture was the Neanderthal. The hand-ax persisted and is found all the way up through the higher levels of Neanderthal deposit. The hand-ax, however, was waning, and the scraper and the point became in time the dominant tools. Bone, too, was employed as a supplementary material for implements. Excavations in Switzerland have revealed that these mighty hunters would break the fibula of the cave bear obliquely about midway of its length. Then by grasping the large knee joint and using this large end of the bone as a handle, the sharply broken point made an excellent skinning tool. The surfaces of these tools which have been recovered are worn smooth, and in some instances are even polished from use. These implements are estimated to have come from deposits left from sixty thousand to a hundred thousand years ago.

[1] From G. G. MacCurdy's *The Coming of Man*. By permission, The University Society, publishers.

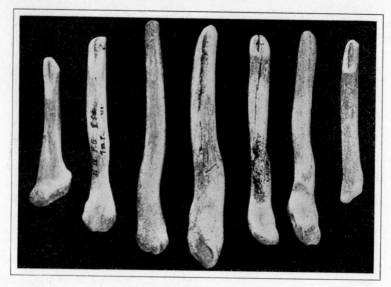

FIG. 214. Mousterian Bone Tools [1]

The *upper Paleolithic* marks a period of great advance from
the more primitive men. This is true both in respect to the
physical type of race that occupied the stage and in respect
to the character of their culture. The people who lived during
this age were the Cro-Magnon type. We have already found
that they belonged to the true *Homo sapiens* group. They
were fine specimens physically and possessed an unusual degree of cultural potentiality.

Up to this time man's kit of tools had been extremely
simple. They consisted largely of ready-at-hand stones and
such artificial tools as he could prepare by crude methods
from flint and bone. But these big-brained Cro-Magnons
were ready for revolutionary advances. They not only made
much better weapons and implements than previous men had,
but their artistic sense carried them far into the field of orna-
mentation and personal adornment.

Flint cores were worked into long, bladelike knives or flakes.

[1] From G. G. MacCurdy's *Human Origins*. By permission, D. Appleton
and Company, publishers.

Gravers were made and scratchers were fashioned. Bone was employed extensively in craftsmanship, as well as horn and ivory. The gravers were used not only to cut the bone and ivory objects which these primitive people manufactured but also to engrave the figures with which they adorned their artifacts. It was during this period that the shapely laurel-leaf points were evolved. These consisted of flints which were about three times as long as their greatest width. Modern European archæologists have likened these to the

FIG. 215. Laurel-Leaf Points

These are typical of the culmination of Paleolithic chipping [1]

laurel leaf and the willow leaf to designate two general types of chipped stone blades characteristic of the late Paleolithic culture. They were deftly chipped on both faces, and in this manner worked down to a remarkable degree of thinness.

Bone was worked into highly polished needles, which were pointed at one end and slit at the other. Javelin points and harpoons were made from reindeer horn and show various stages of development

FIG. 216. Bone Point with Cleft Base

A dart of Paleolithic origin [2]

from the crude forms to the more artistic and effective types.

Mural art in the caves, too, was carried to a remarkable degree of development. Animal drawings, some in three or four colors, adorned the walls. Bison, deer, hairy mammoths, horses, woolly rhinoceroses, and reindeer are represented. Some of these drawings are crude and imperfect; others

[1] From G. G. MacCurdy's *The Coming of Man*. By permission, The University Society, publishers. [2] Ibid.

reach a surprising degree of artistic merit. The figures were often engraved, but they also include many polychrome paintings. The colors apparently were prepared from the mineral pigments which were found at hand. Sculpture, too, found its beginnings here. Animals modeled out of clay have been found in the Cro-Magnon habitations. They were shaped by using sticks or the tips of the fingers. These primitive peoples must have known how to produce artificial light by burning, as a sort of torch, the fat of animals which they killed in the chase. Many of their productions are on the walls of deep, dark recesses, and could scarcely have been produced without the aid of some kind of artificial illumination. Among the animal forms discovered the female was much more often portrayed.

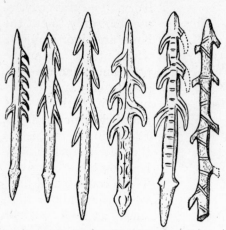

Fig. 217. **Harpoons of Reindeer Horn**

Used by Cro-Magnon man during the Magdalenian epoch [1]

The advantages of personal adornment had risen to the level of consciousness, too, but it is not known which sex was most interested. The materials used for this purpose were largely teeth, bone, and shells. They were pierced with pointed tools, were often highly polished, and were then worked up into necklaces, pendants, and other ornaments. Sometimes these objects had a more or less extensive series of notches cut into them. It is believed that these were used as tally cards to indicate the number of animals killed in the hunt or for the purpose of keeping other numerical records. In this way these objects served utilitarian as well as ornamental functions.

[1] From G. G. MacCurdy's *The Coming of Man*. By permission, The University Society, publishers.

Fig. 218. A Bison as painted by Cro-Magnon Man on the Ceiling of the Cavern
of Altamira [1]

The depth of the deposits in various caves and grottoes
shows conclusively that the period of the upper Paleolithic
was, like the middle Paleolithic, a very long one. Extensive

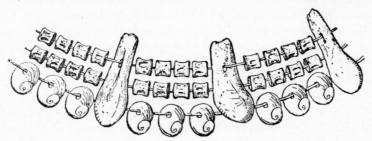

Fig. 219. Upper Paleolithic Necklace

Composed of canines of the stag, fish vertebræ, and the shells of *Nassa neritea* [2]

examples of upper Paleolithic art are found in France, Switzerland, and Spain. Germany, Italy, and almost all other countries of Europe are represented to some degree. Northern Africa

[1] From G. G. MacCurdy's *Human Origins*. By permission, D. Appleton
and Company, publishers.   [2] Ibid.

also furnishes evidence of an industry contemporaneous with parts of the European upper Paleolithic. This culture seems to have been concurrent with the last great glacial advance and to have survived some time after the last glacial retreat. Altogether the men of this period were very remarkable people and advanced the torch of culture to a surprising degree. It can well be questioned as to whether the men of our own day, if placed under the same circumstances, would or could have made greater utilitarian and artistic advancement.

The Mesolithic period is characterized by a halt in cultural development. If the upper Paleolithic represents the Augustan Age of prehistoric development, the Mesolithic represents, as one writer has put it, the "Dark Age" of prehistoric culture. For some reason the forces that raised the former peoples to the zenith of their achievement had become ineffective, and there seemed nothing to take their place. Dr. MacCurdy thinks that a change in climate was partly responsible and that different blood, together with racial shifts, may have completed the picture of the factors concerned. At any rate, the superb art of the Cro-Magnon disappeared and nothing rose to mark further important advancement.

Microlithic (Greek *micros*, "small") tools, such as blades of flint, disk-shaped scrapers and pebbles, which had served as chisels or as paring knives, have been found. One object that has baffled the interpretative genius of anthropologists is the painted pebble unearthed in the deposits of that age. More than two hundred of these pebbles have been found. One inference that has been drawn is that the painted figures represent the early beginnings of a form of character-writing. Another is that these pebbles were gaming-pieces, but of course no one will ever know.

Remains of an attempt at crude pottery work have been excavated, and a form of what is believed to be a rough millstone has been discovered. These latter objects are not intrinsically very important. Their form is too crude. They are believed, however, to indicate that man was turning more persistently to plant life as a source of food and was depending less upon the products of the chase.

Kitchen middens were characteristic of the people of this transitional period. These middens appear to have been piles of kitchen refuse from ancient dwellings. Other periods have had their kitchen middens also, but they differed from those of this period largely in that they were composed mostly of bone piles. The middens of the Mesolithic epoch, however, consist chiefly of huge piles of seashells. This form of aquatic life evidently constituted a large part of the dietary, and the shells represented the discarded portion of the animal body. It is within these heaps that many of the diminutive stone implements, evidently used for purposes of extraction, are found.

Traces of the Mesolithic culture have been found in three continents. In Europe it extended as far north as Denmark and Russia, in Asia as far east as India, and these early people seem to have roamed at will over North Africa.

The Neolithic period was characterized by the introduction of agriculture and domesticated animals. The torch of cultural evolution, once it was lighted, seems never to have become entirely extinguished. It has burned, however, with a very uneven light. Sometimes it flared up brightly and marked a period of rapid and extensive advancement; then its brilliance in a following period died down to the mere uncertain flicker of a candle-like flame. The upper Paleolithic we found to be a period of great refulgence which not only marked a rapid advance in utilitarian art but also in decorative art. Then followed the Mesolithic age, when the cultural torch burned low. It did scarcely more than furnish a transition light to the succeeding period, the Neolithic (Greek *neos*, "new," and *lithos*, "stone"), in which the torch flared high and shone with a steady, penetrating radiance.

The factors which so rapidly advanced the whole aspect of this early civilization were the cultivation of plants, the domestication of animals, the invention of pottery, and the development of the art of weaving. Primitive cattle, dogs, hogs, sheep, and goats seem to have been tamed and to have contributed bountifully in various ways to man's necessity.

FIG. 220. Pottery Vases of the Neolithic Period[1]

Cultivated plants, like domesticated animals, were doubt-less introduced into Europe from Asia. It seems that at least three varieties of wheat and two of barley were grown, in addition to millet, peas, and other fruit and crop plants.

But successful plant culture, even in those remote days of limited yields, required a place of storage. This need seems to have led definitely to the ceramic (Greek *keramos*, "potter's clay") arts and perhaps to the textile arts as well. The first granaries appear to have been pottery jars, which, when filled with grain, could be stored in sheltered places away from rain, birds, and rodents. It seems probable that sack-like containers too might have been made from fibers as well as from skins. The principal fiber plant raised was a narrow-leaved flax which may still be found growing wild in Mediterranean countries. Wool also was used extensively as a weaving fiber.

Back in the Paleolithic period men began to fish in order to increase their food supply. This is known by the engrav-ings of fish found decorating the walls of their caves. At first the fishhook used was a straight sliver of bone about an inch to an inch and a half long, pointed at both ends, and with the cord attached at the middle; but in Neolithic times it had become a well-formed, recurved hook with a groove near the

[1] From G. G. MacCurdy's *Human Origins*. By permission, D. Appleton and Company, publishers.

base for attaching the cord. Remnants of linen fish nets with even mesh have been found also in habitations near the lakes.

In the hands of these intelligent people the making of stone implements and tools became a high art. Flints were deftly chipped into beautifully shaped poniards, knives, and axes. Other stones were picked and polished for implements, and most of the axes were hafted. Daggers, deerhorn picks, and chalk lamps as well as many other objects have been found. Millstones for grinding were also abundant. The efforts of these early people at craftsmanship were not confined to stone alone; other materials, such as bone, ivory, and horn, were likewise employed. The remains of bone flutes have been recovered, showing an awakening in the art of music."

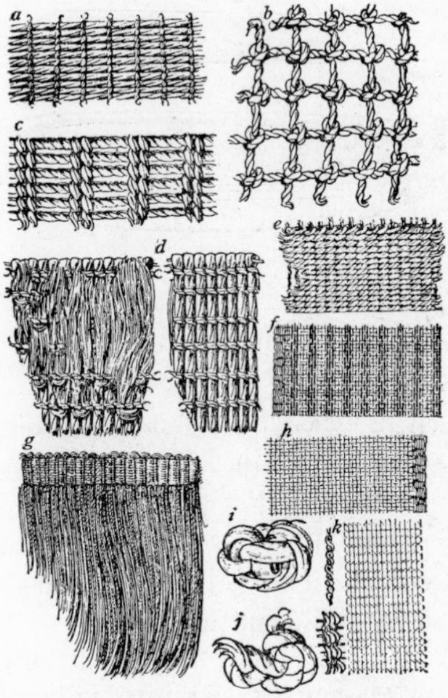

FIG. 221. Examples of Neolithic Weaving

The material is all flax. *a, b, c, d, e*, handwork. Note the structure and the accuracy of workmanship. *f, g, h*, made by some weaving apparatus [1]

Burial practices advanced rapidly. Stone and rockinclosed shelters and rooms for purposes of interment were built. These were widely distributed and show a progressive development from roughly constructed structures to stone cists and compartments with carved and decorated walls.

Flint of good quality was mined commercially and was transported from place to place in primitive boats called dugouts. These boats were hewn out of the trunks of trees and are the earliest type of boat known to have been used by man.

[1] From G. G. MacCurdy's *Human Origins*. By permission, D. Appleton and Company, publishers.

This water commerce carried on by means of boats greatly influenced the type and location of many Neolithic habitations.

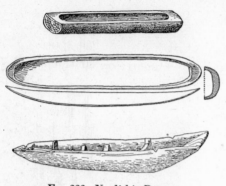

Fig. 222. Neolithic Dugouts

The earliest boats known to have been used by man[1]

The Neolithic house was often, where the environment favored it, a pile dwelling. Such houses were elevated on pilings that extended out into the water along the shores of lakes, streams, and swamps. Boats could be moored to the pilings — a matter of great convenience. The location of the houses offered considerable protection from invaders and made the problem of sewage disposal a very easy one. Their location also doubtless facilitated fishing, since fish and other forms of aquatic life used as food would be attracted to these sites by the refuse cast into the water. The houses were often aggregated into villages, and the pile-village mode of life was widely distributed over Europe, especially during the latter part of the Neolithic period. It persisted also into the Age of Metals, which followed.

If the Neolithic culture had a center, it is believed to have been located not far from the region where Asia, Africa, and Europe meet. Some anthropologists believe it to have begun in Persia at least twenty thousand years ago and to have reached Crete, in the Mediterranean, some six thousand years later. It probably extended westward at a still later date and varied as to time in different places.

The Bronze Age marked the use of metals by man in his cultural activities. As has been aptly said, "Conservatism is deeply ingrained in man's make up." He accepts the distinctly new hesitatingly and oftentimes with great misgivings.

[1] From G. G. MacCurdy's *Human Origins*. By permission, D. Appleton and Company, publishers.

Fig. 223. Neolithic Pile Dwellings (Reconstructed)

Communal house (left); private house (right). The railings did not exist in the original [1]

Especially was this true of primitive man. Some slight attention may have been given to the metals, even in the Neolithic period, but the employment of these new materials in his social pursuits came by degrees and marked a distinct advance in culture. It is a well-known fact that the extensive use of stone persisted far into the Age of Metals; but once the value and superiority of metals came to be recognized, the death knell of stone as a dominant factor in human progress had been sounded. Stone lingered; but thenceforth no further advances on a grand scale, either in artistic design or in the method of fashioning it, were made. New ways of extracting metals were discovered and more extensive uses were found for them, but the stone industry became stagnant and gradually disappeared. Of this transition Dr. MacCurdy says, "The change from the Stone Age to the Age of Metals was the most revolutionary step ever taken by man."

[1] From G. G. MacCurdy's *The Coming of Man*. By permission, The University Society, publishers.

It is probable that early man's attention was first directed to the metals by the small amounts of them which he found free in nature. Copper in the free condition occurs in limited quantities in certain localities, as do gold and some of the other metals. Upon examination and trial man found these new materials to be ductile, fusible, tenacious, and relatively easy to work into desired forms. Copper is believed to have been used in Egypt as early as 5000 B.C.; in the Orient its use antedates the Christian Era by some four thousand years. For a long period, in fact, copper and gold were practically the only metals employed. Most of the objects designed from them were ornaments of different kinds. Copper, however, came to be used for the manufacture of more utilitarian weapons and implements. Ancient objects of pure copper have been found in Ireland and Hungary, as well as in Saxony, Bohemia, Switzerland, and France. In prehistoric objects of gold, Ireland takes first rank. Because of these discoveries some authors have suggested the recognition of an Age of Copper, but the tendency now is to recognize the work in pure copper as merely a preliminary stage of the Age of Bronze.

As long as these early peoples were restricted in their metal work to the small amounts of free copper and gold available, not much advancement in this field could be made. The supply was entirely too limited. Fortunately (and, it seems probable, by accident) more abundant supplies were discovered. This enlarged supply came through the discovery of the art of reducing, or extracting, metals from their ores. It seems probable that a lump of copper carbonate, or of hematite, or of tinstone was inadvertently used in the circle of stones surrounding the open hearth, or fireplace. Becoming embedded in the deep hot embers, the ore was reduced and gave up its metal in the more or less free form. The curiosity of these early people now being aroused by the unusual product of this accidental smelting process, they experimented. Progress was made. Soon they were constructing crude furnaces for purposes of extraction, and the supply of metals grew.

It seems likely, too, that the practice of blending copper and tin to form the alloy called bronze came about through

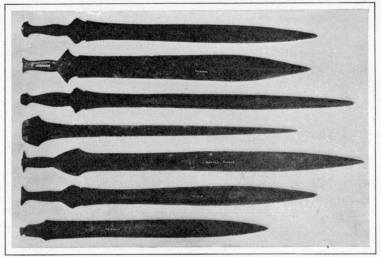

British Museum

FIG. 224. Swords of the Types used in the Bronze Age

accident at these hearth fires. Tinstone and chunks of copper carbonate may have been used simultaneously as hearth-stones, and their extracted metals running together formed a crude bronze product. This alloy proved to be a harder metal than either of its ingredients, and for this reason was more desirable in the constructive arts. The new discovery was followed up, and the Bronze Age was ushered in.

The Age of Bronze is often subdivided into four epochs, depending upon the objects made and the fineness of the art employed; but for purposes of brevity we shall deal with this period of something like two thousand years as being a single period.

As has been made clear, the metal ages were ushered in gradually. For centuries the people continued to live in pile villages. They constructed dolmens and other huge stone monuments. Plant and animal husbandry were not only continued but enlarged and expanded. Nevertheless prehistoric man "had reached the parting of the ways and had wisely chosen the one which led to the dawn of history and to our time."

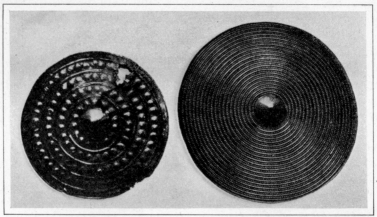

Fig. 225. Bronze-Age Shields from England (Left) and from Wales

The principal weapons fashioned during the Bronze Age were the poniard, the sword, and the lance. Lance heads were made in large numbers, and bronze arrowheads to some extent supplanted the flint points of the previous ages. As the swords were lengthened, bronze helmets and shields for protection were devised. Armor-like plates were prepared for the horse as well as for his rider. These protective devices were made of other more perishable materials, such as wood and leather.

Among the tools and utensils, hafted axes and adzes played a dominant rôle. Sickles were fashioned, and bronze razors were very common. Large musical trumpets which were probably used in warfare and on festival occasions have been found in a perfect state of preservation.

The art of making jewelry was carried far. Huge ornate belt buckles, combs, bracelets, finger rings, and fibulæ, or safety pins, have been recovered in large numbers. Bronze tweezers, probably used to remove superfluous hair from the face, are found associated with combs. This would suggest that the ideas of these ancient peoples as to the toilet were in many respects not so different from ours today. Amber-bead necklaces of intricate design and surprising beauty, as well as many other amber trinkets, were used to adorn the body.

Specimens exhumed from coffins and sepultures reveal a high development of the pattern and beauty of dress.

The art of pottery-making continued to flourish. Urns and jars of both clay and bronze were made in large numbers. These vessels were of very great utilitarian value, serving for storage purposes of all kinds. Some of them measured a meter in diameter. Many of them were characteristically decorated with a variety of engraved designs. The presence of perforated pottery is believed to point to the use of these objects in cheese-making. A peculiar, horizontally oblong-shaped bottle with an opening at the top for filling and another at one end for emptying is regarded by at least one authority as being a nursing bottle.

Relics of the Bronze Age are often found as isolated specimens and in tombs; but by far the greatest number are recovered in caches, where they were presumably deposited as hidden treasure. It is difficult to determine the origin of these caches. They may have been foundry warehouses or traders' hoards, or they may have concealed private collections or contained offerings used in connection with religious vows. Their size is in some instances astonishing. While workmen were constructing a sewer at Bologna, Italy, in 1877 they unearthed an enormous pottery vase which contained nearly fifteen thousand bronze objects and their fragments. Included were axes of various types, knives, chisels, lances, gouges, sickles, harness trappings, and ornaments. The peat bogs of Denmark have yielded large numbers of remarkably well-preserved specimens of the Bronze Age.

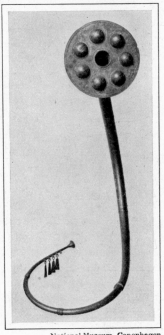

National Museum, Copenhagen

FIG. 226. **Bronze-Age Trumpet**

Found in a peat bog near Copenhagen

This age was represented in Asia, Africa, and practically throughout Europe. It probably had its origin in the region of the eastern Mediterranean, from whence it spread in all directions. The word for copper in most European languages is derived from the term *aes Cyprium*, applied to Cyprus by the Romans. Cyprus today has vast mounds of slag which mark the sites of prehistoric smelting operations. "Bronze" is a term which came from the city of Brundusium, in Italy, where, according to some authorities, highly prized mirrors of bronze were made.

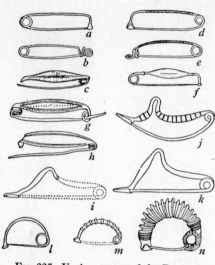

FIG. 227. Various types of the Bronze-Age Safety Pins

a, b, c, from Greece ; *d, e, f*, from Italy ; *h*, from Germany ; *g, i, m*, from France ; *j, k*, from Sicily ; *l, n*, from Switzerland [1]

The Iron Age superseded the Bronze Age. The Iron Age superseded the Age of Bronze, but there was no distinct line of demarcation between the two. It is very difficult to say where the latter left off and the former became dominant. Indeed, even in these advanced days man has not wholly forsaken bronze in favor of iron. Many bronze statues are still cast. Bronze, being much more resistant to the rusting and corroding influences of the atmosphere, is highly preferable for casting exposed figures.

It would be difficult to say where iron was first employed by man as a fabricating material. The first traces of its use doubtless extend far back into the preceding age. Elliot states that iron vessels were used in Egypt from 1500 B.C. to 1450 B.C. and very likely at a much earlier date in Assyria.

[1] From G. G. MacCurdy's *Human Origins*. By permission, D. Appleton and Company, publishers.

Fig. 228. Part of a Cache of Bronze Objects found in a Peat Bog in Denmark

a, two vases; *b*, bronze decoration for a belt; *c*, torque; *d*, bracelets; *e*, chisel;
f, axes with end sockets [1]

MacCurdy believes the significant beginnings of the Iron Age
to have occurred about the same time, both in the Euphrates
and Tigris valleys and in Egypt. The probable date is thought

[1] From G. G. MacCurdy's *Human Origins*. By permission, D. Appleton
and Company, publishers.

FIG. 229. Iron Ingot from Mosul, Assyria [1]

to be from 1300 B.C. to 1200 B.C. It seems more than probable that iron was first accidentally smelted from its ores and obtained as a metal while prehistoric man was purposely smelting copper and tin to make bronze. A piece of hematite or similar iron ore was possibly included in the crude smelter along with the ores of other metals, and the discovery of iron followed. An enormous cache of iron was discovered in 1867 near the ruins of ancient Nineveh. The iron in this cache consisted for the most part of spindle-shaped ingots with a hole in one end for suspension. Each ingot weighed between 4.4 and 8.8 pounds.

Blinkenberg considers that the extraction of iron from its ores was begun by the Hittite kings about 1300 B.C. on the southeast shores of the Black Sea. From this eastern-Mediterranean region as a center of origin its use evidently extended in all directions, especially westward, because the waves of successive emigrations from Asia proceeded in that general direction. In Greece the first epoch of the Iron Age occurred somewhere between 1200 B.C. and 800 B.C. In Italy it is thought to have been about 1000 B.C.

In central and western Europe the Iron Age is divided into two epochs — the Hallstatt and the La Tène. The latter occurred about the time the Gauls occupied the territory now represented by France and Switzerland and were contending with the invading armies of the Romans. The Hallstatt is

[1] From G. G. MacCurdy's *Human Origins*. By permission, D. Appleton and Company, publishers.

British Museum

FIG. 230. Shield of Enameled Bronze

named after a famous cemetery in Hungary which yielded an abundance of type specimens. The La Tène (meaning "shallow") type station is at the northeast end of Lake Neuchâtel, in Switzerland. In the following discussion, however, sharp distinctions as to time will not be made.

The weapons of this age were of various kinds. Many of them were still of bronze, but an increasing number of them were of iron. Long iron swords were abundant, as were shorter ones. The short sword finally developed into a sort of dagger. Swords and daggers were carried in scabbards, which were made first of wood and leather and later of metal. For the first time knives with articulating blades were made which closed similarly to the pocketknife of modern times. Lances (to be held in the hand) and javelins (to be hurled) were effective weapons of both offense and defense. Defensive armor in the form of shields was very common. These were sometimes made of wood, with a metal knob, but the best ones were of iron or bronze. The circular-shaped shield seems to have yielded to the ellipsoidal form, doubtless because the latter, being longer than it was wide, afforded better protection to the whole body. The art of enameling had been discovered, and the shield insignia and decorations were often finished in this manner. One shield, a veritable work of art, has been recovered from the Thames at Battersea, London. Warriors marched to battle with insignia mounted on poles. The wheel and the wild boar were favorite symbols. The advance of the warriors in the Gallic wars was likewise accompanied by loud shouts and noises made by beating their shields with their lances and by the blowing of trumpets.

Among the tools and utensils the iron ax with sockets similar to the modern ax made its appearance. The ax was often used as a weapon also. Anvils and hammers for working iron and other materials did not differ much from those used today in certain regions. Andirons made their appearance early in the Hallstatt epoch, together with other hearth utensils, such as pokers, pothooks, spits, cauldrons, and large forks. Hand mills with revolving millstones also came into use. Harness was made very largely of wood, and the parts were fastened together with metal hooks and rivets. Sickles, scythes, saws, pruning hooks, and shears similar to modern sheep shears were in common use.

Pottery-making and the making of metal dishes flourished. This art was greatly facilitated by the introduction of the potter's wheel from the East. The earthen vessels were frequently highly decorated in colors, and the metal ones were extensively ornamented with figured engravings.

The idea of coinage seems to have originated among the people of the eastern-Mediterranean region and was later introduced among the Gallic people from Greece and Rome. The barbaric coins were at first largely imitations of the classical forms and were made of gold or silver or of an alloy of these two metals, called *electrum*. As time went on, the monetary designs of the northwest European peoples departed farther and farther from the Greek and Roman prototypes and became more distinctly Gallic.

Articles of personal adornment multiplied in the Iron Age and became more varied both in use and in design. Fibulæ, or safety pins, varied as to form and often highly ornamented, are found abundantly in the ruins. Earrings, bracelets, torques, finger rings, belts, tweezers, as well as many other decorative and toilet articles, abounded. Certain curious

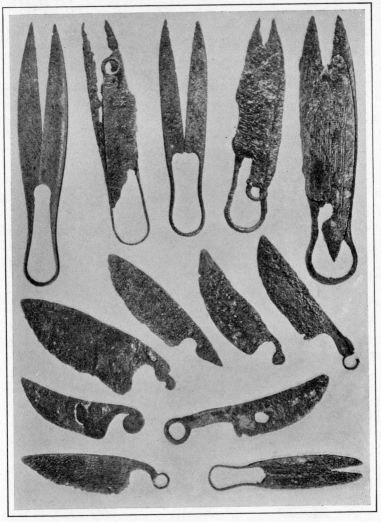

FIG. 231. Scissors and Razors of the La Tène Period[1]

[1] From G. G. MacCurdy's *Human Origins*. By permission, D. Appleton and Company, publishers.

Fig. 232. La Tène Pottery chiefly from the Marne

short-handled bronze spoons have been recovered. These are believed to have been employed in the preparation of paints and pomades, which were used as cosmetics. Glass, often colored with cobalt and other pigments, was made into bracelets, beads, and pendants. Combs of ivory and bone, some of which had double rows of teeth, were in common use. As one reads the list of toilet and personal articles recovered from habitations and sepultures he is led to see that the attempts at personal embellishment are very old practices. In all its ramifications and varied forms in modern times, it still seems to be but the mere extension and refinement of an art begun far back among primitive peoples.

The making of wine, particularly in southern Europe, was engaged in on an extensive scale. The Gauls imported their best wines from south of the Alps. They made pitchers, utensils, and other objects which were used privately and on festival occasions, but the best of these articles were also imported from Italy and Greece.

The age of metals greatly facilitated war as well as the arts. The Iron Age, it will be recognized, continues up to the very present. However, since more attention is to be given in later pages to the present aspects of civilization, we shall pass over them for the moment. Nevertheless, in closing this chapter we wish to point out the relation between the discovery and use of metals and warfare.

Elliot states that it is doubtful whether "serious war was known in Europe until the Bronze Age was established." During the Neolithic period he finds abundant evidence of migrations, colonizations, shifting in populations, and the opening up of new ground. There was a quick growth of civilization, attended by many refinements. Doubtless there was some local discord which led to "squabbles between villages." It is more than probable, too, that savage raids were made to steal cattle and other property, if not, indeed, to carry off the more beautiful maidens and women of neighboring tribes. But this author finds no evidence of serious war.

With the advent of the metal arts, on the other hand, the whole scene was changed. Resistant, ductile materials were now available with which to make weapons of offense and defense, to make chariots and armor, and to manufacture vehicles and other conveyances for the transportation of food and supplies. With these advantages man's ambition grew. He was quick to avenge real or fancied wrongs, and he coveted his neighbor's landed resources and goods and found justification for employing violent means, if necessary, to seize them. He wandered far from the place of his birth, and these migrations often brought him into conflict with other peoples. Consequently, from the Bronze Age on, wars have been numerous and extensive.

So, as we sum up the influence of the metal arts on civilization, we find vast advantages accruing to society. Certainly great advancement such as characterizes the cultured peoples of today would have been quite impossible without them, but their contribution to the waging of war has made the use of metals a somewhat qualified cultural advantage.

QUESTIONS FOR STUDY

1. What is anthropology, and what do we learn from its study?

2. How is prehistoric man believed to have secured his food?

3. What influence would permanent habitations have on man's cultural development?

4. How is knowledge of man's cultural development obtained?

5. How did man discover the use of metals, and what effects did they have on his social and economic life?

6. How does the development of art, sculpture, and agriculture aid us in securing information about the peoples living in the various ages?

7. What effects did the domestication of plants and animals have upon man's mode of living?

8. What cultural advances were made by prehistoric man?

9. Do you think the cultural development of man will continue or has he reached the highest stage of culture which it is possible for him to achieve? Give reasons for your conclusion.

10. Make a chart listing the cultural characteristics of each period (see page 501).

11. Show how different materials have contributed to man's cultural development.

12. Show what effect transportation probably had upon Neolithic culture.

13. Make a list of items in Neolithic culture that in modified form have passed down to the present.

REFERENCES

ELLIOT, G. F. S. Prehistoric Man and His Story.
MacCurdy, G. G. Human Origins, Vols. I and II.
MacCurdy, G. G. Prehistoric Man.
MacCurdy, G. G. The Coming of Man.
PEAKE, H., and FLEURE, H. J. Hunters and Artists.
PEAKE, H., and FLEURE, H. J. Peasants and Potters.
PEAKE, H., and FLEURE, H. J. Priests and Kings.
WELLS, H. G. The Outline of History, pp. 55–116.

CHAPTER XVIII · Many Social Habits and Institutions begun in Prehistoric Periods continue in Modified Form to the Present

Introduction. One cannot separate the cultural history of man into sharply delimited cross sections as a woodsman saws a log into blocks. Yet in order to get some idea of social chronology it is necessary to subdivide the whole into periods. This is imperative if one is to write or think about it conveniently. Perhaps the best basis for such a division is one based upon the chief materials which man used at different times in his fabricating industries. This was the plan used in the preceding chapter. How faulty this scheme of classification was has been perfectly evident. At best it is but an arbitrary division; for we have found that the use of stone extended far into the Age of Bronze, and that the roots of the Bronze Age found their beginnings in the Age of Stone. So it was for the Iron Age with reference to the period that preceded it.

When one leaves the more material aspects of the stream of culture, the attempt to make arbitrary divisions is an even harder one. Somewhere back in the Stone Age man began to wear clothing, and in civilized communities he continues to wear clothing today. Only the styles, the materials used, and the workmanship have changed. Transportation likewise began in the Age of Stone and continues today in an amazingly larger volume. Only the conveyances and the methods have changed. For this reason, then, there are certain aspects of cultural history that should be looked at as a whole, without any attempt to separate them into arbitrary divisions.

Man has extended his food supply and made it more dependable. At the outset man was dependent upon the wilds for his food. His chief food supply was the flesh of animals supplemented to some extent by nuts, bark, seeds, fruits, and other nutritious parts of plants. Without effective weapons wild game was difficult to obtain. As a consequence doubtless primitive man often went supperless to bed. Today, among some of the less intelligent tribes of Africa and other countries the problem of getting enough food to eke out a mere existence is a difficult one. Before the white man brought the horse to the American Indian, he is said to have been compelled to hunt the bison in groups and, at times, to have followed the herd for days before bringing down a single unwary victim.

The ruins of habitations indicate that in prehistoric ages man often concentrated very largely on one species of animal for sustenance. At one time it was the cave bear, at another the woolly elephant, and at a later period the wild horse. During the Mesolithic period the kitchen-middens, as we have seen, indicate that many people of that age lived largely on shellfish.

That the matter of getting food was a pressing problem at times is thought to be revealed by primitive art. The animals depicted are largely game animals. Many of them are shown to have been killed by clubs, javelins, or darts. The hunt must have been one of early man's most thrilling as well as most momentous experiences. Among the animals represented the female form was preponderant as to numbers. This is taken as a prayer that the game animals might be fruitful in reproduction to insure an adequate supply of food.

During the Neolithic period a great advance was achieved — one which, perhaps, more than any other made cultured civilization possible. Man did not entirely forsake the hunt, but he did domesticate plants and animals. Which one of these was first selected for his protective care, or whether they were both tamed about the same time, is not known. As a result the food supply became more certain, variable, and dependable. Moreover, instead of leading a nomadic life, roaming about to follow the wild herds, man could now settle down and build permanent homes. This change in his food habits and resources eventually resulted in many of the constructive advances which have come with permanent habitations and a more or less fixed community life.

Among the plants cultivated by Neolithic man were wheat, rye, barley, the pear, the grape, the apple, the strawberry, the pea, and the lentil. The Bronze Age added oats and a few other species. The Iron Age brought the turnip and the onion. The leading animals domesticated were the dog, hog, ox, sheep, and goat. The remains of the horse are abundant, even in Paleolithic times; but the evidence that it had been domesticated is not conclusive until the Bronze Age, when bridle bits and harness trappings made their appearance.

Man's habits of dress began early and have changed in response to an evolving society. At what stage man forsook his nude ways for clothing it is impossible to say, but it was doubtless early in prehistoric culture. There are different theories as to the origin of clothing, but the glacial advances of the Ice Age probably made some sort of protective covering for the body imperative. As the early aborigines were engaged in skinning the animals which they used for food, it would not have taken a very high grade of mental development to perceive that the skins might be used for protecting the body. Far back in the Paleolithic period, perhaps, indeed, as early as the Eolithic period, flint scrapers have been found. It is believed that the chief use to which these tools were put was to clean and prepare animal skins for clothing.

At first the skins were probably draped over the shoulders without fastenings. Convenience soon required that they be held together in some fashion, and the idea of binding them on the body or of sewing the pieces together with a rawhide thong was introduced. That the Cro-Magnon race in the upper Paleolithic wore clothing is perfectly evident by the numerous bone and ivory needles which they left. The textile art, invented by the Neolithic people, constituted a marked advance and probably meant that woven cloths also were adopted as materials for making garments. However, very little is known definitely about the clothing habits of these people. Textiles are very perishable; consequently, even if the people of that period did have abundant clothing made of these fabrics, the forces of decay acting in the remains of their habitations have largely obliterated the evidence. Nevertheless, the ruins of Neolithic dwellings in Switzerland show that the arts of spinning, netting, knitting, weaving, and embroidery were practiced. It seems perfectly reasonable, then, to suppose that clothing made from cloth was in vogue. Loom weights of stone and clay, spindle whorls, bundles of raw flax fiber, and specimens of both twill and taffeta have been recovered.

From the Neolithic period on, development in clothing seems to

FIG. 233. Male (*a*) and Female (*b*) Costume of the Bronze Age[1]

have been largely a matter of growing diversity of style in response to ever-widening needs and increased refinement. Oak coffins from the Bronze Age exhumed in Denmark have revealed garments of surprisingly advanced patterns and beauty. The sleeves of the female costume were of elbow length, the edges of the garment were either finished with buttonhole stitches or faced with strips of cloth, and two woven belts ending in artistic tassels were apparently used to secure the dress about the waist. A nicely woven headdress completed the attire. The male costume was not so elaborate, but it was very durable, convenient, and serviceable. Fibulæ were used to secure these garments at certain points, and this doubtless accounts for the large number of these primitive safety pins recovered from this age and the succeeding age.

[1] From G. G. MacCurdy's *Human Origins*. By permission, D. Appleton and Company, publishers.

Protective habitations were a primal need of man. When man, through mutation, departed far enough from his furry ancestors to lose the hair from his body, he needed a habitation for protection. The presence of rapacious wild animals of various kinds made this need all the more imperative. Like primitive man in other respects, he met this new need not by purposeful design but by utilizing what nature itself had provided. He may also have resorted to crudely constructed temporary shelters if emergency demanded. The natural shelters were overhanging rocks, cavelike recesses, and grottoes in which man took up his abode and became a cave-dweller.

FIG. 234. Female Costume of the La Tène Epoch[1]

This form of dwelling continued up through Neanderthal life and was wholly dominant during the entire Cro-Magnon epoch. It was the decided and persistent preference of this great race for cave life that has preserved to us so many objects indicative of their culture and advancement. In such caves these primitive people ate, lived, slept, and carried on their artistic endeavors. The Mesolithic peoples for the most part seem likewise to have lived in natural caves and rock shelters, although certain moor villages of primitive type are known. The habit of living in caves continued far into the Neolithic period; but, as would naturally be expected among

---

[1] From G. G. MacCurdy's *The Coming of Man*. By permission, The University Society, publishers.

a people so enterprising as these, a great advance came in the form of artificially constructed homes. Some of these habitations existed as moor villages. These villages consisted of a number of rounded huts sheltering a shallow circular pit. Huts were sometimes so made as to differentiate between sleeping room and kitchen. The walls were made of poles and branches and were coated with a plaster of clay. A more advanced type of architecture was the rectangular-shaped hut with walls of staves or slabs and strengthened by a matlike weft that extended horizontally between the walls.

The custom of building houses over the water front along lakes and streams began in the Neolithic period and persisted far up through both the Bronze Age and the Iron Age. The best-known examples of these were the pile villages found along the lake fronts, particularly in Switzerland (Fig. 223). Erected on pilings and extending out over the water's edge, these homes afforded the maximum of protection and utility. The inhabitants were much less likely to be endangered by wild animals or marauding tribes. Waste disposal took care of itself, and the pile-supported platforms upon which the houses stood afforded convenient docks for fishing or for the incipient commerce in flint carried on by means of man's first crudely constructed boats.

From this point on, among cultured people artificially constructed homes of more or less permanent design came to be the rule. These stood either isolated or, more frequently, aggregated into villages and towns and later into cities. In historic times types of architecture and construction have shown a constant evolutionary advancement.

The knowledge of fire and its control has contributed greatly to man's cultural development. Man's knowledge of fire is perhaps as old as the race itself. The most primitive and lowly of men doubtless saw fires set by lightning in both forest and grassland. Volcanoes and lava flows may have been another source of primeval fires. Such fires, as they swept over large areas, would often prove to be very devastating. Many animals would naturally perish in their wake. Hungry savages who escaped suffocation or burning would feast upon the bodies of parched and half-roasted carcasses. In this way prehistoric man may have learned of the greater tenderness and palatability of cooked flesh as contrasted with raw flesh. At the same time he came to appreciate the efficacy of fire in keeping himself warm. During extreme cold weather such as attended the advance of the glaciers, artificial heat must have been of tremendous advantage. By adding fuel to smoldering logs or peaty deposits passed over by natural timber and grassland fires, men of the Old Stone Age probably learned to prolong the life of fire. With such genial warmth and protection as was afforded by these artificially protracted fires, it is little wonder that primitive peoples came to deify fire and to worship it. But man was still far from an effective control of fire. He could foster a blaze by adding fuel, but he could not restore it once it had died out.

How man first discovered that a fire could be kindled by friction or that a spark could be struck from a piece of flint by percussion is a secret buried in prehistoric antiquity. It is probable that he got his clue at this point too by accidental observations. Theobald tells us that forests in southern India are occasionally set on fire by friction produced by the rubbing of one bamboo branch against another. Once such an observation was made, the inference that fire could be produced artificially by rubbing two dry sticks together would have been an easy one. The Negritos of the Zambales mountains in the Philippines, Elliot tells us, still make fires by "rubbing one bamboo across a nick in another." Fire drills are in common use today among the aborigines of the Malaysian countries and Australia. It would seem too that the Stone Age peoples could not have worked in flint very long without discovering that sparks which might be fanned into a blaze could be derived from this source.

Burnt flints have been recovered from the lower Paleolithic levels, and well-defined hearths appear among the ruins of the Neanderthal race. The oldest known fire-making devices which have been preserved appear in the upper Paleolithic. These consist of flints and lumps of pyrites used as "strike-a-lights." Cro-Magnon lamps, probably used to furnish light for the embellishment of cave walls, have been found. A chemical analysis of scrapings from the bottom of one of these lamps has shown the presence of animal fats.

The use of fire was greatly extended during the succeeding ages. Without fire the ceramic arts could not have reached the high mark of perfection which they attained. Both the Bronze Age and the Iron Age were thoroughly dependent upon artificial heat; for without a high degree of temperature the extensive reducing operations demanded to furnish the requisite quantity of metals would have been quite impossible.

In more recent times frictional and percussion methods of striking fire have given place to matches as a quick, safe, and convenient invention for this purpose. With the advent of matches man's control of fire may be said to have become complete.

Language has made cultural development possible. Language when defined as any means of communication between living beings must have begun at the very dawn of the most primitive races. The question as to whether even lower animals have something that can be compared to human language is still an open one. There can be no doubt that animals do communicate their comparatively rudimentary mental states and emotions to one another.

Language in its more dignified form, however, is thought of as a human achievement essentially basic to the development of intellectual and artistic culture. In its essence language is nothing more than the adoption of arbitrarily spoken symbols to represent things, ideas, actions, or feelings. The symbols used — words, sounds, and signs — are by mutual consent common in their meaning and signify the same to all individuals of the group. By this means the ideas and sentiments which one person achieves may be faithfully conveyed to the fellow members of his tribe or community.

The evolution of language laid the foundation for cultural advancement. The individual animal, aside from instinctive action, learns everything by experience. When that particular animal, however old he may be, dies, all his attainments perish with him. Each succeeding animal, so far as we can see, begins at the very point where his ancestors did — from scratch. With mankind it is entirely different. The discoveries, artistic ideas, knowledge, and experience of each generation can all be preserved. They were first kept and transmitted by tradition. Later they were expressed in written symbols of different kinds and left as historical records of the past. With this advantage each child does not, like the animal, start from scratch. On the contrary, the experiences and achievements of the race are cumulative, and each child may begin where his ancestors left off; each generation, through the advantages of language may, so to speak, stand on the shoulders of the preceding generation. A very large part of all educational endeavor is an attempt to acquaint the individual student with the worth-while discoveries and achievements of the past, in order that he may profit by this knowledge when he takes his place as a constructive member of society. When an electrical engineer starts his career today he does not begin at the level of Volta or Henry or Faraday: he has the advantage of an acquaintance with all the significant discoveries that have been made since the time of these pioneer masters.

The origin of language is shrouded in the uncertainties of the past. Different theories have been advanced, but for lack of well-preserved records the exact truth will probably never be known. Some believe language to have arisen through the imitation of animal sounds, others believe it to be a slow evolution from exclamations uttered in response to emotions of pain, joy, and the like, and still others believe that it possibly grew up from vocal expressions emitted in common by the group as they engaged in certain activities in unison. Whatever its origin was, when the development was once begun language continued to expand in a manner which paralleled man's growing intelligence. Even savage tribes have a surprisingly large vocabulary. A missionary in Tierra del Fuego was able to compile a dictionary of some 30,000 words in the native tongue. Professor Babbitt estimates that people with a common-school education know from 25,000 to 35,000 words, and American college students reported something less than 60,000 words.

The art of writing, like language, is also very old and has reached its present state of development through a long period of evolution. Men of all ages have needed to convey messages to other men with whom they could not talk in person. Consequently it became necessary to invent some objective vehicle of expression.

Man's first attempts in the field of writing were extremely crude and primitive. They seem to have consisted of roughly drawn and often grotesque pictures of animals and other objects. These are known as *pictographs*. It is possible that most products of Cro-Magnon art were designed for the purpose of religious ceremonials or, as some prefer to believe, they were an expression of art for art's sake. At different points in the Rocky Mountains the pictographic writing of primitive Indians inscribed on rocky walls may still be seen.

It is perfectly evident that this type of written expression would have severe limitations. Aside from being extremely slow and laborious in its execution, it would be very difficult, if not quite impossible, to express various shades of meaning. Among these would be the cases of nouns, verbs, and tenses of verbs, exclamations, and interjections. It would seem quite impossible to tell a connected story in this fashion without the extreme likelihood of misinterpretation.

Writing as an art is believed to have come to the greater part of mankind through the ingenuity of the Egyptians. Thot, the god of wealth, is given credit for this invention. The Egyptian system was largely pictographic at the outset. Gradually the form of the picture was abbreviated, so that a part stood for the whole. A horse's head might represent the whole animal. As time went on, these elements would change with usage and finally came to be mere symbols which were interpreted as standing for that particular object. At length many of these and other devised symbols became phonetic syllables and letters which could be connected up to express thought. Written language, then, came to consist of a heterogeneous combination of pictures, abbreviated pictures, and other symbols which were used as letters and syllables. This scheme of writing was known as the hieroglyphic system. The word *hieroglyphic* means "sacred carvings" and is employed because this form of expression was

FIG. 235. Hieroglyphic Writing[1]

used so extensively in religious rites and ceremonies and in preparing inscriptions for the dead. As Maspero declares, it was "a most complicated system, in which syllables and ideograms were mingled with letters properly so called. There is a little of everything in an Egyptian phrase, sometimes even in a word." The hieroglyphic form was so difficult that it "demanded a serious effort of memory and long years of study; indeed, many people never completely mastered it."

A few centuries later the Phœnicians used a cuneiform system of writing which is neither "picturesque nor decorative." Contrasting this with the Egyptian hieroglyphs, Maspero continues:

It does not offer that delightful assemblage of birds and snakes, of men and quadrupeds, of heads and limbs, of tools, weapons, stars, trees, and boats, which succeed each other in perplexing order on

[1] From H. S. Williams's *The History of the Art of Writing.* Merrill and Baker, publishers.

the Egyptian monuments, to give permanence to the glory of Pharaoh and the greatness of his gods. Cuneiform writing is essentially composed of thin short lines, placed in juxtaposition or crossing each other in a somewhat clumsy fashion; it has the appearance of numbers of nails scattered about at haphazard, and its angular configuration, and its stiff and spiny appearance gives the inscriptions a dull and forbidding aspect which no artifice of the engraver can overcome.[1]

FIG. 236. A Baked Clay Tablet of Cuneiform Writing[2]

The Phœnician system, like the Egyptian, had gradually evolved from an earlier pictographic form of writing. All valuable documents were written with a stylus on clay tablets which were afterwards baked to secure permanency.

Picture writing is still practiced by barbarous and semi-civilized peoples of today. Certain Indian tribes as well as bushmen communicate in this way. Even the present Chinese system of writing is but a very peculiar and complex system of sign writing. Some authorities believe that it is probably the complexity of the Chinese writing more than anything else that has kept these people in such a state of retarded cultural development.

Wells believes that the most important thing that happened in the six or seven thousand years that preceded the Christian

[1] G. Maspero, *The Dawn of Civilization* (third edition). By permission, D. Appleton and Company, publishers.

[2] From H. S. Williams's *The History of the Art of Writing*. Merrill and Baker, publishers.

Era was the invention of writing and its gradual advance to importance in human affairs. Once a system of writing had been adopted by progressive peoples, a constant change in the direction of refinement and perfection of form followed. Even the form of modern writing is changing with usage. The script used in the United States a half-century ago is different in many respects from the script letters used today. This change will doubtless continue to evolve in response to practice as society moves on.

Religion has been a dominant force in man's social evolution. The origin of the religious attitude in man's social development is a difficult matter to trace. Much of prehistoric man's physical activity, as we have seen, has been preserved to us through the abundance of artifacts, mural decorations, and remnants of his material industries which he has left. But religion is a wholly mental and emotional thing. No direct, tangible elements of such an institution can be preserved to posterity. What has been learned of prehistoric religion, then, has been inferred indirectly from the more objective aspects of primitive man's culture. Picturesque and ornamental inscriptions and designs, sepultures ranging from the most crude to the large and elaborate, amulets of different kinds, and charms have all been examined for evidence. In addition the religion, magic, and superstitions of the barbaric tribes living today have been observed and studied to find what light they shed on the problem. Naturally there has been disagreement. Some eminent students in this field believe that religion had its origin in *animism*, or the belief in spirits; others, in *magic*, or the possession by objects of hidden supernatural powers; still others, in *totemism, fetishism*, and so on. Moreover, there are some of the more orthodox students who hold that certain religions as to origin are the product of special, direct revelation. The whole problem is so difficult that the facts may never be known with certainty. It is even extremely hard to define religion, because it evidently means different things to different people. It is also probable that religion, in its broad sense, may have arisen from different sources among different tribes and races. In our discussion religion is taken to be man's conception of, and relation to, a power superior to himself which is believed to control the course of nature and of human life. In the discussion to follow we shall touch upon the animistic and magic bases only because they seem to be the most plausible theories that have been advanced in this field.

Animism, according to Sir Edward Tylor, characterizes tribes very low in the scale of humanity and then, in a constantly improved and more intellectual form, ascends high into the levels of modern culture. Animism holds to a belief "in souls and a future state, in controlling deities and subordinate spirits," and naturally leads to worship and conciliation. In its simplest forms animism is extremely crude. Savage philosophers seem to have regarded every man to have had two things belonging to him, namely, a life and a phantom, which came to be combined in what we call a spirit. Spirit embodiment was almost universal. Not only were men and animals conceived to have a spirit, but even plants and apparently inanimate objects, such as stones, weapons, boats, and ornaments, as well.

In the crude form of animism primitive men treat rivers, stones, trees, and weapons as if they were intelligent beings. They talk to them, appease them, or attempt to punish them for the harm they do. Savages of Borneo hold feasts to retain the "spirit of the paddy" within their rice crop lest it spoil in that hot, unfavorable climate. When, for any reason, the rice plants grow sickly these simple-minded people believe the spirit of the rice has departed, and they hold special ceremonies with supposedly appropriate incantations to call it back. The North American Indian will talk to and reason with his horse as if the horse were perfectly rational. Some will spare the rattlesnake, fearing the return and vengeance of its spirit if the reptile is killed. Formerly, among the Kayans of Borneo, whenever one of their great men died, slaves were killed, too, in order that their spirits might follow the spirit of the deceased and attend it. Before they were dispatched the slaves were often besieged and implored by the relatives of the dead man to take good care of his spirit by keeping always near it and obeying all its requests. In Fiji the wives of a great man were strangled in order that their spirits might accompany his spirit into the state beyond.

It seems probable that beliefs similar to these must have inspired the burial customs traced far back into the Stone Age. Food and weapons were interred with the dead, probably to supply their spirits on their journey into the next world. The Egyptians sought by the highly developed art of embalming to preserve the physical body from decay, since its preservation was thought necessary to insure the welfare of the soul in the future life. It was regarded as needing food and drink and everything else that the individual required in this life; so the graves and tombs were supplied with these things. The recovery of these articles in modern times from the sepultures of the ancient pharaohs has thrown much light on ancient Egyptian civilization.

Dr. Tylor believes animism to have been deeply ingrained in primitive man's nature. "The animism of savages stands for and by itself; it explains its own origin." Out of these roots a higher form of animism is believed to have come. As intelligence and culture grew, the more childish superstitious elements were discarded. New conceptions were added, until finally animism evolved into the enlightened religions of today.

The exponents of magic as a source of religion would start farther back to find the roots of religious belief. At the outset magic is considered to have existed without any notion or idea of supernatural deities. According to Sir James Frazer, in its unadulterated form the primitive magician "supplicates no higher power; he sues the favor of no fickle and wayward being; he abases himself before no lawful deity." In a world little understood by primitive man magic took the place of what now, among cultured peoples, is science. It was a sort of cause-and-effect conception. The magician believed that the same causes would always produce the same results; hence the use of the proper tool or instrument, the performance of the appropriate ceremony, or the casting of the correct spell would achieve the desired ends. The objective sought might be a successful hunt, or rain for the growing crops, or recovery from disease. The only thing in his thinking that could thwart the realization of the expected result, when the proper magic was employed, was the more potent charm of some other magician.

The "fatal flaw" in magic lay not in its reliance upon law as a determining force in the universe, but in its complete misconception as to the cause-and-effect relation involved in natural law. Ideas associated by similarity were thought to have a cause-and-effect relation, as were also those connected by time. For instance, the ancient doctors regarded a group of small, lobed plants to be efficacious in the treatment of liver diseases because their lobed form was similar to that of the liver. Paleolithic man fashioned his javelin hurler after the form of some animal, not only that it might be artistic, but in order that it might be more effective in bringing down game. The superstition that warty hands are caused by handling a toad is a more modern example. Instances of the error in cause-and-effect relations as a result of time association are seen in the belief that people might catch a dangerous disease or become insane as the result of contact with a witch. A more familiar case is the superstition that the hind foot of a rabbit carried in the pocket makes one lucky. All magic is false and unreliable; for whatever is true ceases to be magic and rises to the level of dependable science, for it now becomes truth established either by observation or by experimentation.

Magic by this school is believed to have been followed or accompanied by some degree of animism which led to the recognition of spirits. In time magic and religion came to be connected, and man through magic could even influence the gods. The magicians of ancient Egypt claimed the power of compelling even the highest gods to follow their commands and actually threatened them with dire destruction in case of disobedience. Whatever transcended primitive man's ability to understand natural forces came to be looked upon as supernatural and to reside in spirits. On a higher and more abstruse plane spirits and supernatural forces themselves were conceived to be controlled by magic. Soothsayers were simply individuals who understood the more abstruse magic, pertaining to both the natural and the supernatural. It is probable that the priesthood had its origin in this very group.

The question naturally arises at this juncture as to how man came to abandon magic. Much of the evidence is obscure; but, as Dr. Frazer says:

I would suggest that a tardy recognition of the inherent falsehood and barrenness of magic set the more thoughtful part of mankind to cast about for a truer theory of nature and a more fruitful method of turning her resources to account. The shrewder intelligences must in time have come to perceive that magical ceremonies and incantations did not really effect the results which they were designed to produce, and which the majority of their simpler fellows still believed that they did actually produce. . . . The discovery amounted to this, that men for the first time recognized their inability to manipulate at pleasure certain natural forces which hitherto they had believed to be completely within their control. It was a confession of human ignorance and weakness. Man saw that he had taken for causes what were no causes, and that all his efforts to work by means of these imaginary causes had been vain. His painful toil had been wasted, his curious ingenuity had been squandered to no purpose.[1]

Natural events occurred as usual, but they were not now believed to be produced by man himself.

The rain still fell on the thirsty ground: the sun still pursued his daily, and the moon her nightly journey across the sky: the silent procession of the seasons still moved in light and shadow, in cloud and sunshine across the earth: men were still born to labour and sorrow, and still, after a brief sojourn here, were gathered to their fathers in the long home hereafter. All things indeed went on as before, yet all seemed different to him from whose eyes the old scales had fallen. For he could no longer cherish the pleasing illusion that it was he who guided the earth and the heaven in their courses, and that they would cease to perform their great revolutions were he to take his feeble hand from the wheel. . . . Thus cut adrift from his ancient moorings and left to toss on a troubled sea of doubt and uncertainty, his old happy confidence in himself and his powers rudely shaken, our primitive philosopher must have been sadly perplexed and agitated till he came to rest, as in a quiet haven after a tempestuous voyage.

[1] This and the passages following are from Sir James Frazer's *The Golden Bough*. By permission of The Macmillan Company, publishers.

For a dependable understanding of nature he began to observe, investigate, and think. With the advent of these habits of mind, science began. For religious satisfaction and consolation he turned to a higher faith and practice.

If the great world went on its way without the help of him or his fellows, it must surely be because there were other beings, like himself, but far stronger, who, unseen themselves, directed its course and brought about all the varied series of events which he had hitherto believed to be dependent on his own magic. . . . who gave the fowls of the air their meat and the wild beasts of the desert their prey; who bade the fruitful land to bring forth in abundance, the high hills to be clothed with forests, . . . who breathed into man's nostrils and made him live, or turned him to destruction by famine and pestilence and war. To these mighty beings, whose handiwork he traced in all the gorgeous and varied pageantry of nature, man now addressed himself, humbly confessing his dependence on their invisible power, and beseeching them of their mercy to furnish him with all good things, to defend him from the perils and dangers by which our mortal life is compassed about on every hand, and finally to bring his immortal spirit, freed from the burden of the body, to some happier world, beyond the reach of pain and sorrow, where he might rest with them and with the spirits of good men in joy and felicity forever.

By some such mental processes and experiences as these it is believed that man came to be religious.

QUESTIONS FOR STUDY

1. What factors seem to have been responsible for the development of the art of making and wearing clothing?

2. What factors have been responsible for the development of community life?

3. Compare the personal adornments used by modern man with those employed by prehistoric man.

4. How has the art of writing contributed to man's progress?

5. What are some theories as to how language was developed?

6. How has the art of language contributed toward man's social development?

7. How does the animistic theory as to the origin of religion differ from the theory which traces religion to magic?

8. How have the discovery and increased knowledge of fire aided in man's progress?

9. How did the art of cooking probably originate, and what influence has it had upon man's manner of living?

REFERENCES

FRAZER, SIR JAMES. The Golden Bough(one-volume *abridged* edition).
JUDD, C. H. The Psychology of Social Institutions, pp. 160–172, 263–274.
MacCURDY, G. G. Human Origins, Vols. I and II.
MacCURDY, G. G. Prehistoric Man.
MacCURDY, G. G. The Coming of Man, pp. 91–144.
PEAKE, H., and FLEURE, H. J. Priests and Kings.
TYLOR, SIR EDWARD B. Primitive Culture.
WELLS, H. G. The Outline of History, pp. 94–97, 168–176.

CHAPTER XIX · The Age of Power has greatly influenced Man's Social Development

Introduction. In Chapter XVII we found that during his cultural history man had experienced three ages, namely, those of stone, bronze, and iron. On the basis of the chief material used in his fabricating endeavors, there can be no question that we are still deeply immersed in the Iron Age. Since man first smelted iron in his crude outdoor hearths thirty-five hundred to four thousand years ago, the use of this metal, disregarding certain temporary interruptions due to financial depressions, has constantly increased. In the United States, for example, we used almost twice as much pig iron in 1930 as we did in 1921. Other countries show similar, if not such marked, tendencies. So, on the basis of the raw material most extensively used in industrial construction, we are increasingly iron consumers.

About a hundred and fifty years ago, however, a new force of great significance was brought into civilized society. That force was *industrial power*. It has not supplanted iron; neither has it disowned it. But power has dethroned iron and made it its servant. Iron and steel are the chief agents through which it exercises its dominion. It would be impossible to have an Age of Power without iron and steel machines through which power could be made effective on a large scale.

The rule of power has often been referred to as the *Industrial Age*, the *Machine Age*, the *Age of Energy*, and so on. In this brief chapter, however, we prefer to call it the Age of Power and to think of it not as an age that has succeeded the metals, but as an age that has subjugated them and made them the chief instrument of its own rule.

Society entered the Age of Power about a hundred and fifty years ago. Stewart Chase states that culture officially entered the Age of Power when James Watt made his steam engine in 1775. We have often heard that Watt invented the steam engine. That, however, is not exactly true. We have also been thrilled by picturing a keen-minded boy grasping the bright idea that led up to his invention as he sat by his mother's kitchen stove watching the lid of the boiling teakettle bob up and down. That is also not true. A crude sort of steam engine had been designed by Hero of Alexandria as early as 200 B.C. However, history shows that the first practical application of steam as a source of power is attributed to Thomas Newcomen. Even Newcomen adopted earlier ideas when he made an engine with a piston which worked in a cylinder closed at one end and open at the other. Steam was admitted at the closed end of the cylinder. This forced the piston along. When the piston reached the opposite end of its stroke, the steam was shut off, and the cylinder was cooled with cold water to condense the steam. Atmospheric pressure working through the open end of the cylinder now forced the piston back to its original position, when the steam could again be applied. This crude engine was extremely slow in its action and was noisy and very wasteful of power.

Watt himself was a clever tinker. He could repair all sorts of mechanical devices. One of the Newcomen engines installed at the University of Glasgow did not perform well. Watt, having worked with John Black, the great physicist who discovered latent heat, was called in to fix it. While Watt was working on this engine one Sunday afternoon the idea occurred to him that the cylinder might be closed at both ends and that the steam might be both admitted and released at the ends by automatic valves. This idea marked a tremendous advance. Steam could now be used to push the piston both backward and forward. Watt took out his initial patent in 1769; but his first successful engine, as a power unit, was not in operation until 1776. It was pounding away, making fourteen strokes per minute, about the time our forefathers were attaching their names to the Declaration of Independence.

The steam engine was rapidly employed in industry. By 1785 Watt's engines were being introduced into factories. When his patent expired in 1800 they were revolutionizing industry. They were found in textile mills, paper mills, foundries, and mines, and were being experimented with as a source of power for carriages and boats.

Many other factors entered into the Industrial Revolution, to be sure; but the dominant one was power. Up to that time man, animals, and, to a very limited extent, water power had been the chief sources of energy. Now reservoirs of chemical energy stored up by the plant in prehistoric times could be tapped and made to do work.

The textile industry was revolutionized by mechanical inventions and the application of power. The Industrial Revolution began in England, but quickly spread to other countries. In 1700 the method of spinning and weaving all kinds of fiber materials into cloth was practically unchanged from those that prevailed among Neolithic men five thousand years B.C. In 1738 Kay invented a flying shuttle for weaving cloth. Previously two men, one standing on either side of the piece of cloth being woven, had to pass the shuttle back and forth by hand. Kay's innovation was a wide departure from primitive practice and marked the advent of a new era in the weaving industry.

This invention was rapidly adopted throughout England. Soon the facilities for spinning fell behind the demands for weaving. The hand-driven spinning wheels could not produce thread fast enough to supply the new flying-shuttle looms. In 1764 Hargreaves brought out the spinning jenny, which made one wheel turn eight spindles. In a short time the jenny was improved to make the wheel operate a hundred spindles. Arkwright next brought out the power-driven roller spinning frame, which was followed by Crompton's spinning mule in 1779. This machine could twist fine, hard, smooth yarn from raw fiber all in one process.

Now, with these highly perfected machines for manufacturing cotton fiber into finished cloth, the demand for raw cotton rose to unprecedented heights almost overnight. Cotton fibers had to be separated from the seed slowly and laboriously by hand. Under this handicap the demand presently far exceeded

2

the supply. The ingenious mind of Eli Whitney tackled the difficulty and produced the cotton gin. This device could clean the seeds from the fiber more than a hundred times faster than Negro slaves could pick them out. With this invention the chain was complete, and the processes involved in the cotton industry from picked fiber to finished cloth were speeded up immeasurably. From time to time improved patents have been added, machines have been enlarged, and new forms of driving power have been applied. The result has been a highly specialized textile industry with enormous capacity and an almost endless variety of products.

The electric generator and the electric motor have played an important rôle in the Age of Power. In 1821 when Michael Faraday made a magnetized needle rotate under the influence of a current-bearing wire, a new epoch in industry was initiated. That incident was the beginning of the electric motor. The motor had to pass through many stages of development, but the principle remains exactly the same as was used by Faraday in his London laboratory.

In 1831 the principle of electrical induction was embodied in a simple generator. The credit for this achievement also belongs to Faraday. With the subsequent improvement of the generator the two conditions required for a wide application of electrical power to industry had been realized. Almost unlimited quantities of electricity could now be generated, and this electrical energy, through the agency of the motor, could be converted into mechanical power to drive anything from a toy train to a huge electric locomotive.

With the electric generator and the electric motor as practical realities, industrial development received a great impetus. Electrical energy to furnish thousands of horse power could be generated at one point where the coal supply or running water made it advantageous to do so. The current could then be carried hundreds of miles on transmission lines to furnish power to the individual operator, to small companies, to great corporations, and to private homes. In 1928, in the United States alone, electric motors supplied mechanical energy at the rate of fifty million horse power.

In the hands of such ingenious men as Edison, Westinghouse, and others the applications of electricity knew almost no bounds. To hundreds of electrical toys, tools, and household contrivances may be added such machines and appliances as the electric furnace, the electromagnet, the telegraph, the telephone, the linotype, the trolley car, the rotary press, the locomotive, the electrotype machine, the incandescent lamp, the automatic adding machine, the radio, the teletypewriter, the electric drill, the X-ray tube, television apparatus, and the electrically driven sea-going vessel. This enumeration is, moreover, scarcely a beginning of the total enumeration that might be made.

Railroads and steamships have added greatly to the mechanization of modern life. We have found that in 1770 Joseph Cugnot of France built a steam carriage to transport artillery. In 1825 Gurney, an English inventor, constructed a steam vehicle which was successfully used to transport passengers. But the first man to make practical application of steam to the railway locomotive was George Stephenson in 1829. His locomotive was a tiny thing with cylinders set on a slant along the sides and with their power applied directly to the rear wheels. It went puffing and wheezing across the country at the terrific pace of fifteen miles an hour. But this infant was the progenitor of huge locomotives with enough power to drag after them serpentine trains of loaded box cars a mile long.

The first steamboat was a little tug called the *Charlotte Dundas*, built by William Symington in 1802. The paddle wheel was in the rear, and a Watt engine furnished the power. In 1807 Robert Fulton built the steam-driven *Clermont*, which terrified the natives at night as it went roaring up the Hudson, throwing a shower of sparks heavenward. In 1837 the *Savannah* was the first steamship to make a successful voyage across the Atlantic. It took many days to make the trip. In these times the great ocean greyhounds cleave the distance across the Atlantic under favorable conditions in less than five days. Their capacity has grown from a handful of people to thousands. Great sea-going freighters carry literally mountains of freight on a single voyage.

Fig. 237. Power Plant and Operating Equipment in a Burlington Two-Mile-a-
Minute Train

The eight-in-line Diesel 600-horse-power oil-burning Winton that will drive the
Zephyr along steel rails with the speed of an airplane. Generators, an integral part
of the motor, actually translate the power into electricity for operation and
illumination

**The internal-combustion engine has taken a significant part in
the Age of Power.** In 1859, after almost going bankrupt in the
adventure, Colonel Drake brought in the first oil well in the
United States, near Titusville, Pennsylvania. Cap rock was
pierced at sixty-nine feet, the "golden liquid half filling the
hole." On pumping, the well yielded a thousand gallons of
petroleum a day and instituted an industry that has become
the basis of some of the most colossal private fortunes in
America. That tiny pump-thrown stream grew, and at present it has swollen to a total in this country alone of something
like eight hundred million barrels a year.

With petroleum came the internal-combustion engine —

Fig. 238. The *Zephyr* is a New Type of Transportation Unit designed to meet Modern Needs

The entire train weighs less than one standard Pullman coach

first the gasoline motor, then the Diesel engine. This power applied to the automobile has revolutionized modern life and speeded up the movement of private and public business many fold. The remote rancher today is almost as close to the city as the country dweller who lived a dozen miles away a half-century ago. With the advent of the airplane, men in these days literally fly to their appointments.

The Diesel engine, a motor much more economical to operate, now bids fair to displace the gas engine in automobiles and airplanes and to supplant steam locomotives on the railroads as well. The economies and time-saving involved will further accelerate industry and make the reign of power complete.

One might continue for many pages to point out and enumerate the application of power to the enterprises of modern life. But this is not necessary. The examples already given will suffice to make the picture clear. The installed horse power in the United States has grown from 70,000,000 in 1900 to 1,026,000,000 in 1928. The Age of Power is upon us, and the civilized nations if not already industrialized are rapidly becoming so.

The Age of Power has brought far-reaching social and economic changes. The change from handicrafts to machine-made goods has brought an entirely new social epoch with its attendant problems. Before the introduction of machinery most of the manufactured goods were made either at home or in relatively small guilds. The hum of the spinning wheel could be heard in almost every country home. The products of these individual efforts were then gathered up by traders and sold on the market. Our word "spinster" is a relic of those days. A "spinster" was a woman who did not marry but stayed at home and spun. Leather was tanned either in the home or in privately owned tanyards. Shoes, from sole to upper, were made by one man. Weavers in small groups or often in their own homes produced cloth for the market. The blacksmith made hoes and spades, and the private wagon-maker turned out a vehicle under his own trade name.

The introduction of power and machinery changed all this. Now industrial power, instead of hand energy, was to be employed. At first it was either water power or steam which could not be transmitted but must be used on the site. This necessitated the abandonment of their former homes by the workers and their congregating in the vicinity of the factory. For a time private workers and manufacturers tried to hold out against the machine and to continue production in their more or less isolated and scattered homes. But they were waging a losing fight. The strength of one's arm, or the draft of a horse, or the treadmill of an individual shop was no match for the mighty water wheel or the prodigious power generated by a huge stationary boiler.

The development of railroads and steamship lines greatly facilitated the congestion of the workers about the point of manufacture. Before the days of rapid transportation goods could not be sold very far from the place in which they were made. Means of transfer were too slow and the cost was too great. Consequently every region had to be more or less self-sufficient. Within certain limits the people of a community had both to make the goods and to use them. Railroads and ocean-going freighters changed the situation. Goods could now be made at one point in large quantities and could then be sent rapidly and cheaply to points hundreds of miles away. Great factories employing hundreds of workers grew up, and crowded cities became a reality.

With the aggregation of workers into cities and towns the evils of the slum and tenement districts developed. Moreover, at the outset, working conditions were in most cases terrible. Long hours were the rule, the factories had little sanitation and were poorly lighted and ventilated, and any provision for recreation and rest was practically unknown. In certain parts of England at the outset of the Industrial Revolution, conditions were downright inhuman and revolting. The workers often were women and children who were powerless to protect themselves. Oftentimes from dire necessity the children of the pauper class were apprenticed to the mill owners, who worked them like beasts of burden. The following testimony given by a pauper father before the factory commission in 1833 furnishes a typical example:

My two sons (one ten, the other thirteen) work at the Milnes's factory at Lenton. They go at half past five in the morning; don't stop at breakfast or tea time. They stop at dinner half an hour. Come home at a quarter before ten. They used to work till ten, sometimes eleven, sometimes twelve. They earn between them 6s. 2d. per week. One of them, the eldest, worked at Wilson's for two years at 2s. 3d. per week. He left because the overlooker beat him and loosened a tooth for him. I complained, and they turned him away for it. They have been gone to work sixteen hours now; they will be very tired when they come home at half past nine. I have a deal of trouble to get 'em up in the morning. I have been obliged to beat 'em with a strap in their shirts, and to pinch 'em, in order to get them well awake. It made me cry to be obliged to do it.[1]

Conditions such as these, to be sure, no longer exist. In these days an enlightened public opinion would permit no such inhuman treatment. Moreover, workers have banded

[1] Robinson and Beard, *Readings in Modern European History*, Vol. II, p. 283. Ginn and Company.

themselves together into labor unions and have made demands as to the length of the working day, sanitary conditions, child labor, wages, rest periods, and many other desirable needs. To the credit of the mill owners and operators, too, it must be said that no one in this enlightened age would attempt to enforce even the best of codes which prevailed at that time in England. Good business would dictate otherwise. No employer can treat his workers as he might treat animals and get their best service in return.

But in the matter of the relationship that exists between labor and capital we are, even today, far from Utopia. Some industrial leaders are thoroughly human in their point of view and make the comfort and welfare of their workers as much an objective as they do the financial returns from their business. But, alas, many do not. Concessions are made grudgingly and then only at the compelling demand of a labor organization or a government regulation. Many industrial operators still deny their laborers the right to organize for mutual advantage. Wages in some industries and at many places in all industries are far too low. Child labor has gone only in response to the stern command of law. Unsanitary working conditions still may be found. Workers are compelled to live closely crowded in small ugly company houses, playground facilities are lacking for the children, and the workers are compelled to buy provisions from company stores and too frequently at exorbitant prices.

So the struggle between the worker and the mine owner and the mill owner still goes on. We seem to be moving slowly toward social and industrial justice; but the going is difficult and needs the intelligent, enlightened coöperation of every citizen.

An Age of Power has meant mass production. An Age of Power has come to mean mass production. When goods were made by manual labor, only limited quantities of them could be produced. In most cases each man made either the whole article or a large part of it himself. He took great pride in the quality of product which he turned out, but the labor output was low and the price relatively high.

When industrialized factories were built, however, this slow, individualized method of producing quickly became obsolete. Parts of a manufactured article were turned out separately and many of them at a time. If trousers were being made, many layers of cloth were piled on one another, and the cutter fashioned dozens of trouser legs at one operation. Sheet metal to be punched for rivet holes was piled up, and one stroke of the massive mechanical punch penetrated many separate sheets. Moreover, labor came to be largely specialized, and each man did just one small part.

In assembling plants the contrivance or machine being set up often moves along on an endless belt or chain. Each man does his own particular task, which may be no more than to put a bolt or two in place. It is reported that a worker from an automobile plant, when applying for another job, claimed to be a skilled workman. When pressed for an answer as to the nature of his skill, he finally admitted that he screwed nut 479 on its bolt. In one automobile plant which is a model of extreme specialization and industrialization the separate parts never go into the storage bins at all. On the contrary, they start for the assembling line at once, being carried by special conveyors. As the cars move along, the parts are fed in from each side by these conveyors, and each man in the assembling line does just one thing and nothing more. At the end of the line the cars are complete, have the radiator filled, the gas tank supplied, and are all ready to go. While making one model the daily output, if pressed to capacity, was said to be a line of cars eighteen miles long.

Throughout industry manufactured articles are emitted from the factory in a steady stream when demand warrants. Carpet sweepers, toasters, washing machines, automobiles, trucks, tractors, plows, clothing, motion-picture films, and radios, to name but a few of the many objects, pour forth. What is the effect of industry and of such an array of modern conveniences and pleasure-giving devices on society?

Opinions as to the social effect of mass production vary. Some socially minded persons view the scene with alarm; others, with hope and optimism. Critics say that man himself is being mechanized. Because of highly specialized tasks it is claimed that employees today lose the pride which the old worker experienced in the product of his hand. Since formerly the craftsman, as we have found, made the whole article himself or pretty much the whole of it, he was proud of the object and felt a thrill at its completion. This gave him a self-respect, a stimulation, an urge toward originality, an alertness, and a standing among his fellows that is denied the modern factory worker. An industrial worker today is rapidly becoming a dull, complying, mechanized individual.

The more hopeful turn to the other side of the picture. It is said that only between 5 and 10 per cent of our population are engaged in highly specialized industrial tasks. The result is a rich supply of the necessities of life, of household conveniences that supplant dull drudgery, and of pleasure-giving and leisure-satisfying devices such as the ancient world never dreamed of. Even the highly industrialized individual himself has so many more advantages — the product of science and industry — that life for him is more interesting than it was for the average worker a century ago.

Which one of these groups is right it is hard to say. It is probable that both contentions are in part correct. It seems reasonable to believe that industry brings dangers to the individual worker as well as advantages, and that society in the future must seek to mitigate the evils and to multiply the blessings of an Age of Power.

The question of international relations has become a pressing one. Before the adoption of power on a wide scale nations were more or less isolated. Only those whose territory was contiguous, or lay very close to each other, had serious problems of international relationship. Sometimes a monarch grew imperialistic in his desires and waged wars of conquest; or the religious zeal of one group caused them to invade the territory of another group. But the acute problems of an international character that are so perplexing today have risen largely because industrial power has done so much to annihilate space and produce a community of interests.

Most of the nations that have a supply of raw materials or

can get it have become manufacturing peoples. In seeking world markets their products come into competition, and this is conducive to jealousies and animosities. Electrical communication is bringing the nations of different continents closer together today in respect to incidents and events than the people of two adjoining counties were a hundred years ago. Just now Admiral Byrd is on an expedition of exploration to Antarctica. On Wednesday, January 3, 1934, he made a three-hour reconnoitering airplane flight from his flagship. The entire distance covered was about five or six hundred miles. That evening newspapers in all parts of the country apprised their readers of the event and gave details as to the weather encountered, the behavior of the plane, and the difficulties involved in the flight. Not very many months ago, October 11, 1932, the Postal Telegraph Company was observing the hundredth anniversary of the inception of the idea of telegraphy by its inventor, S. F. B. Morse. As a part of the exercises it was planned to see how quickly a message could be made to span the globe. The message "What hath God wrought," which was the original one sent over Morse's first line and also the first round-the-world message sent by President Theodore Roosevelt in 1903, was transmitted from the Broadway studio by the president of the company, Mr. Clarence Mackay. In just four minutes and forty-five seconds the message returned to the room from which it started, having circumscribed the earth. Fast ocean liners now cross the Atlantic in less than five days. In 1933 Wiley Post made a round-the-world airplane voyage in a little less than five days' actual flying time.

With these products of modern science and an Age of Power at hand, the leading nations of the earth have all become neighbors and keen business competitors. Population in each country is becoming denser, and the people are turning more and more from agriculture and pastoral pursuits to manufacturing and commerce as a means of livelihood. The constantly increasing volume of goods seeking markets in even the remotest corners of the earth naturally brings the various nations into financial and commercial conflict. Trade rivalry and economic struggle is a continuing condition. Naturally out of these situations there emanates a permanent possibility, if not a probability, of war.

In the past when international relations became too strained resort to force was the only appeal known. But now war is too costly and deadly. No nation can expect to carry on a prolonged major conflict under modern conditions without facing the possibility of bankruptcy. Besides, the atrocious potentialities of future wars are becoming terrifying. We were appalled by the cruelties of the World War; but competent judges tell us that "the destructive power of large-scale air attacks is a hundred times greater now than it was in the World War, and can be carried out in a fraction of the time needed then." Bombing fleets of two hundred *unmanned* airplanes controlled by ten manned planes far up in the sky can be sent over a city. The bombs sent down in showers may be loaded with high explosives, deadly gases, or cultures of germs. War, then, will not be fought to a draw. It will mean either victory or annihilation.

Besides, international agreements reached by force are never satisfactory or permanent. They usually mean only a breathing spell until the temporarily vanquished can improve their resources and acquire other allies. For this reason war as a means of obtaining justice and a state of peace and fair dealing among nations must be discarded. Reason, a sense of right, fair-mindedness, and humanitarian motives, be the nations large or small, demand it. Some agency, then, for reaching international settlements on this basis must be established. This may take the form of a world court or a league of nations, — more successful than the past attempts have been, — or it may be achieved through some other means. The problem remains for coming generations to solve.

The industrialization of society has brought serious problems of unemployment. The application of power to manufacturing processes has greatly increased the amount of goods an individual can produce. Mr. T. T. Read, a reputable statistician, has attempted to calculate the relative output of work per person in different nations due to mechanization. If the output in China, where industrial power is employed to a small degree only, is taken as 1, it jumps to 8.25 in France, to 12 in Germany, to 18 in Great Britain, to 20 in Canada, and to 30 in the United States; in other words, a single worker in the United States has the same capacity for producing goods that thirty laborers have in China. It took 100,000 men twenty years to build the Great Pyramid; but a couple of years ago a force of 4500 men, using power-driven stone-cutters, steam railways for transportation, and electric cranes and hoists, constructed the great Empire State Building from foundation to flagpole in less than a year. This gigantic building is the loftiest structure ever erected by man; its uppermost floor stands a hundred and two stories above the bustling street.

On the authority of the League of Nations man's individual productive power within the decade 1919–1929 has been increased at a rate twenty times as fast as during the twenty years preceding 1919. According to the *Monthly Labor Review* of the United States government the increased output of a worker in certain industries in 1927 as compared with 1914 is as follows: in slaughterhouses 26 per cent; in blast furnaces 103 per cent; in automobile factories 178 per cent; and in tire factories 292 per cent. Agricultural production also has been greatly increased through scientific methods and the application of power to farm machinery.

There was a time when it was feared that the population of civilized countries would increase to the point where insufficient food could be produced to feed the people. Malthus saw in this the specter of want stalking the land; and even as late as 1899 Sir William Crookes predicted that there would be a dearth of wheat within a generation. But scientific invention and power have changed the whole outlook. Because of these advances man is no longer threatened with a shortage of commodities — boots, shoes, sugar, clothing, automobiles, rice, and potatoes — but with an excess of them. The individual's capacity to produce has become so marked that the nation today is seriously threatened with the problem of unemployment; the fear is that in the future only a portion of the available workers will be required to supply all the goods the people need, and that the remainder will find themselves without employment and hence with no money to purchase the necessities of life.

There are those who tell us, however, that the unemployment bugaboo is overworked; that, though automatic machinery may cut down the number of men in a factory, the very introduction of such new machinery makes an increased and compensating need for labor at other points. For instance, if automatic machinery is installed, that machinery has to be made before it can be used, and this requires men. The defenders of the Machine Age also claim that although one scientific invention may displace workers in one field, it will at the same time provide many new jobs in new, related services. The automobile, for instance, assuredly did throw many carriage and wagon makers, harness makers, and horseshoers out of a job; but at the same time it created a vast number of other kinds of positions to take up the labor slack — automobile salesmen, garage mechanics, oil and gas producers, and service-station attendants. The recent report of the President's Research Committee on Social Trends states that in spite of our heavy growth in population, from 1900 to 1930 the *percentage* of our population engaged in gainful occupations slightly increased. Thus, in 1900 the percentage of the population in occupations was 38.3; in 1910 it was 40.6, in 1920 it was 39.6, and in 1930 it was 39.8. As these figures stand they would indicate that, barring temporary depressions and the "hard times" incident thereto, our people will be as busily occupied in the future as they have been in the past.

It seems probable, however, that these figures cannot be reliably interpreted in any such fashion. There are several other factors that must be taken into consideration. Child-labor laws have raised the legal age at which a youth may become an industrial worker, school-attendance laws require the child to stay in school until he has reached a stated age, and (probably of most importance) the daily hours of labor have been greatly shortened. The report cited in the preceding paragraph states that while no definite figures are available, it is probable that during the past fifty years the normal working week has been shortened twenty hours. Twenty-five years ago the common workday was ten hours long, and in some industries (for example, in steel manufacturing) it was twelve hours long. Now few laborers work over eight hours a day, and many of them but five and one-half days a week. This tendency to shorten hours of labor has doubtless made available many additional jobs, and these have swelled the ranks of the employed.

During the past business depression it has been estimated that the peak of unemployment saw from twelve million to fifteen million people out of work. Moreover, owing to the further rapid mechanization of industry from 1929 to 1933 to cut down overhead, many economists believe it probable that large numbers of these workers never will find profitable employment again. If not, then serious problems in this field also must confront society in the future.

Industry has made possible the accumulation of vast private fortunes. Greatly increased manufacturing capacity through the use of power has proved to be very profitable. Wages have advanced in many industries, but these have not been commensurate with the profits. The result has been the accumulation of huge fortunes in the hands of a relatively few individuals. The situation in the United States in this respect is becoming more accentuated. In 1921 one per cent of the population owned 13.16 per cent of the nation's total taxable wealth; in 1929 slightly less than one per cent of the population possessed 24.27 per cent of the taxable wealth. Because of the power that wealth gives, this tendency for the property of the country to become amassed in the hands of a relatively very small minority is considered to be very undesirable. The workers whose efforts help to create this wealth are in many instances underpaid and consequently deprived of many of the necessities and legitimate luxuries of life. Apparently some of the motive for profit must be taken out of business. Some way must be contrived to insure to the workers a more just and equal distribution of the profits of their labor. The need for such distribution is imperative, but how it is to be done without destroying individual enterprise and initiative remains a most perplexing problem.

In the future the schools must educate for leisure. We have found that, as a result of the Age of Power and many conditions growing out of it, the future is likely to see a reduced demand for labor.

It is probable that the length of the working day will be further shortened, and it is not improbable that the working week will be reduced to five days instead of six. If it is, society is faced with the pressing problem of effectively teaching boys and girls and young men and women how to use their leisure time profitably; how to get legitimate recreation and pleasure; how to recuperate the powers of mind and body; how to appreciate, to the full, nature and books; how to enjoy cultured social intercourse; in short, how to use the hours of respite from labor to live more satisfyingly and abundantly. Moreover, these opportunities must be provided partly at public expense and, it is likely, through the coöperative effort of the individual himself.

One should have no delusions about education for prolonged and permanent leisure. No individual who is distressed and unhappy over unemployment or who has an unprovided-for sick wife at home and a starving family is in any mood to appreciate the most elaborate provisions for leisure. Any self-respecting man must have employment — an opportunity to support himself and his family by the fruits of his own labor. If he does not have these rights fulfilled and sees no way in the future by which he can attain that end, then he surely becomes a menace to society. If competent self-government means anything, it means a self-supporting job and a fair return to labor for its contribution to industry. These considerations are basic and must be provided by every social organization that expects to live and perpetuate itself. If the realization of this condition means that the factory owner and the industrialist must recognize that there are higher human values than profits and capital, then this knowledge must be forthcoming at any price. If it means that our present form of economic organization must be entirely torn down and radically changed and rebuilt, then that must come too. In the light of past history whatever changes are necessary to provide the economic basis for an increasingly fuller life will come. They may be retarded and delayed; but come they will. The slow but certain evolution of a democratic society will, we believe, insure that result.

But in this section we are not discussing these fundamental considerations. The provision of these necessities of life is the *sine qua non* of a self-perpetuating free society. They are the basic things and must be provided before we can begin to talk about the type of leisure to which we are referring. After the physical needs of the body have been met for the worker and his family, then society must offer to them the opportunity to enjoy more spiritual and recreational experiences during their leisure hours. It must provide adequate parks, animal and plant preserves, music halls, libraries, clean picture shows, theaters, playgrounds, baseball parks, swimming pools, social centers, and the like. Moreover, in its educational program it must develop the ability on the part of each individual to enjoy and appreciate these agencies and through them to live a life in which growing leisure will bring deeper, richer, and more satisfying social experiences.

There are many other perplexing social and cultural problems that grow out of the change from private production to industrialized economic life. Old-age pensions; retirement incomes; public health; crime; the elimination of the eugenically unfit by humane methods, — these and a thousand other vexatious questions press for settlement by the adoption of rational intelligent policies. Volumes have been written, and others will follow. In a brief chapter such as this only a few of what the writers conceive to be the major problems of an Age of Power could be presented.

The progress of man has been from a stage of development characterized by a furious struggle with the forces of nature for a mere existence, to a period in which the forces of nature are, to a large degree, being brought under control.

Energy and materials from natural resources are used to

meet man's needs. The progress which he has made has been progress in the use of intelligence. Through the use of his intellect he has built up a mighty social and industrial system, the operation of which makes increasing demands for intelligence. Will the demands of the Age of Power outrun man's intellectual capacity? Will civilization destroy itself, and will its destruction carry man back to the level of the primitive ages? That disturbing question demands the attention and the deepest concern of every intelligent individual. Never before in the history of man has the challenge to intellectual activity been so keen as it is today.

QUESTIONS FOR STUDY

1. Trace the developments that have led to the Age of Power.

2. How has the utilization of power modified our manner of living?

3. What effect may the general distribution of electrical power have upon congested populations?

4. How has mass production in industry affected modern social life?

5. In an industrialized economic life what changes seem necessary to insure a living wage to all workers?

6. What influence has the advance in communication had upon the farmer and the city dweller and upon our international relations?

7. List the factors in modern life that demand a better agency for adjusting international disagreements.

REFERENCES

BARNES, H. E. Living in the Twentieth Century, pp. 62–150.

BEARD, C. A. The Rise of American Civilization, pp. 713–800.

BEARD, C. A. (Editor). Whither Mankind, pp. 65–82.

CHASE, STEWART. A New Deal.

CHASE, STEWART. Men and Machines.

KIMBALL, D. S. "The Social Effects of Mass Production," Science, Vol. 77, pp. 1–7.

MILLIKAN, R. A. Science and the New Civilization, pp. 1–31, 52–86.

RUSSELL, W. F. "Leisure and National Security," Recreation, Vol. 26, pp. 171–174.

President's conference on unemployment, Washington, D. C., 1921. Committee, on Recent Economic Changes, Vol. II, pp. 757–839.

President's Research Committee on Social Trends, Vols. I and II.

GLOSSARY

aborigines: primitive peoples; the original inhabitants of a country.

abortive: brought forth prematurely; hence imperfectly developed.

abscission layer: the layer of cells formed at the place of attachment of a leaf to a stem, where the leaf separates from the stem when it falls naturally.

absorption: the passage of liquids into the cells of the body.

adaptation: the process of becoming fitted to an environment.

adipose tissue: a form of connective tissue containing many cells in which a large portion of each cell body has been replaced by fat.

adventitious: occurring away from the ordinary place.

affinity: a relationship between species or groups depending on likeness of structure or of chemical constitution.

agnostic: one who professes ignorance of the existence of God.

algæ: a group of lower aquatic plants, as, pond scum and kelp.

alimentary canal: the digestive tract through which food is conveyed and through which it is absorbed by the body.

altruism: regard for and devotion to the interests of others; opposed to selfishness.

amalgamate: to mix or combine so as to become one.

amœboid: a term applied to cell movements which resemble the movement of *Amœba*.

amino-acid: one of a number of organic acids in which one hydrogen atom is replaced by the amino radicle (NH_2); the end product of protein digestion. These acids are the building stones of proteins.

anæmia: a disease characterized by a deficiency of blood or of red blood corpuscles.

anatomy: the study of the structure of living things.

animate: possessing life; living.

anterior: situated in front; relatively nearer the head.

anthrax: a contagious and malignant disease of cattle and sheep, sometimes affecting man.

anthropology: the science of man in general.

archæology: the science of antiquities, concerned with the systematic investigation of the relics of man and his industries.

armadillo: a burrowing mammal having an upper covering, or carapace, of bony plates.

artifact: anything made or modified by human art.

asexual reproduction: a type of reproduction in which no germ cells, or gametes, are involved.

assimilation: the conversion of food materials into protoplasmic substances.

atrophy: a wasting or withering of the body or any of its parts.

auditory: pertaining to the sense of hearing.

auricle: one of the receiving chambers of the heart.

auto-intoxication: the poisoning of the body or some part of the body by toxic matter generated therein.

bacteria: an important and widely distributed group of microscopic, one-celled plants.

barnacle: a marine crustacean usually covered by a hard shell and found on rocks, piles, and ship bottoms.

basal metabolism: the energy required to keep the body alive. Basal metabolism is determined when the patient is lying quiet and relaxed in a room of comfortable temperature, from twelve to eighteen hours after he has eaten a meal.

bile: the digestive fluid secreted by the liver of vertebrates.

bolus: a rounded mass of any material.

botany: the science treating of plants.

bulb: an underground, nearly spherical stem formed by the overlapping of thick fleshy leaves; a means of vegetative reproduction.

calorimeter: any apparatus for measuring the quantity of heat generated in a body or emitted by it.

calyx: the leaflike parts of a flower which surround the petals.

capillaries: the minute vessels of the blood-vascular system which connect the arteries and the veins.

capitulate: to sum up.

capon: a castrated cock.

carbohydrates: a class of organic substances composed of carbon, hydrogen, and oxygen, and usually with two atoms of hydrogen to one of oxygen.

carnivorous: eating or living on flesh.

cartilage: an elastic tissue which composes a large part of the skeleton of most young vertebrate animals.

catalyzer: a substance which changes the speed of a chemical reaction but does not itself enter into the reaction.

cellulose: a carbohydrate used in the formation of the walls of plant cells.

centrosome: the structure which forms the focus of the spindle fibers in mitotic cell division in animal cells.

ceramic: pertaining to pottery.

chlorophyll: the green coloring matter found in plant cells.

chloroplasts: a plastid containing chlorophyll developed in plant cells exposed to the rays of the sun.

chromatin: a deeply staining protoplasmic substance characteristic of the nucleus.

cilia: delicate whiplike protoplasmic projections from the surface of a cell.

cinema: a motion picture.

clinical thermometer: a thermometer for determining the temperature of the body.

coagulate: to change a liquid into a clot or a jelly by heat, by chemical action, or by a ferment.

coalesce: to grow or come together so as to form one body.

cochineal bug: the insect from which the scarlet dye *cochineal* is obtained.

cocoon: a sac, or envelope, in which the larvæ of certain animals pass through the pupa stage.

colloid: a state of matter in which particles larger than molecules are held in suspension in a fluid or semifluid medium.

congenital: born with one; existing from or before birth.

conifer: a plant of the pine family.

connective tissue: a tissue composed of cells and materials excreted by cells, which in its simpler forms binds other tissues or organs together.

corm: a bulbiform, non-scaly underground stem produced by some plants, such as the gladiolus, as a propagative organ.

corolla: the petals of a flower, usually colored.

crustacea: a large class of invertebrate animals with jointed bodies, usually hard-shelled, and with a varying number of jointed legs.

cuneiform: wedge-shaped characters used in an ancient form of writing.

cutin: a waxy substance in the epidermis of plants which serves the purpose of reducing the evaporation of water from inner tissues.

deciduous plants: plants bearing leaves all of which fall every year.

defecate: to discharge excrement from the bowels.

deglutition: the act, process, or power of swallowing.

desiccate: to free from moisture.

diarrhea: an abnormal condition of the intestines characterized by a morbid frequency and fluidity of the evacuations.

diatom: one of a group of microscopic green algæ.

dicotyledon: a plant whose seed contains two seed leaves, or cotyledons.

digestion: the power of converting food into substances that can be absorbed and assimilated.

dilemma: a situation in which one must make a difficult choice between two or more alternatives.

dinosaur: an extinct animal related to the modern reptile.

dicecious: having the male and female organs borne by different individuals.

dominant character: a marked parental character exhibited by a crossbred organism and its descendants.

dropsy: an abnormal accumulation of serous fluid in some cavity of the body or a diffusion of such fluid through the cellular tissues.

ductile: capable of being drawn out, as into a fine wire.

duodenum: the first portion of the small intestine.

ebullition: the boiling, or bubbling, of any liquid caused by the formation of gaseous bubbles in its mass.

ecology: the study that treats of plants and animals with reference to their environment; that branch of biology dealing with the relation of plants and animals to their environment.

ellipsoidal: having the shape of an ellipse.

embryo: the early stage of development of an organism.

emulsoid: a colloidal solution in which the particles in suspension are of a liquid not soluble in the suspending liquid.

enameling: the process of covering with enamel; decorating with enamel.

endosperm: a tissue within the embryo sac in seed plants which contains food materials for the developing embryo.

environment: all the external circumstances which influence an organism.

enzyme: an organic substance, usually secreted by a gland, which has the power to bring about or to hasten a chemical reaction but which is not consumed in the process.

epidemic: a widespread occurrence of a disease in a certain region.

epidermis: the outer layer of skin; the outer layer of cells in plants.

epilepsy: a chronic nervous disease characterized by paroxysms recurrent at uncertain intervals, attended by loss of consciousness and sensation, facial distortion, foaming at the mouth, convulsions of the limbs, and difficult, stertorous breathing.

epithelium: a layer of cells at the center or inner surface of an organ or passage.

esophagus: the tube which conveys the food from the mouth to the stomach.

ethics: the principles of right conduct.

evolution: the process of change.

excretion: the process of elimination of the dissolved waste products of metabolism; the waste products.

exudate: a fluid or semifluid deposited in the tissues or in a cavity as a result of some vital process.

Fallopian tubes: the ducts which serve to convey the egg cells from the ovaries to the uterus.

fatigue: a condition of organs or cells caused by excessive exertion.

fauna: the native animals of any given area.

feral: (1) wild by nature, untamed or undomesticated; (2) escaped from domestication.

ferment: an agent capable of producing fermentation.

fermentation: the decomposition produced in an organic substance by the action of living organisms, such as certain fungi or bacteria.

fertilization: the union of a pair of gametes.

fetishism: the belief in or worship of fetishes; rites used in fetish worship.

fibula: the outer of the two bones forming skeleton of the lower leg.

filamentous: like, consisting of, or bearing threads.

fission: the splitting of a cell into two parts; a form of reproduction in unicellular organisms.

flint: a subvitreous, dull-colored variety of quartz.

flora: the native plants of a particular region or period of the earth's history.

fortuitous: occurring by chance; pertaining to events that are without any cause.

fulcrum: the support on or against which a lever rests, or the point or pivot about which it turns.

fusible: capable of being fused, or melted, by heat.

fusiform: spindle-shaped; tapering toward each end.

gastronomy: the art of preparing and serving appetizing food; the discriminating appreciation of such food.

gene: a factor or element in a chromosome which conditions a character of an organism.

genesis: the act or process of producing or originating.

genetics: the study of heredity and its laws.

germination: the beginning of growth.

gland: an organ which secretes or excretes some special substance.

glycerin: chemically an alcohol formed by the decomposition of oils, fats, or molasses.

grain rust: a diseased condition in cereals which is produced by a parasitic fungus.

graver: an engraver's burin; a sculptor's chisel.

grotto: a small cavern; a cavern-like retreat.

guild: a corporation or association of persons engaged in kindred pursuits for mutual protection or aid or for coöperation.

hæmoglobin: the red coloring matter of the red blood corpuscles of vertebrates.

hematite: one of the most important ores of iron.

hemorrhage: the escape of blood from a blood vessel.

herbaceous: having the character of, or similar to, an herb.

herbarium: a collection of dried plants for study, usually scientifically classified.

herbivorous: feeding on plants.

heredity: the transmission of characteristics from generation to generation.

homologous: corresponding; similar or alike.

hormone: a substance manufactured by ductless glands, which assists the regulation of bodily activities.

horticulture: the science of cultivating gardens or orchards.

humus: decayed or decaying organic material in the soil.

husbandry: agriculture.

hybrid: the offspring of parents unlike in one or more characters.

hydrolysis: a chemical reaction produced by the decomposition of a compound, its elements taking up those of water.

ideograms: graphic representations of ideas by symbolic characters.

imbecile: a person of feeble mind, either by nature or by old age.

inanimate: without life.

inbreed: to follow a course of breeding from nearly related animals.

incipient: beginning to exist; belonging to the first stages.

indican: a compound contained in the blood and urine of certain animals, including man.

inhibit: to hold back, check, or restrain.

innervation: the nervous stimulation of an organ necessary for its proper functioning.

inorganic: designating matter which is not and has not been alive; without physical structure which was produced by life processes.

instinct: a natural spontaneous impulse in animals, moving them without reason toward actions essential to their existence, preservation, and development.

integument: the natural covering of an animal or vegetable body, as the human skin, the shell of a lobster, or the rind, or husk, of seeds.

intercellular: between cells.

internal combustion: a burning within; in an engine, the burning or explosion which takes place within a cylinder.

irritability: the ability to respond to stimuli.

juxtaposition: a placing close together.

kelp: any of the large brown seaweeds.

kitchen middens: the heaps of kitchen refuse of ancient dwellings.

lacteals: the lymphatic vessels of the small intestine.

larva: the immature, often wormlike form of an insect that undergoes metamorphosis in the stage between the egg and the pupa.

leucocyte: a white blood corpuscle.

ligament: a tough band of fibrous connective tissue which binds bones together.

lymph: the liquid part of the blood and the white blood corpuscles, which have passed through the capillary walls into spaces between the cells.

lymphatics: vessels which contain lymph.

medulla: the inner portion of an organ.

mesentery: a membrane that attaches the intestine to the body wall.

microbe: a microscopic organism, especially a bacterium instrumental in producing disease.

micropyle: the opening in a seed through which the pollen tube enters the ovule.

migration: the moving from one region to another.

modification: a change in a living organism resulting from its own activity and not transmitted to descendants.

monocotyledon: a plant with one cotyledon, or seed leaf.

monolith: any structure in stone formed of a single piece.

mutant: an organism which exhibits new characters unlike the usual characters of the species.

narcotic: a drug that relieves pain and produces sleep or, in sufficiently large doses, stupor or complete insensibility.

neuritis: inflammation of a nerve.

nitrogenous: designating any compound in which nitrogen is an important constituent.

nocturnal: active at night.

nomadic: having no fixed abode; roaming.

nostrum: a medicine the formula of which is a secret; a quack medicine.

nymph: a stage in the metamorphosis of some insects in which the young resembles the adult.

obstetrics: the branch of medical science concerned with the treatment and care of women during pregnancy and labor.

olfactory: pertaining to the sense of smell.

oracle: the medium by which a god reveals the divine purpose or makes known hidden knowledge.

organ: a group of tissues performing a special function.

organic: in biology, pertaining to or derived from anything that has life; in chemistry, applied to that branch of chemistry which deals with compounds of carbon.

organism: a living being, either plant or animal.

orifice: a mouth, or small opening, into a cavity.

osmosis: the diffusion of a solution through a semipermeable membrane.

ossification: the process by which soft animal tissue becomes bone.

ossify: to convert into bone.

ovipositor: a special organ which certain insects possess for depositing eggs.

ovule: an immature seed.

oxygenate: to combine or impregnate with oxygen.

oxyhæmoglobin: a loose combination of oxygen and hæmoglobin found in the red blood corpuscles of the pulmonary capillaries.

pampas: great treeless plains in South America.

panacea: a remedy or cure for all ills.

Paramecium: a one-celled, more or less slipper-shaped, ciliated animal, common in stagnant pools.

parasite: an organism that lives in or on another (host) at the expense of the latter.

pathogenic: productive of disease.

pébrine: a destructive epidemic disease of silkworms caused by a bacterium.

pedigree: a list or table of descent and relationship.

perennial: lasting more than one year; a plant that lives from year to year.

peristalsis: a wavelike muscular contraction occurring in some tubular structures, especially in the alimentary canal, and serving to propel the contents onward.

pharynx: the upper part of the alimentary canal which connects the mouth with the esophagus.

phylogeny: the history or development of a race, type, or species.

pictograph: a picture representing an idea.

pipette: a small tube, often graduated, for removing small portions of a fluid.

pistil: the part of a flower which contains the ovules.

plasma: the fluid part of the blood; blood from which all solid or semisolid ingredients have been removed.

plastid: a small granule of specialized protoplasm in the cytoplasm of certain cells.

pollen: the male spores of seed-bearing plants.

pollination: the process of transferring pollen grains from the stamen to the pistil.

polyp: a simple invertebrate animal consisting of a sac with, at one end, a mouth surrounded by tentacles and, at the other end, a means of attachment.

posterior: situated behind or toward the hinder part; opposed to anterior.

prehensile: adapted for grasping or holding.

prickly pear: any of a genus of cactuses bearing a pear-shaped fruit.

primordial: the earliest formed.

proboscis: a tubular prolongation of the head or mouth region of an animal.

progenitor: an ancestor in the direct line; a forefather or parent.

protoplasm: the essential material of all plant and animal cells; the physical basis of all life.

Protozoa: the division of animals into which all the one-celled animals are grouped.

pseudopodium: a blunt finger-like projection used by *Amœba* and its relatives for locomotion and feeding.

pupa: the quiescent stage in the metamorphosis of insects during which the larva changes to the adult.

pylorus: the opening between the stomach and the small intestine.

pyrite: fool's gold; a pale metallic brass-yellow opaque iron disulfide.

quartz: a natural form of crystallized silica.

recapitulate: to repeat briefly what has already been said at length.

recessive: with reference to a gene, such as does not express itself in the soma in the presence of a contrasting dominant gene.

reflex: an automatic subconscious response to a specific stimulus.

regeneration: the replacing, or forming anew, of lost parts by lower animals and plants.

renal: of or pertaining to the kidney.

retrograde: a moving backward or downward; declining.

rhizome: an elongated underground stem which produces roots below and leaves above.

rodent: any of the gnawing animals having strong incisor teeth.

ruminating animal: an animal that chews the cud.

saprophyte: an organism that lives upon dead organic matter.

sedge: any of a large family of grasslike herbs growing in damp places.

sequoia: either of two evergreen trees of California which grow to an immense size — the big tree and the redwood.

serum: the clear fluid of the blood that is left after the blood has clotted and the clot is removed.

sexual reproduction: a type of reproduction in which two kinds of germ cells are present — the sperm and the ovum.

siblings: two or more children having one or both parents in common, but not of the same birth.

sloth: an insectivorous mammal of South and Central America.

species: a group of animals or plants having certain common characteristics which clearly distinguish the group from other groups; the subdivision next smaller than a genus.

sperms: the reproductive cells of plants and animals which are produced by the male.

spiracle: an aperture, or orifice, for the passage of air or water in respiration.

spore: a variable, asexual structure capable under proper conditions of developing into a new individual.

stamen: the pollen-bearing part of the flower.

steelyard: a simple device for weighing.

stimulus: any condition which produces a response in an organism.

subcutaneous: situated, found, or existing beneath the skin.

sweet wort: unfermented beer.

synthesis: the uniting or building up of substances into a compound.

taurine: relating to or designating the group of bovine ruminants that includes the domesticated ox and the zebu.

tenacious: having great cohesiveness of parts; tough.

tendon: the inelastic tissue which binds muscles to bones, to other muscles, or to organs of the body.

tentacles: flexible appendages of many lower animals used as feelers and as organs of motion and attachment.

thorax: the chest cavity in man and the other higher animals.

tinstone: a mineral, brown or black in color, occurring in crystals of brilliant luster and also in massive forms; it is the chief source of metallic tin.

tissue: a group of similar cells of an animal or plant body, forming a continuous mass or layer; as, muscular or nervous tissue.

T. N. T. (trinitrotoluene): a powerful explosive.

totemism: the system of dividing a tribe into clans according to their totems.

toxin: a poisonous substance produced as a secretion in animal or vegetable tissue.

tuber: a short thickened portion of an underground stem.

tumor: a local swelling on or in any part of the body, especially from some morbid growth.

tundra: a treeless plain consisting of black, mucky soil with a permanently frozen subsoil, but supporting a dense growth of mosses and lichens.

turgor: a condition of normal rigidity of living plant cells, produced by the pressure of water upon the cell walls.

udder: the milk gland of animals.

urea: the main waste product of protein matter in the bodies of animals, a result of the process of katabolism.

ureter: the tube, or duct, through which urine passes from a kidney to the bladder.

vacuole: a space, or cavity, formed in the protoplasm of a cell.

variation: a deviation in form, structure, or function from the parent form.

ventricle: a muscular pumping chamber of the heart; a small cavity in the body of an animal.

vertebrates: the group of animals characterized by having a vertebral column.

vestigial: having become small or degenerate, as a structure which was once more complete in functional activity, but which has become more or less nonfunctional.

virus: the poison from any infectious disease.

viscera: the internal organs of animals.

vitiate: to injure the substance or qualities of a thing so as to impair or spoil its use or value.

INDEX